Springer Undergraduate Mathematics Series

T0225994

For further volumes:
www.springer.com/series/3423

Luis Barreira · Claudia Valls

Complex Analysis and Differential Equations

 Springer

Luis Barreira
Departamento de Matemática
Instituto Superior Técnico
Lisboa, Portugal

Claudia Valls
Departamento de Matemática
Instituto Superior Técnico
Lisboa, Portugal

Based on translations from the Portuguese language editions:

Análise Complexa e Equações Diferenciais by Luis Barreira
Copyright © IST Press 2009, Instituto Superior Técnico

Exercícios de Análise Complexa e Equações Diferenciais by Luís Barreira and Clàudia Valls
Copyright © IST Press 2010, Instituto Superior Técnico
All Rights Reserved

ISSN 1615-2085   Springer Undergraduate Mathematics Series
ISBN 978-1-4471-4007-8   ISBN 978-1-4471-4008-5 (eBook)
DOI 10.1007/978-1-4471-4008-5
Springer London Heidelberg New York Dordrecht

Library of Congress Control Number: 2012936978

Mathematics Subject Classification: 30-01, 34-01, 35-01, 42-01

Printed on acid-free paper

Springer is part of Springer Science+Business Media (www.springer.com)

Preface

This book is essentially two books in one. Namely, it is an introduction to two large areas of mathematics—*complex analysis* and *differential equations*—and the material is naturally divided into two parts. This includes holomorphic functions, analytic functions, ordinary differential equations, Fourier series, and partial differential equations. Moreover, half of the book consists of approximately 200 worked-out problems plus 200 exercises of variable level of difficulty. The worked-out problems fill the gap between the theory and the exercises.

To a considerable extent, the parts of complex analysis and differential equations can be read independently. In the second part, some special emphasis is given to the applications of complex analysis to differential equations. On the other hand, the material is still developed with sufficient detail in order that the book contains an ample introduction to differential equations, and not strictly related to complex analysis.

The text is tailored to any course giving a first introduction to complex analysis or to differential equations, assuming as prerequisite only a basic knowledge of linear algebra and of differential and integral calculus. But it can also be used for independent study. In particular, the book contains a large number of examples illustrating the new concepts and results. Moreover, the worked-out problems, carefully prepared for each part of the theory, make this the ideal book for independent study, allowing the student to actually see how the theory applies, before solving the exercises.

Lisbon, Portugal   Luis Barreira and Claudia Valls

Contents

Part I
Complex Analysis

1
Basic Notions

In this chapter we introduce the set of complex numbers, as well as some basic notions. In particular, we describe the operations of addition and multiplication, as well as the powers and roots of complex numbers. We also introduce various complex functions that are natural extensions of corresponding functions in the real case, such as the exponential, the cosine, the sine, and the logarithm.

1.1 Complex Numbers

We first introduce the set of complex numbers as the set of pairs of real numbers equipped with operations of addition and multiplication.

Definition 1.1

The set $\mathbb{C}$ of *complex numbers* is the set $\mathbb{R}^2$ of pairs of real numbers equipped with the operations

$$(a, b) + (c, d) = (a + c, b + d) \tag{1.1}$$

and

$$(a, b) \cdot (c, d) = (ac - bd, ad + bc) \tag{1.2}$$

for each $(a, b), (c, d) \in \mathbb{R}^2$.

L. Barreira, C. Valls, *Complex Analysis and Differential Equations*,
Springer Undergraduate Mathematics Series,
DOI 10.1007/978-1-4471-4008-5_1, © Springer-Verlag London 2012

One can easily verify that the operations of addition and multiplication in (1.1) and (1.2) are commutative, that is,

$$(a, b) + (c, d) = (c, d) + (a, b)$$

and

$$(a, b) \cdot (c, d) = (c, d) \cdot (a, b)$$

for every $(a, b), (c, d) \in \mathbb{R}^2$.

Example 1.2

For example, we have

$$(5, 4) + (3, 2) = (8, 6)$$

and

$$(2, 1) \cdot (-1, 6) = \big(2 \cdot (-1) - 1 \cdot 6, 2 \cdot 6 + 1 \cdot (-1)\big) = (-8, 11).$$

For simplicity of notation, we always write

$$(a, 0) = a,$$

thus identifying the pair $(a, 0) \in \mathbb{R}^2$ with the real number a (see Figure 1.1). We define the *imaginary unit* by

$$(0, 1) = i$$

(see Figure 1.1).

Proposition 1.3

We have $i^2 = -1$ and $a + ib = (a, b)$ for every $a, b \in \mathbb{R}$.

Proof

Indeed,

$$i^2 = (0, 1) \cdot (0, 1) = (-1, 0) = -1,$$

and

$$a + ib = (a, 0) + (0, 1) \cdot (b, 0)$$
$$= (a, 0) + (0, b) = (a, b),$$

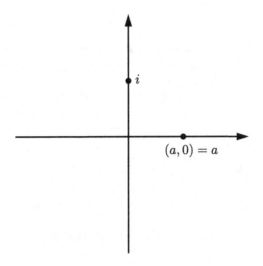

Figure 1.1 Real number a and imaginary unit i

which yields the desired statement. □

We thus have

$$\mathbb{C} = \{a + ib : a, b \in \mathbb{R}\}.$$

Now we introduce some basic notions.

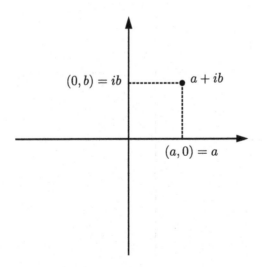

Figure 1.2 Real part and imaginary part

Definition 1.4

Given $z = a + ib \in \mathbb{C}$, the real number a is called the *real part* of z and the real number b is called the *imaginary part* of z (see Figure 1.2). We also write

$$a = \operatorname{Re} z \quad \text{and} \quad b = \operatorname{Im} z.$$

Example 1.5

If $z = 2 + i3$, then $\operatorname{Re} z = 2$ and $\operatorname{Im} z = 3$.

Two complex numbers $z_1, z_2 \in \mathbb{C}$ are equal if and only if

$$\operatorname{Re} z_1 = \operatorname{Re} z_2 \quad \text{and} \quad \operatorname{Im} z_1 = \operatorname{Im} z_2.$$

Definition 1.6

Given $z \in \mathbb{C}$ in the form

$$z = r \cos \theta + i r \sin \theta, \tag{1.3}$$

with $r \geq 0$ and $\theta \in \mathbb{R}$, the number r is called the *modulus* of z and the number θ is called an *argument* of z (see Figure 1.3). We also write

$$r = |z| \quad \text{and} \quad \theta = \arg z.$$

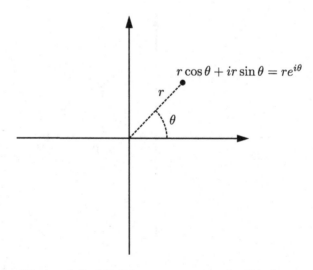

Figure 1.3 Modulus, argument and polar form

We emphasize that the number θ in (1.3) is not unique. Indeed, if identity (1.3) holds, then

$$z = r\cos(\theta + 2k\pi) + ir\sin(\theta + 2k\pi) \quad \text{for } k \in \mathbb{Z}.$$

One can easily establish the following result.

Proposition 1.7

If $z = a + ib \in \mathbb{C}$, then

$$|z| = \sqrt{a^2 + b^2} \tag{1.4}$$

and

$$\arg z = \begin{cases} \tan^{-1}(b/a) & \text{if } a > 0, \\ \pi/2 & \text{if } a = 0 \text{ and } b > 0, \\ \tan^{-1}(b/a) + \pi & \text{if } a < 0, \\ -\pi/2 & \text{if } a = 0 \text{ and } b < 0, \end{cases} \tag{1.5}$$

where $\tan^{-1}$ is the inverse of the tangent with values in the interval $(-\pi/2, \pi/2)$.

It follows from (1.4) that

$$|\operatorname{Re} z| \leq |z| \quad \text{and} \quad |\operatorname{Im} z| \leq |z|. \tag{1.6}$$

Example 1.8

If $z = 2 + i2\sqrt{3}$, then

$$|z| = \sqrt{2^2 + 2^2 \cdot 3} = \sqrt{16} = 4,$$

and using the first branch in (1.5), we obtain

$$\arg z = \tan^{-1} \frac{2\sqrt{3}}{2} = \tan^{-1} \sqrt{3} = \frac{\pi}{3}.$$

The following result is a simple consequence of Definition 1.6.

Proposition 1.9

Two complex numbers $z, w \in \mathbb{C}$ are equal if and only if $|z| = |w|$ and

$$\arg z - \arg w = 2k\pi \quad \text{for some } k \in \mathbb{Z}.$$

1.2 Polar Form

It is often useful to write a complex number in the form (1.3) or also in the following alternative form.

Definition 1.10

Given $z \in \mathbb{C}$ in the form $z = r\cos\theta + ir\sin\theta$, with $r \geq 0$ and $\theta \in \mathbb{R}$, we write

$$z = re^{i\theta} = |z|e^{i\arg z}.$$

We say that $z = a + ib$ is the *Cartesian form* of z and that $z = re^{i\theta}$ is the *polar form* of z.

Example 1.11

If $z = 1 + i$, then

$$|z| = \sqrt{2} \quad \text{and} \quad \arg z = \tan^{-1} 1 = \pi/4.$$

Hence, the polar form of z is $\sqrt{2}e^{i\pi/4}$.

Now we describe the product and the quotient of complex numbers in terms of the polar form.

Proposition 1.12

If $z_1 = r_1 e^{i\theta_1}$ and $z_2 = r_2 e^{i\theta_2}$, then

$$z_1 z_2 = r_1 r_2 e^{i(\theta_1 + \theta_2)} \tag{1.7}$$

and

$$\frac{z_1}{z_2} = \frac{r_1}{r_2} e^{i(\theta_1 - \theta_2)} \quad \text{for} \quad z_2 \neq 0.$$

Proof

For the product, by (1.3) we have

$$z_1 z_2 = (r_1 \cos\theta_1 + ir_1 \sin\theta_1)(r_2 \cos\theta_2 + ir_2 \sin\theta_2),$$

and thus,

$$z_1 z_2 = r_1 r_2 (\cos\theta_1 + i\sin\theta_1)(\cos\theta_2 + i\sin\theta_2)$$

$$= r_1 r_2 (\cos\theta_1 \cos\theta_2 - \sin\theta_1 \sin\theta_2)$$

$$+ i r_1 r_2 (\cos\theta_1 \sin\theta_2 + \sin\theta_1 \cos\theta_2)$$

$$= r_1 r_2 \cos(\theta_1 + \theta_2) + i r_1 r_2 \sin(\theta_1 + \theta_2)$$

$$= r_1 r_2 e^{i(\theta_1 + \theta_2)}. \tag{1.8}$$

For the quotient, we note that if $w = \rho e^{i\alpha}$ is a complex number satisfying $w z_2 = z_1$, then it follows from (1.8) that

$$w z_2 = \rho r_2 e^{i(\alpha + \theta_2)} = r_1 e^{i\theta_1}.$$

By Proposition 1.9, we obtain

$$\rho r_2 = r_1 \quad \text{and} \quad \alpha + \theta_2 - \theta_1 = 2k\pi$$

for some $k \in \mathbb{Z}$. Therefore,

$$\frac{z_1}{z_2} = w = \rho e^{i\alpha} = \frac{r_1}{r_2} e^{i(\theta_2 - \theta_1 + 2k\pi)} = \frac{r_1}{r_2} e^{i(\theta_2 - \theta_1)}$$

for $z_2 \neq 0$, which yields the desired statement. $\square$

Now we consider the powers and the roots of complex numbers, also expressed in terms of the polar form. For the powers, the following result is an immediate consequence of (1.7).

Proposition 1.13

If $z = re^{i\theta}$ and $k \in \mathbb{N}$, then $z^k = r^k e^{ik\theta}$.

The roots of complex numbers require some extra care.

Proposition 1.14

If $z = re^{i\theta}$ and $k \in \mathbb{N}$, then the complex numbers w such that $w^k = z$ are given by

$$w = r^{1/k} e^{i(\theta + 2\pi j)/k}, \quad j = 0, 1, \dots, k-1. \tag{1.9}$$

Proof

If $w = \rho e^{i\alpha}$ satisfies $w^k = z$, then it follows from Proposition 1.13 that

$$w^k = \rho^k e^{ik\alpha} = re^{i\theta}.$$

By Proposition 1.9, we obtain $\rho^k = r$ and $k\alpha - \theta = 2\pi j$ for some $j \in \mathbb{Z}$. Therefore,

$$w = \rho e^{i\alpha} = r^{1/k} e^{i(\theta + 2\pi j)/k},$$

and the distinct values of $e^{i(\theta + 2\pi j)/k}$ are obtained for $j \in \{0, 1, \ldots, k-1\}$. $\square$

We note that the roots in (1.9) of the complex number z are uniformly distributed on the circle of radius $r^{1/k}$ centered at the origin.

Example 1.15

For $k = 5$ the roots of 1 are

$$w = 1^{1/5} e^{i(0 + 2\pi j)/5} = e^{i2\pi j/5}, \quad j = 0, 1, 2, 3, 4$$

(see Figure 1.4).

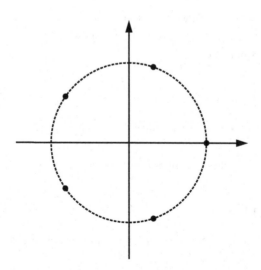

Figure 1.4 Roots of 1 for $k = 5$

1.3 Conjugate

Now we introduce the notion of the conjugate of a complex number.

Definition 1.16

Given $z = a + ib \in \mathbb{C}$, the complex number $\bar{z} = a - ib$ is called the *conjugate* of z (see Figure 1.5).

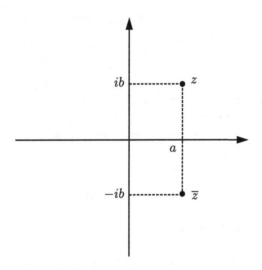

Figure 1.5 $\bar{z}$ is the conjugate of z

Clearly, $\bar{\bar{z}} = z$. Moreover, if $z = re^{i\theta}$, then

$$\bar{z} = \overline{r\cos\theta + ir\sin\theta}$$
$$= r\cos\theta - ir\sin\theta$$
$$= r\cos(-\theta) + ir\sin(-\theta) = re^{-i\theta}.$$

Proposition 1.17

For every $z \in \mathbb{C}$, we have $z\bar{z} = |z|^2$.

Proof

Given a complex number $z = re^{i\theta}$, we have

$$z\bar{z} = re^{i\theta}re^{-i\theta} = r^2 e^{i0} = |z|^2.$$

This yields the desired identity.   □

Proposition 1.18

For every $z, w \in \mathbb{C}$, we have

$$\overline{z + w} = \bar{z} + \bar{w} \quad \text{and} \quad \overline{zw} = \bar{z}\,\bar{w}.$$

Proof

Let $z = a + ib$ and $w = c + id$, with $a, b, c, d \in \mathbb{R}$. Taking conjugates, we obtain

$$\bar{z} = a - ib \quad \text{and} \quad \bar{w} = c - id.$$

Therefore,

$$\bar{z} + \bar{w} = (a + c) - i(b + d) \tag{1.10}$$

(see Figure 1.6). On the other hand,

$$z + w = (a + c) + i(b + d),$$

and thus,

$$\overline{z + w} = (a + c) - i(b + d). \tag{1.11}$$

The identity $\overline{z + w} = \bar{z} + \bar{w}$ now follows readily from (1.10) and (1.11).
Moreover, if $z = re^{i\theta}$ and $w = \rho e^{i\alpha}$, then

$$zw = r\rho e^{i(\theta + \alpha)},$$

and thus,

$$\overline{zw} = r\rho e^{-i(\theta + \alpha)} = re^{-i\theta}\rho e^{-i\alpha} = \bar{z}\,\bar{w}.$$

This completes the proof of the proposition.   □

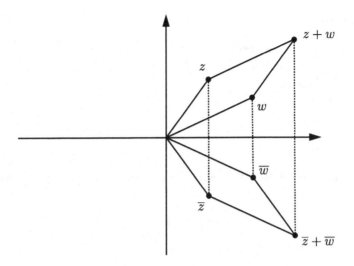

Figure 1.6 Points z, w, $z+w$ and their conjugates

Example 1.19

Let us consider the polynomial

$$p(z) = \sum_{k=0}^{n} a_k z^k$$

for some real numbers $a_k \in \mathbb{R}$. We have $\overline{a_k} = a_k$ for each k, and thus,

$$\overline{p(z)} = \sum_{k=0}^{n} \overline{a_k z^k} = \sum_{k=0}^{n} \overline{a_k}\, \overline{z}^k$$

$$= \sum_{k=0}^{n} a_k \overline{z}^k = p(\overline{z}).$$

In particular, if $p(z) = 0$ for some $z \in \mathbb{C}$, then

$$p(\overline{z}) = \overline{p(z)} = \overline{0} = 0.$$

This implies that the nonreal roots of p occur in pairs of conjugates.

We also use the notion of conjugate to establish the following result.

Proposition 1.20

For every $z, w \in \mathbb{C}$, we have:
1. $|z| \geq 0$, and $|z| = 0$ if and only if $z = 0$;
2. $|zw| = |z| \cdot |w|$;
3. $|z + w| \leq |z| + |w|$.

Proof

The first property follows immediately from (1.4). For the second property, we note that

$$|zw|^2 = zw\overline{zw} = zw\overline{z}\,\overline{w}$$

$$= z\overline{z}w\overline{w} = |z|^2|w|^2 \tag{1.12}$$

for every $z, w \in \mathbb{C}$. Finally, for the third property, we observe that

$$|z + w|^2 = (z + w)(\overline{z + w})$$

$$= (z + w)(\overline{z} + \overline{w})$$

$$= z\overline{z} + z\overline{w} + w\overline{z} + w\overline{w}$$

$$= |z|^2 + |w|^2 + 2\operatorname{Re}(z\overline{w}).$$

It follows from (1.6) and (1.12) that

$$\operatorname{Re}(z\overline{w}) \leq |z\overline{w}| = |z| \cdot |\overline{w}| = |z| \cdot |w|,$$

and hence,

$$|z + w|^2 \leq |z|^2 + |w|^2 + 2|z| \cdot |w|$$

$$= \big(|z| + |w|\big)^2.$$

This completes the proof of the proposition.   □

1.4 Complex Functions

In this section we consider complex-valued functions of a complex variable. Given a set $\Omega \subset \mathbb{C}$, a function $f \colon \Omega \to \mathbb{C}$ can be written in the form

$$f(x + iy) = u(x, y) + iv(x, y),$$

with $u(x,y), v(x,y) \in \mathbb{R}$ for each $x + iy \in \Omega$. In fact, since the set of complex numbers $\mathbb{C}$ is identified with $\mathbb{R}^2$, we obtain functions $u, v \colon \Omega \to \mathbb{R}$.

Definition 1.21

The function u is called the *real part* of f and the function v is called the *imaginary part* of f.

Example 1.22

For $f(z) = z^2$, we have

$$f(x + iy) = (x + iy)^2 = x^2 - y^2 + i2xy,$$

and hence,

$$u(x, y) = x^2 - y^2 \quad \text{and} \quad v(x, y) = 2xy.$$

Example 1.23

For $f(z) = z^3$, we have

$$f(x + iy) = (x + iy)^3 = x^3 - 3xy^2 + i(3x^2y - y^3),$$

and hence,

$$u(x, y) = x^3 - 3xy^2 \quad \text{and} \quad v(x, y) = 3x^2y - y^3.$$

Now we introduce various complex functions.

Definition 1.24

We define the *exponential* of the complex number $z = x + iy$ by

$$e^z = e^x(\cos y + i \sin y).$$

Example 1.25

For each $z = x + i0 \in \mathbb{R}$, we have

$$e^z = e^x(\cos 0 + i \sin 0) = e^x(1 + i0) = e^x.$$

Hence, the exponential of a real number x coincides with the exponential of x when this is seen as a complex number.

Example 1.26

For $z = i\pi$, we have

$$e^{i\pi} = e^{0+i\pi} = e^0(\cos\pi + i\sin\pi) = 1(-1 + i0) = -1.$$

We also describe several properties of the exponential.

Proposition 1.27

For every $z, w \in \mathbb{C}$ and $k \in \mathbb{Z}$, we have:
1. $e^{z+w} = e^z e^w$ and $1/e^z = e^{-z}$;
2. $\overline{e^z} = e^{\overline{z}}$;
3. $(e^z)^k = e^{kz}$;
4. $e^{z+i2k\pi} = e^z$.

Proof

Given $z = x + iy$ and $w = x' + iy'$, we have

$$e^{z+w} = e^{(x+x')+i(y+y')}$$

$$= e^{x+x'}\left[\cos(y+y') + i\sin(y+y')\right]$$

$$= e^x e^{x'}\left[(\cos y\cos y' - \sin y\sin y') + i(\sin y\cos y' + \sin y'\cos y)\right]$$

$$= e^x e^{x'}(\cos y + i\sin y)(\cos y' + i\sin y')$$

$$= e^x(\cos y + i\sin y)e^{x'}(\cos y' + i\sin y')$$

$$= e^z e^w.$$

In particular,

$$e^z e^{-z} = e^{z-z} = e^0 = 1,$$

and thus $1/e^z = e^{-z}$. This establishes the first property. For the second, we note that

$$\overline{e^z} = \overline{e^x\cos y + ie^x\sin y}$$

$$= e^x\cos y - ie^x\sin y = e^x(\cos y - i\sin y)$$

$$= e^x\left[\cos(-y) + i\sin(-y)\right] = e^{x-iy} = e^{\overline{z}}.$$

The third property follows from the first one by induction, and for the fourth we note that

$$e^{z+i2\pi k} = e^{x+i(y+2k\pi)}$$

$$= e^x \left[\cos(y + 2k\pi) + i\sin(y + 2k\pi) \right]$$

$$= e^x (\cos y + i\sin y) = e^z.$$

This completes the proof of the proposition. □

Now we consider the trigonometric functions.

Definition 1.28

The *cosine* and the *sine* of $z \in \mathbb{C}$ are defined respectively by

$$\cos z = \frac{e^{iz} + e^{-iz}}{2}$$

and

$$\sin z = \frac{e^{iz} - e^{-iz}}{2i}.$$

Example 1.29

For $z = x + i0 \in \mathbb{R}$, we have

$$\cos z = \frac{e^{ix} + e^{-ix}}{2}$$

$$= \frac{1}{2}(\cos x + i\sin x + \cos x - i\sin x) = \cos x$$

and

$$\sin z = \frac{e^{ix} - e^{-ix}}{2i}$$

$$= \frac{1}{2i}(\cos x + i\sin x - \cos x + i\sin x) = \sin x.$$

Hence, the cosine and the sine of a real number x coincide respectively with the cosine and the sine of x when this is seen as a complex number.

Example 1.30

For $z = iy$, we have

$$\cos(iy) = \frac{e^{-y} + e^{y}}{2}.$$

In particular, the cosine is not a bounded function in $\mathbb{C}$, in contrast to what happens in $\mathbb{R}$. One can show in a similar manner that the sine is also unbounded in $\mathbb{C}$.

Example 1.31

Let us solve the equation $\cos z = 1$, that is,

$$\frac{e^{iz} + e^{-iz}}{2} = 1.$$

For $w = e^{iz}$, we have $1/w = e^{-iz}$, and thus,

$$w + \frac{1}{w} = 2,$$

that is, $w^2 - 2w + 1 = 0$. This yields $w = 1$, which is the same as $e^{iz} = 1$. Writing $z = x + iy$, with $x, y \in \mathbb{R}$, we obtain

$$e^{iz} = e^{i(x+iy)} = e^{-y+ix}$$

$$= e^{-y} \cos x + i e^{-y} \sin x,$$

and it follows from $e^{iz} = 1 + i0$ that

$$e^{-y} \cos x = 1 \quad \text{and} \quad e^{-y} \sin x = 0.$$

Since $e^{-y} \neq 0$, we obtain $\sin x = 0$. Together with the identity $\cos^2 x + \sin^2 x = 1$, this yields $\cos x = \pm 1$. But since $e^{-y} > 0$, it follows from $e^{-y} \cos x = 1$ that $\cos x = 1$, and hence, $e^{-y} = 1$. Therefore, $x = 2k\pi$, with $k \in \mathbb{Z}$, and $y = 0$. The solution of the equation $\cos z = 1$ is thus $z = 2k\pi$, with $k \in \mathbb{Z}$.

The following result is an immediate consequence of Proposition 1.27.

Proposition 1.32

For every $z \in \mathbb{C}$ and $k \in \mathbb{Z}$, we have

$$\cos(z + 2k\pi) = \cos z \quad \text{and} \quad \sin(z + 2k\pi) = \sin z.$$

We also introduce the logarithm of a complex number.

Definition 1.33

We define the (*principal value of the*) *logarithm* of $z \in \mathbb{C} \setminus \{0\}$ by

$$\log z = \log |z| + i \arg z, \tag{1.13}$$

taking $\arg z \in (-\pi, \pi]$.

It follows from

$$e^{\log z} = e^{\log |z| + i \arg z}$$
$$= e^{\log |z|} e^{i \arg z}$$
$$= |z| e^{i \arg z} = z \tag{1.14}$$

that the (principal value of the) logarithm is a (right) inverse of the exponential.

Example 1.34

For each $z = -x + i0$ with $x > 0$, we have

$$\log(-x) = \log|-x| + i \arg(-x) = \log x + i\pi.$$

For $z = i$, we have

$$\log i = \log |i| + i\frac{\pi}{2} = \log 1 + i\frac{\pi}{2} = i\frac{\pi}{2}.$$

Example 1.35

For each $z = x + iy$ with $x > 0$, by (1.5) we have

$$\arg z = \tan^{-1} \frac{y}{x},$$

where $\tan^{-1}$ is the inverse of the tangent with values in the interval $(-\pi/2, \pi/2)$, so that $\arg z \in (-\pi, \pi)$. Therefore,

$$\log z = \log |z| + i \arg z$$
$$= \frac{1}{2} \log(x^2 + y^2) + i \tan^{-1} \frac{y}{x},$$

and the functions

$$u(x,y) = \frac{1}{2}\log(x^2 + y^2) \quad \text{and} \quad v(x,y) = \tan^{-1}\frac{y}{x}$$

are respectively the real and imaginary parts of $\log z$.

One can use the logarithm to define powers with a complex exponent.

Definition 1.36

Given $z \in \mathbb{C} \setminus \{0\}$ and $w \in \mathbb{C}$, we define

$$z^w = e^{w \log z}, \tag{1.15}$$

where $\log z$ is the principal value of the logarithm.

We note that $z^0 = e^{0 \log z} = 1$ for every $z \in \mathbb{C} \setminus \{0\}$.

Example 1.37

We have

$$2^i = e^{i \log 2} = \cos \log 2 + i \sin \log 2.$$

Since $\log i = i\pi/2$, we have

$$i^i = e^{i \log i} = e^{i(i\pi/2)} = e^{-\pi/2}.$$

Example 1.38

We have

$$(-1)^{2i} = e^{2i \log(-1)} = e^{2i(\log 1 + i\pi)} = e^{-2\pi}.$$

Incidentally, we note that

$$\left[(-1)^2\right]^i = 1^i = e^{i \log 1} = e^0 = 1,$$

while

$$\left[(-1)^i\right]^2 = \left(e^{i \log(-1)}\right)^2 = \left(e^{i(\log 1 + i\pi)}\right)^2 = \left(e^{-\pi}\right)^2 = e^{-2\pi}.$$

This shows that, in general, the numbers $(z^{w_1})^{w_2}$ and $z^{w_1 w_2}$ do not coincide.

1.5 Solved Problems and Exercises

Problem 1.1

Compute $(2 + 3i) + (5 - i)$ and $(2 + 4i)(3 - i)$.

Solution

We have

$$(2 + 3i) + (5 - i) = (2 + 5) + (3 - 1)i = 7 + 2i$$

and

$$(2 + 4i)(3 - i) = \big(2 \cdot 3 - 4 \cdot (-1)\big) + \big(2 \cdot (-1) + 4 \cdot 3\big)i = 10 + 10i.$$

Problem 1.2

Find the real and imaginary parts of $(2 + i)/(3 - i)$.

Solution

Multiplying the numerator and the denominator of $(2 + i)/(3 - i)$ by the conjugate of $3 - i$, we obtain

$$\frac{2 + i}{3 - i} = \frac{(2 + i)(3 + i)}{(3 - i)(3 + i)} = \frac{5 + 5i}{10} = \frac{1}{2} + \frac{1}{2}i.$$

Therefore,

$$\operatorname{Re}\frac{2 + i}{3 - i} = \frac{1}{2} \quad \text{and} \quad \operatorname{Im}\frac{2 + i}{3 - i} = \frac{1}{2}$$

(see Figure 1.7).

Problem 1.3

Find the modulus and the argument of $i^3/(2 + i)$.

Solution

Since

$$\frac{i^3}{2 + i} = \frac{-i(2 - i)}{(2 + i)(2 - i)} = \frac{-1 - 2i}{5},$$

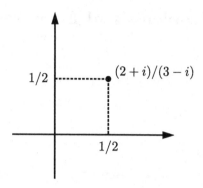

Figure 1.7 Real part and imaginary part of $(2+i)/(3-i)$

we have

$$\left|\frac{i^3}{2+i}\right| = \sqrt{\frac{1}{5^2} + \frac{2^2}{5^2}} = \sqrt{\frac{5}{5^2}} = \frac{1}{\sqrt{5}},$$

and, by (1.5),

$$\arg\frac{i^3}{2+i} = \pi + \tan^{-1}\frac{-2/5}{-1/5} = \pi + \tan^{-1}2,$$

where $\tan^{-1}$ is the inverse of the tangent with values in the interval $(-\pi/2, \pi/2)$.

Problem 1.4

Write the complex number $z = \sqrt{2} - \sqrt{2}i$ in polar form and compute z^5.

Solution

We have $|z| = \sqrt{2+2} = 2$ and

$$\arg z = \tan^{-1}\frac{-\sqrt{2}}{\sqrt{2}} = \tan^{-1}(-1) = -\frac{\pi}{4}.$$

Hence, $z = 2e^{-i\pi/4}$, and by Proposition 1.13, we obtain

$$z^5 = 2^5 e^{-i5\pi/4} = 32e^{-i5\pi/4}.$$

Problem 1.5

Find the cube roots of -4.

Solution

Let $z = -4$. Since $|z| = 4$ and $\arg z = \pi$, we have $z = 4e^{i\pi}$, and hence, by Proposition 1.14, the cube roots of -4 are

$$w_j = \sqrt[3]{4}e^{i(\pi + 2\pi j)/3} = \sqrt[3]{4}e^{i\pi(1+2j)/3}, \quad j = 0, 1, 2.$$

More precisely,

$$w_0 = \frac{1 + i\sqrt{3}}{\sqrt[3]{2}}, \quad w_1 = -\sqrt[3]{4} \quad \text{and} \quad w_2 = \frac{1 - i\sqrt{3}}{\sqrt[3]{2}}$$

(see Figure 1.8).

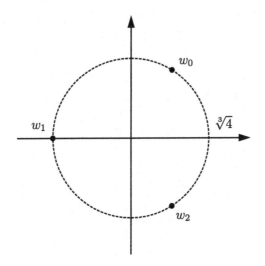

Figure 1.8   Cube roots of -4

Problem 1.6

Compute $\log(-3)$ and $\log(2 + 2i)$.

Solution

Let $z = -3$. We have $|z| = 3$ and $\arg z = \pi$. Therefore, by (1.13),

$$\log(-3) = \log 3 + i\pi.$$

Now let $z = 2 + 2i$. We have

$$|z| = \sqrt{2^2 + 2^2} = \sqrt{8} = 2\sqrt{2}$$

and

$$\arg z = \tan^{-1} \frac{2}{2} = \tan^{-1} 1 = \frac{\pi}{4}.$$

Therefore,

$$\log z = \log(2\sqrt{2}) + i\frac{\pi}{4} = \frac{3}{2}\log 2 + i\frac{\pi}{4}.$$

Problem 1.7

Compute $(2i)^{2i}$ and $(-1)^i$.

Solution

Since $2i = 2e^{i\pi/2}$, we have

$$\log(2i) = \log 2 + i\frac{\pi}{2},$$

and thus, by (1.15),

$$\begin{aligned}
(2i)^{2i} &= e^{2i\log(2i)} \\
&= e^{2i(\log 2 + i\pi/2)} \\
&= e^{i2\log 2}e^{-\pi} \\
&= e^{-\pi}\big[\cos(2\log 2) + i\sin(2\log 2)\big].
\end{aligned}$$

On the other hand, since $-1 = 1e^{i\pi}$, we have

$$\log(-1) = \log 1 + i\pi = i\pi,$$

and hence,

$$(-1)^i = e^{i\log(-1)} = e^{i(i\pi)} = e^{-\pi}.$$

Problem 1.8

Use the Cartesian form of z to verify that $z\bar{z} = |z|^2$.

Solution

We write $z = a + ib$, with $a, b \in \mathbb{R}$. Then $\bar{z} = a - ib$, and hence,

$$z\bar{z} = (a + ib)(a - ib)$$
$$= (a^2 + b^2) + i(a(-b) + ba)$$
$$= a^2 + b^2 = |z|^2.$$

Problem 1.9

Determine the set of points $z \in \mathbb{C}$ such that $2|z| \leq |z - 4|$.

Solution

Since $|z|$ and $|z - 4|$ are nonnegative, the condition $2|z| \leq |z - 4|$ is equivalent to $4|z|^2 \leq |z - 4|^2$. Writing $z = x + iy$, with $x, y \in \mathbb{R}$, we obtain

$$4|z|^2 = 4(x^2 + y^2)$$

and

$$|z - 4|^2 = (x - 4)^2 + y^2 = x^2 - 8x + 16 + y^2.$$

Therefore, the condition $4|z|^2 \leq |z - 4|^2$ is equivalent to

$$4(x^2 + y^2) \leq x^2 - 8x + 16 + y^2,$$

which yields

$$3x^2 + 8x + 3y^2 \leq 16. \tag{1.16}$$

Since

$$3x^2 + 8x = 3\left(x + \frac{4}{3}\right)^2 - \frac{16}{3},$$

condition (1.16) is equivalent to

$$\left(x + \frac{4}{3}\right)^2 + y^2 \leq \frac{64}{9}.$$

Therefore, the set of points $z \in \mathbb{C}$ such that $2|z| \leq |z - 4|$ is the closed disk of radius $8/3$ centered at $-4/3$ (see Figure 1.9).

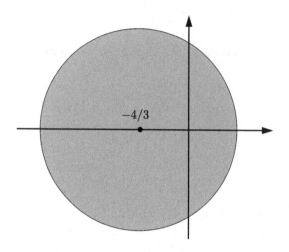

Figure 1.9 Closed disk of radius 8/3 centered at −4/3

Problem 1.10

Determine the set of points $z \in \mathbb{C}$ such that $|z| = z^2$.

Solution

For $z = |z|e^{i\alpha}$, the condition $|z| = z^2$ is equivalent to $|z| = |z|^2 e^{2i\alpha}$. Clearly, $z = 0$ is a solution. For $z \neq 0$ we obtain the equivalent condition $1 = |z|e^{2i\alpha}$, which by Proposition 1.9 yields $|z| = 1$ and $2\alpha = 2k\pi$ with $k \in \mathbb{Z}$, that is, $\alpha = k\pi$ with $k \in \mathbb{Z}$. Hence,

$$z = |z|e^{i\alpha} = 1e^{i0} = 1 \quad \text{or} \quad z = |z|e^{i\alpha} = 1e^{i\pi} = -1.$$

Therefore, the set of points $z \in \mathbb{C}$ such that $|z| = z^2$ is $\{-1, 0, 1\}$.

Problem 1.11

Verify that the function $z^2 - z$ is not one-to-one and find whether it is onto.

Solution

The equation

$$z^2 - z = z(z - 1) = 0$$

has the solutions $z = 0$ and $z = 1$, and hence, the function $z^2 - z$ is not one-to-one. On the other hand, since the equation $z^2 - z = w$ has the solution

$$z = \frac{1}{2}(1 + \sqrt{1 + 4w}),$$

where $\sqrt{1 + 4w}$ is any square root of $1 + 4w$, the function $z^2 - z$ is onto.

Problem 1.12

Solve the equation $\cos z = 2$.

Solution

Let $w = e^{iz}$. We have

$$0 = \cos z - 2$$

$$= \frac{1}{2}\left(e^{iz} + e^{-iz}\right) - 2$$

$$= \frac{1}{2}\left(w + \frac{1}{w}\right) - 2 = \frac{w^2 - 4w + 1}{2w}.$$

Since e^{iz} never vanishes, the equation $\cos z = 2$ is equivalent to $w^2 - 4w + 1 = 0$, which has the solutions

$$w_1 = \frac{4 + \sqrt{16 - 4}}{2} = 2 + \sqrt{3} \quad \text{and} \quad w_2 = \frac{4 - \sqrt{16 - 4}}{2} = 2 - \sqrt{3}.$$

We thus consider the equations

$$e^{iz} = 2 + \sqrt{3} \quad \text{and} \quad e^{iz} = 2 - \sqrt{3}. \tag{1.17}$$

Writing $z = x + iy$, with $x, y \in \mathbb{R}$, we obtain $e^{iz} = e^{-y}e^{ix}$. By Proposition 1.9, the condition

$$e^{-y}e^{ix} = 2 + \sqrt{3}$$

yields $e^{-y} = 2 + \sqrt{3}$ and $x = 2\pi k$ with $k \in \mathbb{Z}$. Similarly, the condition

$$e^{-y}e^{ix} = 2 - \sqrt{3}$$

yields $e^{-y} = 2 - \sqrt{3}$ and $x = 2\pi k$ with $k \in \mathbb{Z}$. Hence, the equations in (1.17) have respectively the solutions

$$z = 2\pi k - i\log(2 + \sqrt{3})$$

and

$$z = 2\pi k - i\log(2 - \sqrt{3}),$$

with $k \in \mathbb{Z}$.

Problem 1.13

Solve the equation $\cos z = \sin z$.

Solution

Let $w = e^{iz}$. We have

$$
\begin{aligned}
0 &= \cos z - \sin z \\
&= \frac{1}{2}\left(e^{iz} + e^{-iz}\right) - \frac{1}{2i}\left(e^{iz} - e^{-iz}\right) \\
&= \frac{1}{2}\left(w + \frac{1}{w}\right) - \frac{1}{2i}\left(w - \frac{1}{w}\right) \\
&= \frac{1}{2iw}\left(iw^2 + i - w^2 + 1\right) \\
&= \frac{1}{2iw}\left((i-1)w^2 + (i+1)\right).
\end{aligned}
$$

Thus, we obtain the equation

$$w^2 = -\frac{i+1}{i-1} = i,$$

which has the solutions

$$w_1 = e^{i\pi/4} \quad \text{and} \quad w_2 = e^{i5\pi/4}.$$

Therefore, we must solve the equations

$$e^{iz} = e^{i\pi/4} \quad \text{and} \quad e^{iz} = e^{i5\pi/4}. \tag{1.18}$$

Writing $z = x + iy$, with $x, y \in \mathbb{R}$, we obtain $e^{iz} = e^{-y}e^{ix}$, and the equations in (1.18) have respectively the solutions

$$z = \frac{\pi}{4} + 2\pi k \quad \text{and} \quad z = \frac{5\pi}{4} + 2\pi k, \quad \text{with } k \in \mathbb{Z}.$$

These can be written together in the form $z = \pi/4 + \pi k$, with $k \in \mathbb{Z}$.

Problem 1.14

Show that if $e^{i\theta} \neq -1$, then

$$\mathrm{Re}\log\left(1 + e^{i\theta}\right) = \log\left|2\cos\frac{\theta}{2}\right|.$$

Solution

Since

$$\log\left(1 + e^{i\theta}\right) = \log\left|1 + e^{i\theta}\right| + i\arg\left(1 + e^{i\theta}\right),$$

we have

$$
\begin{aligned}
\mathrm{Re}\log\left(1 + e^{i\theta}\right) &= \log\left|1 + e^{i\theta}\right| \\
&= \log\left|1 + \cos\theta + i\sin\theta\right| \\
&= \log\sqrt{(1 + \cos\theta)^2 + \sin^2\theta} \\
&= \log\sqrt{2 + 2\cos\theta}.
\end{aligned}
$$

It then follows from the identity

$$\left|\cos\frac{\theta}{2}\right| = \sqrt{\frac{1 + \cos\theta}{2}}$$

that

$$\mathrm{Re}\log\left(1 + e^{i\theta}\right) = \log\left(2\sqrt{\frac{1 + \cos\theta}{2}}\right) = \log\left|2\cos\frac{\theta}{2}\right|.$$

Problem 1.15

For $z = x + iy$, with $x, y \in \mathbb{R}$, show that

$$|\cos z|^2 + |\sin z|^2 = \cosh^2 y + \sinh^2 y.$$

Solution

We have

$$\cos z = \frac{e^{iz} + e^{-iz}}{2} = \frac{e^{ix}e^{-y} + e^{-ix}e^{y}}{2}$$

$$= \frac{1}{2}\left(e^{-y}\cos x + e^{y}\cos x\right) + \frac{i}{2}\left(e^{-y}\sin x - e^{y}\sin x\right)$$

$$= \cos x \cosh y - i \sin x \sinh y$$

and

$$\sin z = \frac{e^{iz} - e^{-iz}}{2i} = \frac{e^{ix}e^{-y} - e^{-ix}e^{y}}{2i}$$

$$= \frac{1}{2}\left(e^{-y}\sin x + e^{y}\sin x\right) - \frac{i}{2}\left(e^{-y}\cos x - e^{y}\cos x\right)$$

$$= \sin x \cosh y + i \cos x \sinh y.$$

Therefore,

$$|\cos z|^2 + |\sin z|^2 = \cos^2 x \cosh^2 y + \sin^2 x \sinh^2 y$$

$$+ \sin^2 x \cosh^2 y + \cos^2 x \sinh^2 y$$

$$= \cosh^2 y + \sinh^2 y.$$

Problem 1.16

Show that

$$|z - w|^2 \le \left(1 + |z|^2\right)\left(1 + |w|^2\right) \quad \text{for } z, w \in \mathbb{C}.$$

Solution

By Proposition 1.20, we have

$$|z - w| \le |z| + |-w| = |z| + |w|,$$

and hence,

$$|z - w|^2 \le \left(|z| + |w|\right)^2 = |z|^2 + 2|z| \cdot |w| + |w|^2.$$

On the other hand,

$$0 \le \left(1 - |z| \cdot |w|\right)^2 = 1 - 2|z| \cdot |w| + |z|^2|w|^2, \tag{1.19}$$

and thus,

$$2|z| \cdot |w| \leq 1 + |z|^2 |w|^2.$$

It then follows from (1.19) that

$$|z - w|^2 \leq |z|^2 + 1 + |z|^2 |w|^2 + |w|^2$$
$$= |z|^2 + 1 + (|z|^2 + 1)|w|^2$$
$$= (1 + |z|^2)(1 + |w|^2).$$

Problem 1.17

Verify that the identity $\log(zw) = \log z + \log w$ is not always satisfied.

Solution

Let $z = r_1 e^{i\theta_1}$ and $w = r_2 e^{i\theta_2}$, with $r_1, r_2 > 0$ and $\theta_1, \theta_2 \in (-\pi, \pi]$. We have

$$\log z = \log r_1 + i\theta_1 \quad \text{and} \quad \log w = \log r_2 + i\theta_2,$$

and thus,

$$\log z + \log w = \log r_1 + \log r_2 + i(\theta_1 + \theta_2)$$
$$= \log(r_1 r_2) + i(\theta_1 + \theta_2).$$

On the other hand, since $zw = r_1 r_2 e^{i(\theta_1 + \theta_2)}$ (see (1.7)), we also have

$$\log(zw) = \log(r_1 r_2) + i(\theta_1 + \theta_2 - 2k\pi),$$

where $k \in \mathbb{Z}$ is the unique integer such that

$$\theta_1 + \theta_2 - 2k\pi \in (-\pi, \pi].$$

In particular, when $\theta_1 + \theta_2$ does not belong to the interval $(-\pi, \pi]$ we have

$$\log(zw) \neq \log z + \log w.$$

For example, if $z = w = -1$, then

$$\log(zw) = \log 1 = 0 \quad \text{and} \quad \log z + \log w = 2\log(-1) = 2i\pi.$$

Problem 1.18

Find all complex numbers $z \in \mathbb{C}$ such that $\log \log z$ is purely imaginary.

Solution

For $z \neq 0$ we have

$$\log z = \log |z| + i \arg z,$$

with $\arg z \in (-\pi, \pi]$. Hence, for $\log z \neq 0$ we obtain

$$\log \log z = \log|\log z| + i \arg \log z$$

$$= \log \sqrt{(\log |z|)^2 + (\arg z)^2} + i \arg \log z,$$

with $\arg \log z \in (-\pi, \pi]$. This implies that $\log \log z$ is purely imaginary if and only if

$$\operatorname{Re} \log \log z = \frac{1}{2} \log \left[(\log |z|)^2 + (\arg z)^2\right] = 0,$$

which is equivalent to

$$(\log |z|)^2 + (\arg z)^2 = 1,$$

with $z \neq 0$. Taking $\alpha \in \mathbb{R}$ such that

$$\log |z| = \cos \alpha \quad \text{and} \quad \arg z = \sin \alpha,$$

we obtain $|z| = e^{\cos \alpha}$, and thus,

$$z = |z| e^{i \arg z} = e^{\cos \alpha} e^{i \sin \alpha} = e^{e^{i\alpha}}.$$

We note that $z \neq 0$ and that

$$|\log z| = \sqrt{(\log |z|)^2 + (\arg z)^2} = 1 \neq 0.$$

EXERCISES

1.1. Find:
   (a) the real part of the imaginary part of z;
   (b) the imaginary part of the real part of z.
1.2. Compute the modulus and the argument of:
   (a) $i^3 + 1$;
   (b) $(5 + i6)(1 - i7)$;
   (c) $(1 + i\sqrt{3})/(1 - i\sqrt{3})$;
   (d) $(5 + i4)/(2 - i2)^2$.

1.3. Find the Cartesian and polar forms of:
  (a) $(1-i)^3$;
  (b) $(5+i4)/(2-i)$;
  (c) $i^5 + i^{20}$;
  (d) $\cos i$;
  (e) $\sinh(2i)$.

1.4. Compute the conjugate of $5(2+i3)^3/(2+i)$.

1.5. Find the square roots of:
  (a) i;
  (b) $1-i$;
  (c) $2+i2$.

1.6. Verify that:
  (a) the cube roots of i are $(\sqrt{3}+i)/2$, $(-\sqrt{3}+i)/2$ and $-i$;
  (b) the 4th roots of i are $e^{i\pi/8}$, $e^{i5\pi/8}$, $e^{i9\pi/8}$ and $e^{i13\pi/8}$;
  (c) a 5th root of 1 is

$$\frac{\sqrt{5}-1}{4} + i\sqrt{\frac{5+\sqrt{5}}{8}}.$$

1.7. Find all complex numbers $z \in \mathbb{C}$ such that $(z^2)^2 = 1$.

1.8. Verify that $1 + e^{i2x} = 2e^{ix}\cos x$ for every $x \in \mathbb{R}$.

1.9. Compute $\log\log i$.

1.10. Find whether $\log\log z$ can be computed for every $z \neq 0$.

1.11. Find all solutions of the equation:
  (a) $(z+1)^2 = (\bar{z}-1)^2$;
  (b) $2z^2 + iz + 4 = 0$;
  (c) $z^4 + z^3 + z^2 + z = 0$.

1.12. Solve the equation:
  (a) $e^z = 3$;
  (b) $\cosh z = i$;
  (c) $e^{e^z} = 1$.

1.13. Solve the equation:
  (a) $\cos z \sin z = 0$;
  (b) $\sin z + \cos z = 1$;
  (c) $\sin z = \sin(2z)$.

1.14. Determine the set of points $(x,y) \in \mathbb{R}^2$ such that:
  (a) $x + iy = |x + iy|$;
  (b) $2|x + iy| \leq |x + iy - 1|$.

1.15. Determine the set of points $z \in \mathbb{C}$ such that:
  (a) $z + \bar{z} = |z - \bar{z}|$;
  (b) $z^2 - z = 1$;

(c) $z - |z| = \bar{z}$;

(d) $3|z| \le |\bar{z} - 2|$.

1.16. Determine the set:

(a) $\{z \in \mathbb{C} : |z|^2 = z^2\}$;

(b) $\{z \in \mathbb{C} : z^{-1} = 4\bar{z}\}$.

1.17. Determine the set $\{zw : z, w \in \mathbb{C}\}$.

1.18. Compute the real and imaginary parts of the function:

(a) $f(z) = (z + 1)^2$;

(b) $f(z) = z^2(z - 3)$;

(c) $f(z) = z/(z - 1)$;

(d) $f(z) = \bar{z}/(z\bar{z} - 1)$.

1.19. Find whether the function is one-to-one:

(a) $2z$;

(b) e^z;

(c) $z^2 + z$;

(d) $\cos z$;

(e) $z^2 - \cos z$.

1.20. Find whether the function is onto:

(a) e^z;

(b) z^3.

1.21. Find whether the function is bijective (one-to-one and onto):

(a) $z^4 - z$;

(b) e^z.

1.22. Identify each statement as true or false.

(a) $\cos^2 z + \sin^2 z = 1$ for every $z \in \mathbb{C}$.

(b) $\operatorname{Re}(iz) = \operatorname{Im}(z)$ for every $z \in \mathbb{C} \cap \mathbb{R}$.

(c) The function e^z is bounded.

(d) The function $\cos z - \sin z$ is bounded.

(e) $\cos \cos z$ is always real.

1.23. Show that:

(a) $\operatorname{Re}(iz) = -\operatorname{Im} z$ and $\operatorname{Im}(iz) = \operatorname{Re} z$;

(b) $\cos(iz) = \cosh z$;

(c) $\sin(iz) = i \sinh z$;

(d) $\overline{\sin z} = \sin \bar{z}$;

(e) $\overline{\cos z} = \cos \bar{z}$.

1.24. Show that:

(a) $\cos(z + w) = \cos z \cos w - \sin z \sin w$;

(b) $\sin(z + w) = \sin z \cos w + \cos z \sin w$;

(c) $\cos(2z) = \cos^2 z - \sin^2 z$;

(d) $\sin(2z) = 2 \sin z \cos z$;

(e) $\sin(x + iy) = \cosh y \sin x + i \cos x \sinh y$.

1.25. Verify that:

(a) $\mathrm{Re}[(1+z)/(1-z)] = (1-|z|^2)/|z-1|^2$;

(b) $\sin[-i\log(iz + \sqrt{1-z^2})] = z$.

1.26. Show that $|z^i| < e^\pi$ for every $z \in \mathbb{C} \setminus \{0\}$.

1.27. Show that

$$|z| \le |\mathrm{Re}\, z| + |\mathrm{Im}\, z| \le \sqrt{2}|z|.$$

1.28. Verify that

$$|z-w|^2 + |z+w|^2 = 2(|z|^2 + |w|^2) \quad \text{for } z, w \in \mathbb{C}.$$

1.29. Show that

$$\big||z| - |w|\big| \le |z+w| \quad \text{for every } z, w \in \mathbb{C}.$$

1.30. Find whether the inequality $|z-w| \le |z+w|$ is always satisfied.

1.31. Show that if $z^n = 1$ and $z \ne 1$, then $1 + z + \cdots + z^{n-1} = 0$.

1.32. Given $x \in \mathbb{R} \setminus \{2k\pi : k \in \mathbb{Z}\}$ and $n \in \mathbb{N}$, show that

$$\sin(2x) + \sin(4x) + \cdots + \sin(2nx) = \frac{\sin(nx)\sin[(n+1)x]}{\sin x}$$

and

$$\cos(2x) + \cos(4x) + \cdots + \cos(2nx) = \frac{\sin(nx)\cos[(n+1)x]}{\sin x}.$$

Hint: compute $e^{i2x} + e^{i4x} + \cdots + e^{i2nx}$.

1.33. Find whether the function is bounded or unbounded:

(a) $f(z) = |z+i|^2$;

(b) $f(z) = |z+i|^2 - |z-i|^2$.

1.34. Compute the limit of:

(a) $|1 + i/n|^n$ when $n \to \infty$;

(b) $|\cos(2 + 3i/n)|^n$ when $n \to \infty$.

Holomorphic Functions

In this chapter we introduce the notion of a differentiable function, or of a holo-morphic function. It turns out that differentiability is characterized by a pair of (partial differential) equations—the Cauchy–Riemann equations. We also introduce the notion of the integral along a path and we study its relation to the notion of a holomorphic function. Finally, we introduce the index of a closed path, we obtain Cauchy's integral formula for a holomorphic function, and we discuss the relation between integrals and homotopy.

2.1 Limits and Continuity

Let $f \colon \Omega \to \mathbb{C}$ be a complex-valued function in a set $\Omega \subset \mathbb{C}$. We first introduce the notion of limit.

Definition 2.1

We say that the *limit* of f at a point $z_0 \in \Omega$ exists, and that it is given by $w \in \mathbb{C}$ if for each $\varepsilon > 0$ there exists $\delta > 0$ such that

$$\bigl|f(z) - w\bigr| < \varepsilon \quad \text{whenever} \quad |z - z_0| < \delta.$$

L. Barreira, C. Valls, *Complex Analysis and Differential Equations*,   37
Springer Undergraduate Mathematics Series,
DOI 10.1007/978-1-4471-4008-5_2, © Springer-Verlag London 2012

In this case we write

$$\lim_{z \to z_0} f(z) = w.$$

Now we introduce the notion of continuity.

Definition 2.2

We say that f is *continuous* at a point $z_0 \in \Omega$ if

$$\lim_{z \to z_0} f(z) = f(z_0).$$

Otherwise, the function f is said to be *discontinuous* at z_0. We also say that f is *continuous* in Ω if it is continuous at all points of Ω.

Example 2.3

For the function $f(z) = |z|$, we have

$$\left| f(z) - f(z_0) \right| = \left| |z| - |z_0| \right| \leq |z - z_0|.$$

This implies that $|f(z) - f(z_0)| < \delta$ whenever $|z - z_0| < \delta$, and hence, the function f is continuous in $\mathbb{C}$.

Example 2.4

For the function $f(z) = z^2$, we have

$$\begin{aligned}
\left| f(z) - f(z_0) \right| &= \left| (z - z_0)(z + z_0) \right| \\
&= |z - z_0| \cdot |z - z_0 + 2z_0| \\
&\leq |z - z_0| \left(|z - z_0| + 2|z_0| \right) \\
&< \delta(\delta + 2|z_0|)
\end{aligned}$$

whenever $|z - z_0| < \delta$. Since $\delta(\delta + 2|z_0|) \to 0$ when $\delta \to 0$, the function f is continuous in $\mathbb{C}$.

Example 2.5

Now we show that the function $f(z) = \log z$ is discontinuous at all points $z = -x + i0$ with $x > 0$. For $w \in \mathbb{C}$ in the second quadrant, we have

$$\log w = \log |w| + i \arg w$$

with $\arg w \in [\pi/2, \pi]$. On the other hand, for $w \in \mathbb{C}$ in the third quadrant and outside the half-line $\mathbb{R}^-$, the same formula holds, but now with $\arg w \in (-\pi, -\pi/2]$. Letting $w \to z$ in the second and third quadrants, we obtain respectively

$$\log w \to \log x + i\pi$$

and

$$\log w \to \log x - i\pi.$$

Since the right-hand sides are different, the logarithm has no limit at points of $\mathbb{R}^-$. Therefore, the function f is discontinuous at all points of $\mathbb{R}^-$. On the other hand, one can show that it is continuous in $\mathbb{C} \setminus \mathbb{R}_0^-$ (see Exercise 2.25).

2.2 Differentiability

Now we consider a function $f \colon \Omega \to \mathbb{C}$ in an open set $\Omega \subset \mathbb{C}$, that is, in an open set $\Omega \subset \mathbb{R}^2$.

Definition 2.6

We say that f is *differentiable* at a point $z_0 \in \Omega$ if the limit

$$f'(z_0) = \lim_{z \to z_0} \frac{f(z) - f(z_0)}{z - z_0}$$

exists. In this case, the number $f'(z_0)$ is called the *derivative* of f at z_0.

We also introduce the notion of a holomorphic function.

Definition 2.7

When f is differentiable at all points of Ω we say that f is *holomorphic* in Ω.

Example 2.8

We show that the function $f(z) = z^2$ is holomorphic in $\mathbb{C}$. Indeed,

$$\lim_{z \to z_0} \frac{z^2 - z_0^2}{z - z_0} = \lim_{z \to z_0} \frac{(z - z_0)(z + z_0)}{z - z_0}$$
$$= \lim_{z \to z_0} (z + z_0) = 2z_0,$$

and thus $(z^2)' = 2z$. One can show by induction that

$$\left(z^n\right)' = nz^{n-1}$$

for every $n \in \mathbb{N}$ (with the convention that $0^0 = 1$).

Example 2.9

Now we consider the function $f(z) = \overline{z}$. Given $h = re^{i\theta}$, we have

$$\frac{f(z+h) - f(z)}{h} = \frac{\overline{z} + \overline{h} - \overline{z}}{h}$$

$$= \frac{\overline{h}}{h} = e^{-2i\theta}. \tag{2.1}$$

Since $e^{-2i\theta}$ varies with θ, one cannot take the limit in (2.1) when $r \to 0$. Hence, the function f is differentiable at no point.

Example 2.10

For the function $f(z) = |z|^2$, given $h = re^{i\theta}$ we have

$$\frac{f(z+h) - f(z)}{h} = \frac{(z+h)(\overline{z} + \overline{h}) - z\overline{z}}{h}$$

$$= \frac{z\overline{h} + \overline{z}h + h\overline{h}}{h}$$

$$= \frac{z\overline{h}}{h} + \overline{z} + \overline{h}$$

$$= \frac{zre^{-i\theta}}{re^{i\theta}} + \overline{z} + re^{-i\theta}$$

$$= ze^{-2i\theta} + \overline{z} + re^{-i\theta} \to ze^{-2i\theta} + \overline{z} \tag{2.2}$$

when $r \to 0$. For $z \neq 0$, since the limit in (2.2) varies with θ, the function f is not differentiable at z. On the other hand,

$$\frac{f(z) - f(0)}{z - 0} = \frac{|z|^2}{z} = \frac{z\overline{z}}{z} = \overline{z} \to 0$$

when $z \to 0$. Therefore, f is only differentiable at the origin, and $f'(0) = 0$.

The following properties are obtained as in $\mathbb{R}$, and thus their proofs are omitted.

Proposition 2.11

Given holomorphic functions $f, g \colon \Omega \to \mathbb{C}$, we have:
1. $(f + g)' = f' + g'$;
2. $(fg)' = f'g + fg'$;
3. $(f/g)' = (f'g - fg')/g^2$ at all points where $g \neq 0$.

Proposition 2.12

Given holomorphic functions $f \colon \Omega \to \mathbb{C}$ and $g \colon \Omega' \to \mathbb{C}$, with $g(\Omega') \subset \Omega$, we have

$$(f \circ g)' = (f' \circ g)g'.$$

Now we show that any differentiable function is continuous.

Proposition 2.13

If f is differentiable at z_0, then f is continuous at z_0.

Proof

For $z \neq z_0$, we have

$$f(z) - f(z_0) = \frac{f(z) - f(z_0)}{z - z_0}(z - z_0),$$

and thus,

$$\lim_{z \to z_0} f(z) = \lim_{z \to z_0} \left[f(z) - f(z_0) \right] + f(z_0)$$

$$= \lim_{z \to z_0} \frac{f(z) - f(z_0)}{z - z_0} \lim_{z \to z_0} (z - z_0) + f(z_0)$$

$$= f'(z_0) \cdot 0 + f(z_0) = f(z_0).$$

This yields the desired property. $\square$

We also describe a necessary condition for the differentiability of a function $f \colon \Omega \to \mathbb{C}$ at a given point. We always write

$$f(x + iy) = u(x, y) + iv(x, y),$$

where u and v are real-valued functions.

Theorem 2.14 (Cauchy–Riemann equations)

If f is differentiable at $z_0 = x_0 + iy_0$, then

$$\frac{\partial u}{\partial x} = \frac{\partial v}{\partial y} \quad \text{and} \quad \frac{\partial u}{\partial y} = -\frac{\partial v}{\partial x} \tag{2.3}$$

at (x_0, y_0). Moreover, the derivative of f at z_0 is given by

$$f'(z_0) = \frac{\partial u}{\partial x}(x_0, y_0) + i\frac{\partial v}{\partial x}(x_0, y_0). \tag{2.4}$$

Proof

Writing $f'(z_0) = a + ib$, we obtain

$$f'(z_0)(z - z_0) = (a + ib)\big[(x - x_0) + i(y - y_0)\big]$$
$$= \big[a(x - x_0) - b(y - y_0)\big] + i\big[b(x - x_0) + ia(y - y_0)\big]$$
$$= C(x - x_0, y - y_0),$$

where

$$C = \begin{pmatrix} a & -b \\ b & a \end{pmatrix},$$

and hence,

$$f(z) - f(z_0) - f'(z_0)(z - z_0) = \big(u(x,y), v(x,y)\big) - \big(u(x_0, y_0), v(x_0, y_0)\big)$$
$$- C(x - x_0, y - y_0).$$

For $z \neq z_0$, we have

$$\frac{f(z) - f(z_0) - f'(z_0)(z - z_0)}{|z - z_0|} = \frac{f(z) - f(z_0) - f'(z_0)(z - z_0)}{z - z_0} \cdot \frac{z - z_0}{|z - z_0|}$$
$$= \left(\frac{f(z) - f(z_0)}{z - z_0} - f'(z_0)\right)\frac{z - z_0}{|z - z_0|},$$

and since

$$\left|\frac{z - z_0}{|z - z_0|}\right| = \frac{|z - z_0|}{|z - z_0|} = 1,$$

we obtain

$$\frac{f(z) - f(z_0) - f'(z_0)(z - z_0)}{|z - z_0|} \to 0$$

when $z \to z_0$. Since

$$|z - z_0| = \|(x - x_0, y - y_0)\|,$$

this is the same as

$$\frac{(u(x,y), v(x,y)) - (u(x_0, x_0), v(x_0, y_0)) - C(x - x_0, y - y_0)}{\|(x - x_0, y - y_0)\|} \to 0$$

when $(x, y) \to (x_0, y_0)$. It thus follows from the notion of differentiability in $\mathbb{R}^2$ that the function $F \colon \Omega \to \mathbb{R}^2$ given by

$$F(x, y) = \big(u(x,y), v(x,y)\big) \tag{2.5}$$

is differentiable at (x_0, y_0), with derivative

$$DF(x_0, y_0) = \begin{pmatrix} \frac{\partial u}{\partial x}(x_0, y_0) & \frac{\partial u}{\partial y}(x_0, y_0) \\ \frac{\partial v}{\partial x}(x_0, y_0) & \frac{\partial v}{\partial y}(x_0, y_0) \end{pmatrix}$$

$$= C = \begin{pmatrix} a & -b \\ b & a \end{pmatrix}.$$

This shows that the identities in (2.3) are satisfied. $\square$

The equations in (2.3) are called the *Cauchy–Riemann equations*.

Example 2.15

Let

$$f(x + iy) = u(x, y) + iv(x, y)$$

be a holomorphic function in $\mathbb{C}$ with $u(x, y) = x^2 - xy - y^2$. By Theorem 2.14, the Cauchy–Riemann equations are satisfied. Since

$$\frac{\partial u}{\partial x} = 2x - y,$$

it follows from the first equation in (2.3) that

$$\frac{\partial v}{\partial y} = 2x - y.$$

Therefore,

$$v(x, y) = 2xy - \frac{y^2}{2} + C(x)$$

for some function C. Taking derivatives, we obtain

$$\frac{\partial u}{\partial y} = -x - 2y \quad \text{and} \quad -\frac{\partial v}{\partial x} = -2y - C'(x).$$

Hence,

$$-x - 2y = -2y - C'(x),$$

and $C'(x) = x$. We conclude that $C(x) = x^2/2 + c$ for some constant $c \in \mathbb{R}$, and hence,

$$v(x, y) = \frac{x^2}{2} + 2xy - \frac{y^2}{2} + c.$$

We thus have

$$f(x + iy) = (x^2 - xy - y^2) + i\left(\frac{x^2}{2} + 2xy - \frac{y^2}{2} + c\right).$$

Rearranging the terms, we obtain

$$f(x + iy) = \left[(x^2 - y^2) + i2xy\right] + \left[-xy + i\left(\frac{x^2}{2} - \frac{y^2}{2}\right)\right] + ic$$

$$= z^2 + \frac{i}{2}\left[(x^2 - y^2) + i2xy\right] + ic$$

$$= z^2 + \frac{i}{2}z^2 + ic = \left(1 + \frac{i}{2}\right)z^2 + ic.$$

In particular, $f'(z) = (2 + i)z$.

Example 2.16

We show that a holomorphic function $f = u + iv$ cannot have $u(x, y) = x^2 + y^2$ as its real part. Otherwise, by the first Cauchy–Riemann equation, we would have

$$\frac{\partial u}{\partial x} = 2x = \frac{\partial v}{\partial y},$$

and thus, $v(x, y) = 2xy + C(x)$ for some function C. But then

$$\frac{\partial u}{\partial x} = 2y \quad \text{and} \quad \frac{\partial v}{\partial x} = 2y + C'(x),$$

and by the second Cauchy–Riemann equation we would also have

$$2y = -(2y + C'(x)).$$

Therefore, $C'(x) = -4y$, but this identity cannot hold for every $x, y \in \mathbb{R}$. For example, taking derivatives with respect to y we would obtain $0 = -4$, which is impossible.

As an illustration of the former concepts, in the remainder of this section we shall describe conditions for a holomorphic function to be constant.

Given a set $A \subset \mathbb{C}$, we denote by $\overline{A}$ the *closure* of A. This is the smallest closed subset of $\mathbb{C} = \mathbb{R}^2$ containing A. It is also the set of points $a \in \mathbb{C}$ such that

$$\{z \in \mathbb{C} : |z - a| < r\} \cap A \neq \emptyset$$

for every $r > 0$. In spite of the notation, the notion of closure should not be confused with the notion of the conjugate of a complex number. Now we recall the notion of a connected set.

Definition 2.17

A set $\Omega \subset \mathbb{C}$ is said to be *disconnected* if there exist nonempty sets $A, B \subset \mathbb{C}$ such that

$$\Omega = A \cup B \quad \text{and} \quad \overline{A} \cap B = A \cap \overline{B} = \emptyset.$$

A set $\Omega \subset \mathbb{C}$ is said to be *connected* if it is not disconnected.

Finally, we introduce the notion of a connected component.

Definition 2.18

Given $\Omega \subset \mathbb{C}$, we say that a connected set $A \subset \Omega$ is a *connected component* of Ω if any connected set $B \subset \Omega$ containing A is equal to A.

We note that if a set $\Omega \subset \mathbb{C}$ is connected, then it is its own unique connected component.

Now we show that in any connected open set, a holomorphic function with zero derivative is constant.

Proposition 2.19

If f is a holomorphic function in a connected open set Ω and $f' = 0$ in Ω, then f is constant in Ω.

Proof

By (2.4), we have

$$f'(x+iy) = \frac{\partial u}{\partial x} + i\frac{\partial v}{\partial x} = 0.$$

Together with the Cauchy–Riemann equations, this yields

$$\frac{\partial u}{\partial x} = \frac{\partial u}{\partial y} = \frac{\partial v}{\partial x} = \frac{\partial v}{\partial y} = 0.$$

Now let us consider points $x + iy$ and $x + iy'$ in Ω such that the line segment between them is contained in Ω. By the Mean value theorem, we obtain

$$u(x,y) - u(x,y') = \frac{\partial u}{\partial y}(x,z)(y-y') = 0,$$

where z is some point between y and y'. Analogously,

$$v(x,y) - v(x,y') = \frac{\partial v}{\partial y}(x,w)(y-y') = 0,$$

where w is some point between y and y'. This shows that

$$f(x+iy) = f(x+iy'). \tag{2.6}$$

One can show in a similar manner that if $x + iy'$ and $x' + iy'$ are points in Ω such that the line segment between them is contained in Ω, then

$$f(x+iy') = f(x'+iy'). \tag{2.7}$$

Now we consider an open rectangle $R \subset \Omega$ with horizontal and vertical sides. Given $x + iy, x' + iy' \in R$, the point $x + iy'$ is also in R, as well as the vertical segment between $x + iy$ and $x + iy'$, and the horizontal segment between $x + iy'$ and $x' + iy'$ (each of these segments can be a single point). It follows from (2.6) and (2.7) that

$$f(x+iy) = f(x+iy') = f(x'+iy').$$

This shows that f is constant in R. Finally, we consider sequences $\mathcal{R}_R = (R_n)_{n\in\mathbb{N}}$ of open rectangles in Ω, with horizontal and vertical sides, such that $R_1 = R$ and $R_n \cap R_{n+1} \neq \emptyset$ for each $n \in \mathbb{N}$. We also consider the set

$$U_R = \bigcup_{\mathcal{R}_R} \bigcup_{n=1}^{\infty} R_n.$$

Clearly, U_R is open (since it is a union of open sets), and f is constant in U_R, since it is constant in each union $\bigcup_{n=1}^{\infty} R_n$.

We show that $U_R = \Omega$. On the contrary, let us assume that $\Omega \setminus U_R \neq \emptyset$. We note that

$$(\overline{U_R} \cap \Omega) \setminus U_R \neq \emptyset,$$

since otherwise $\overline{U_R} \cap \Omega = U_R$, and hence,

$$\Omega = U_R \cup (\Omega \setminus U_R),$$

with

$$\overline{U_R} \cap (\Omega \setminus U_R) = (\overline{U_R} \cap \Omega) \setminus U_R = \emptyset$$

and

$$U_R \cap \overline{\Omega \setminus U_R} = U_R \cap \overline{(\Omega \setminus U_R)} = \emptyset$$

(since U_R is open); that is, Ω would be disconnected. Let us then take $z \in (\overline{U_R} \cap \Omega) \setminus U_R$ and a rectangle $S \subset \Omega$ with horizontal and vertical sides such that $z \in S$. Then $S \cap U_R \neq \emptyset$ and thus, S is an element of some sequence $\mathcal{R}_R$. This implies that $S \subset U_R$ and hence $z \in U_R$, which yields a contradiction. Therefore, $U_R = \Omega$ and f is constant in Ω. $\qquad \square$

We also describe some applications of Proposition 2.19.

Example 2.20

We show that for a holomorphic function $f = u + iv$ in a connected open set, if u is constant or v is constant, then f is also constant. Indeed, if u is constant, then

$$f'(x + iy) = \frac{\partial u}{\partial x} + i\frac{\partial v}{\partial x}$$
$$= \frac{\partial u}{\partial x} - i\frac{\partial u}{\partial y} = 0,$$

and it follows from Proposition 2.19 that f is constant. Similarly, if v is constant, then

$$f'(x + iy) = \frac{\partial u}{\partial x} + i\frac{\partial v}{\partial x}$$
$$= \frac{\partial v}{\partial y} + i\frac{\partial v}{\partial x} = 0,$$

and again it follows from Proposition 2.19 that f is constant.

Example 2.21

Now we show that for a holomorphic function $f = u + iv$ in a connected open set, if $|f|$ is constant, then f is constant. We first note that by hypothesis $|f|^2 = u^2 + v^2$ is also constant. If the constant is zero, then $u = v = 0$ and hence, $f = u + iv = 0$. Now we assume that $|f|^2 = c$ for some constant $c \neq 0$. Then $u^2 + v^2 = c$, and taking derivatives with respect to x and y, we obtain

$$2u\frac{\partial u}{\partial x} + 2v\frac{\partial v}{\partial x} = 0$$

and

$$2u\frac{\partial u}{\partial y} + 2v\frac{\partial v}{\partial y} = 0.$$

Using the Cauchy–Riemann equations, one can rewrite these two identities in the matrix form

$$\begin{pmatrix} u & v \\ v & -u \end{pmatrix} \begin{pmatrix} \frac{\partial u}{\partial x} \\ \frac{\partial v}{\partial x} \end{pmatrix} = 0. \tag{2.8}$$

Since the determinant of the 2×2 matrix in (2.8) is $-(u^2 + v^2) = -c \neq 0$, the unique solution is

$$\frac{\partial u}{\partial x} = \frac{\partial v}{\partial x} = 0,$$

and thus,

$$f'(x + iy) = \frac{\partial u}{\partial x} + i\frac{\partial v}{\partial x} = 0.$$

It follows again from Proposition 2.19 that f is constant.

2.3 Differentiability Condition

The following example shows that for a function f to be differentiable at a given point it is not sufficient that the Cauchy–Riemann equations are satisfied at that point.

Example 2.22

We show that the function $f(x + iy) = \sqrt{|xy|}$ is not differentiable at the origin. Given

$$h = re^{i\theta} = r\cos\theta + ir\sin\theta,$$

we have

$$\frac{f(h) - f(0)}{h - 0} = \frac{\sqrt{|(r\cos\theta)(r\sin\theta)|}}{re^{i\theta}}$$

$$= \frac{r\sqrt{|\cos\theta\sin\theta|}}{re^{i\theta}}$$

$$= \sqrt{|\cos\theta\sin\theta|}\,e^{-i\theta}.$$

Since the last expression depends on θ, one cannot take the limit when $r \to 0$. Therefore, f is not differentiable at the origin. On the other hand, we have

$$\frac{\partial u}{\partial x}(0,0) = \lim_{x \to 0} \frac{u(x,0) - u(0,0)}{x - 0} = 0$$

and

$$\frac{\partial u}{\partial y}(0,0) = \lim_{y \to 0} \frac{u(0,y) - u(0,0)}{y - 0} = 0,$$

as well as

$$\frac{\partial v}{\partial x}(0,0) = \frac{\partial v}{\partial y}(0,0) = 0,$$

since $v = 0$. Hence, the Cauchy–Riemann equations are satisfied at the origin.

Now we give a necessary and sufficient condition for the differentiability of a function f in some open set.

Theorem 2.23

Let $u, v \colon \Omega \to \mathbb{C}$ be C^1 functions in an open set $\Omega \subset \mathbb{C}$. Then the function $f = u + iv$ is holomorphic in Ω if and only if the Cauchy–Riemann equations are satisfied at all points of Ω.

Proof

By Theorem 2.14, if f is holomorphic in Ω, then the Cauchy–Riemann equations are satisfied at all points of Ω.

Now we assume that the Cauchy–Riemann equations are satisfied in Ω. This implies that

$$\begin{pmatrix} \frac{\partial u}{\partial x} & \frac{\partial u}{\partial y} \\ \frac{\partial v}{\partial x} & \frac{\partial v}{\partial y} \end{pmatrix} = \begin{pmatrix} a & -b \\ b & a \end{pmatrix}$$

at every point of Ω, for some constants a and b possibly depending on the point. On the other hand, since u and v are of class C^1, the function $F = (u, v)$ in (2.5) is differentiable in Ω. It follows from the proof of Theorem 2.14 that f is differentiable at z_0, with

$$f'(z_0) = a + ib,$$

if and only if F is differentiable at (x_0, y_0), with

$$DF(x_0, y_0) = \begin{pmatrix} \frac{\partial u}{\partial x}(x_0, y_0) & \frac{\partial u}{\partial y}(x_0, y_0) \\ \frac{\partial v}{\partial x}(x_0, y_0) & \frac{\partial v}{\partial y}(x_0, y_0) \end{pmatrix} = \begin{pmatrix} a & -b \\ b & a \end{pmatrix}.$$

This shows that the function f is differentiable at all points of Ω.   □

Example 2.24

Let us consider the function $f(z) = e^z$. We have

$$u(x, y) = e^x \cos y \quad \text{and} \quad v(x, y) = e^x \sin y,$$

and both functions are of class C^1 in the open set $\mathbb{R}^2 = \mathbb{C}$. Since

$$\frac{\partial u}{\partial x} = e^x \cos y, \qquad \frac{\partial v}{\partial y} = e^x \cos y,$$

and

$$\frac{\partial u}{\partial y} = -e^x \sin y, \qquad -\frac{\partial v}{\partial x} = -e^x \sin y,$$

the Cauchy–Riemann equations are satisfied in $\mathbb{R}^2$. By Theorem 2.23, we conclude that the function f is differentiable in $\mathbb{C}$. Moreover, it follows from (2.4) that

$$f'(z) = \frac{\partial u}{\partial x} + i\frac{\partial v}{\partial x} = e^x \cos y + ie^x \sin y = e^z,$$

that is, $(e^z)' = e^z$.

Example 2.25

For the cosine and sine functions, we have respectively

$$(\cos z)' = \left(\frac{e^{iz} + e^{-iz}}{2} \right)' = \frac{ie^{iz} - ie^{-iz}}{2}$$

$$= -\frac{e^{iz} - e^{-iz}}{2i} = -\sin z.$$

and

$$(\sin z)' = \left(\frac{e^{iz} - e^{-iz}}{2i} \right)' = \frac{ie^{iz} + ie^{-iz}}{2i}$$

$$= \frac{e^{iz} + e^{-iz}}{2} = \cos z.$$

Example 2.26

Now we find all points at which the function

$$f(x + iy) = xy + ixy$$

is differentiable. We first note that

$$u(x, y) = v(x, y) = xy$$

is of class C^1 in $\mathbb{R}^2$. On the other hand, the Cauchy–Riemann equations

$$\frac{\partial u}{\partial x} = \frac{\partial v}{\partial y} \quad \text{and} \quad \frac{\partial u}{\partial y} = -\frac{\partial v}{\partial x},$$

take the form

$$y = x \quad \text{and} \quad x = -y.$$

The unique solution is $x = y = 0$. By Theorem 2.14, we conclude that the function f is differentiable at no point of $\mathbb{C} \setminus \{0\}$. But since $\{0\}$ is not an open set, one cannot apply Theorem 2.23 to decide whether f is differentiable at the origin. Instead, we have to use the definition of derivative, that is, we must verify whether the limit

$$\lim_{(x,y) \to (0,0)} \frac{f(x + iy) - f(0)}{x + iy - 0} = \lim_{(x,y) \to (0,0)} \frac{xy(1 + i)}{x + iy}$$

exists. It follows from (1.6) that

$$\left| \frac{xy(1 + i)}{x + iy} \right| \le \frac{|x| \cdot |y| \sqrt{2}}{|x + iy|} \le \sqrt{2}|x + iy| \to 0$$

when $(x, y) \to (0, 0)$, and hence, f is differentiable at the origin, with $f'(0) = 0$.

Example 2.27

Let us consider the function $\log z$. It follows from (1.14) that

$$\left(e^{\log z} \right)' = 1.$$

Hence, if $\log z$ is differentiable at z, then it follows from the formula for the derivative of a composition in Proposition 2.12 that

$$e^{\log z}(\log z)' = 1.$$

Therefore,

$$(\log z)' = \frac{1}{e^{\log z}} = \frac{1}{z}.$$

Now we show that $\log z$ is differentiable (at least) in the open set $\mathbb{R}^+ \times \mathbb{R}$. For this we recall the formula

$$\log z = \frac{1}{2}\log(x^2 + y^2) + i\tan^{-1}\frac{y}{x}$$

obtained in Example 1.35 for $x > 0$. We note that the functions

$$u(x, y) = \frac{1}{2}\log(x^2 + y^2) \quad \text{and} \quad v(x, y) = \tan^{-1}\frac{y}{x}$$

are of class C^1 in $\mathbb{R}^+ \times \mathbb{R}$. Since

$$\frac{\partial u}{\partial x} = \frac{x}{x^2 + y^2}, \qquad \frac{\partial v}{\partial y} = \frac{1/x}{1 + (y/x)^2} = \frac{x}{x^2 + y^2},$$

and

$$\frac{\partial u}{\partial y} = \frac{y}{x^2 + y^2}, \qquad -\frac{\partial v}{\partial x} = -\frac{-y/x^2}{1 + (y/x)^2} = \frac{y}{x^2 + y^2},$$

it follows from Theorem 2.23 that the function $\log z$ is holomorphic in $\mathbb{R}^+ \times \mathbb{R}$.

2.4 Paths and Integrals

In order to define the integral of a complex function, we first introduce the notion of a path.

Definition 2.28

A continuous function $\gamma \colon [a, b] \to \Omega \subset \mathbb{C}$ is called a *path* in Ω, and its image $\gamma([a, b])$ is called a *curve* in Ω (see Figure 2.1).

We note that the same curve can be the image of several paths.
Now we define two operations. The first is the inverse of a path.

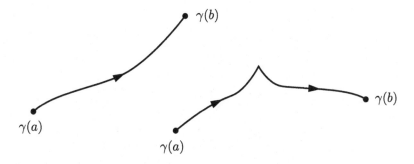

Figure 2.1   Paths and curves

Definition 2.29

Given a path $\gamma\colon [a,b] \to \Omega$, we define the path $-\gamma\colon [a,b] \to \Omega$ by

$$(-\gamma)(t) = \gamma(a+b-t)$$

for each $t \in [a,b]$ (see Figure 2.2).

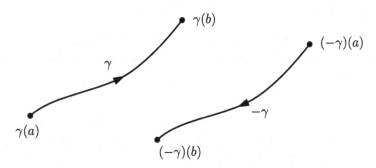

Figure 2.2   Paths γ and $-\gamma$

The second operation is the sum of paths.

Definition 2.30

Given paths $\gamma_1\colon [a_1,b_1] \to \Omega$ and $\gamma_2\colon [a_2,b_2] \to \Omega$ such that $\gamma_1(b_1) = \gamma_2(a_2)$, we define the path $\gamma_1 + \gamma_2\colon [a_1, b_1 + b_2 - a_2] \to \Omega$ by

$$(\gamma_1 + \gamma_2)(t) = \begin{cases} \gamma_1(t) & \text{if } t \in [a_1,b_1], \\ \gamma_2(t - b_1 + a_2) & \text{if } t \in [b_1, b_1 + b_2 - a_2] \end{cases}$$

(see Figure 2.3).

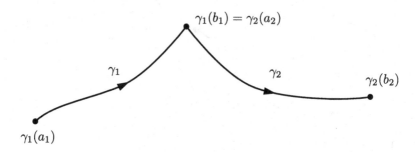

Figure 2.3 Path $\gamma_1 + \gamma_2$

We also consider the notions of a regular path and a piecewise regular path.

Definition 2.31

A path $\gamma\colon [a,b] \to \Omega$ is said to be *regular* if it is of class C^1 and $\gamma'(t) \neq 0$ for every $t \in [a,b]$, taking the right-sided derivative at a and the left-sided derivative at b.

More precisely, the path $\gamma\colon [a,b] \to \Omega$ is regular if there exists a path $\alpha\colon (c,d) \to \Omega$ of class C^1 in some open interval (c,d) containing $[a,b]$ such that $\alpha(t) = \gamma(t)$ and $\alpha'(t) \neq 0$ for every $t \in [a,b]$.

Definition 2.32

A path $\gamma\colon [a,b] \to \Omega$ is said to be *piecewise regular* if there exists a partition of $[a,b]$ into a finite number of subintervals $[a_j,b_j]$ (intersecting at most at their endpoints) such that each path $\gamma_j\colon [a_j,b_j] \to \Omega$ defined by $\gamma_j(t) = \gamma(t)$ for $t \in [a_j,b_j]$ is regular, taking the right-sided derivative at a_j and the left-sided derivative at b_j.

We have the following result.

Proposition 2.33

If the path $\gamma\colon [a,b] \to \mathbb{C}$ is piecewise regular, then

$$L_\gamma := \int_a^b |\gamma'(t)| \, dt < \infty. \qquad (2.9)$$

Proof

Since γ is piecewise regular, the function $t \mapsto |\gamma'(t)|$ is continuous in each interval $[a_j, b_j]$ in Definition 2.32. Therefore, it is Riemann-integrable in each of these intervals, and thus also in their union, which is equal to $[a,b]$. $\qquad \square$

The number L_γ is called the *length* of the path γ.

Example 2.34

Let $\gamma\colon [0,1] \to \mathbb{C}$ be the path given by $\gamma(t) = t(1+i)$ (see Figure 2.4). We have

$$L_\gamma = \int_0^1 |\gamma'(t)| \, dt = \int_0^1 |1+i| \, dt = \sqrt{2}.$$

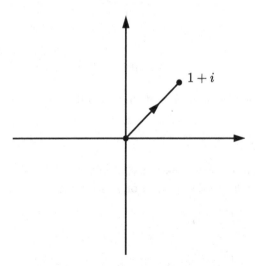

Figure 2.4 The path γ in Example 2.34

Example 2.35

Let $\gamma\colon [0, 2\pi] \to \mathbb{C}$ be the path given by $\gamma(t) = re^{it}$ (see Figure 2.5). We have

$$L_\gamma = \int_0^{2\pi} |\gamma'(t)| \, dt = \int_0^{2\pi} |rie^{it}| \, dt$$

$$= \int_0^{2\pi} r \, dt = 2\pi r,$$

since $|i| = 1$ and

$$|e^{it}| = |\cos t + i \sin t| = \sqrt{\cos^2 t + \sin^2 t} = 1.$$

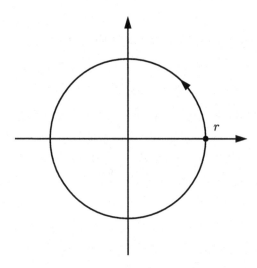

Figure 2.5 The path γ in Example 2.35

Now we introduce the notion of the integral along a path.

Definition 2.36

Let $f\colon \Omega \to \mathbb{C}$ be a continuous function and let $\gamma\colon [a, b] \to \Omega$ be a piecewise regular path. We define the *integral* of f along γ by

$$\int_\gamma f = \int_a^b f(\gamma(t))\gamma'(t) \, dt$$

$$= \int_a^b \operatorname{Re}\big[f(\gamma(t))\gamma'(t)\big] \, dt + i \int_a^b \operatorname{Im}\big[f(\gamma(t))\gamma'(t)\big] \, dt.$$

We also write

$$\int_\gamma f = \int_\gamma f(z)\,dz.$$

We note that under the hypotheses of Definition 2.36, the functions

$$t \mapsto \operatorname{Re}\big[f(\gamma(t))\gamma'(t)\big] \quad \text{and} \quad t \mapsto \operatorname{Im}\big[f(\gamma(t))\gamma'(t)\big]$$

are Riemann-integrable in $[a,b]$, and thus the integral $\int_\gamma f$ is well defined.

Example 2.37

We compute the integral $\int_\gamma \operatorname{Re} z\,dz$ along the paths $\gamma_1, \gamma_2 \colon [0,1] \to \mathbb{C}$ given by

$$\gamma_1(t) = t(1+i) \quad \text{and} \quad \gamma_2(t) = t^2(1+i).$$

We have

$$
\begin{aligned}
\int_{\gamma_1} \operatorname{Re} z\,dz &= \int_0^1 \operatorname{Re}\big[t(1+i)\big] \cdot \big[t(1+i)\big]'\,dt \\
&= \int_0^1 t \cdot (1+i)\,dt \\
&= \frac{t^2}{2}(1+i)\bigg|_{t=0}^{t=1} = \frac{1+i}{2}
\end{aligned}
$$

and

$$
\begin{aligned}
\int_{\gamma_2} \operatorname{Re} z\,dz &= \int_0^1 \operatorname{Re}\big[t^2(1+i)\big] \cdot \big[t^2(1+i)\big]'\,dt \\
&= \int_0^1 t^2 \cdot 2t(1+i)\,dt \\
&= \int_0^1 2t^3(1+i)\,dt \\
&= \frac{t^4}{2}(1+i)\bigg|_{t=0}^{t=1} = \frac{1+i}{2}.
\end{aligned}
$$

Example 2.38

Now we compute the integral $\int_\gamma \operatorname{Im} z \, dz$ along the path $\gamma \colon [0, \pi] \to \mathbb{C}$ given by $\gamma(t) = e^{it}$. We have $\operatorname{Im} \gamma(t) = \sin t$, and hence,

$$
\begin{aligned}
\int_\gamma \operatorname{Im} z \, dz &= \int_0^\pi \sin t \cdot i e^{it} \, dt \\
&= \int_0^\pi \frac{e^{it} - e^{-it}}{2i} i e^{it} \, dt \\
&= \int_0^\pi \frac{1}{2} \left(e^{2it} - 1 \right) dt = \left. \left(\frac{1}{4i} e^{2it} - \frac{1}{2} t \right) \right|_{t=0}^{t=\pi} \\
&= \frac{1}{4i} \left(e^{2\pi i} - 1 \right) - \frac{1}{2} (\pi - 0) = 0 - \frac{\pi}{2} = -\frac{\pi}{2}.
\end{aligned}
$$

The integral has the following properties.

Proposition 2.39

If $f, g \colon \Omega \to \mathbb{C}$ are continuous functions and $\gamma \colon [a, b] \to \Omega$ is a piecewise regular path, then:

1. for any $c, d \in \mathbb{C}$, we have

$$
\int_\gamma (cf + dg) = c \int_\gamma f + d \int_\gamma g;
$$

2.

$$
\int_{-\gamma} f = -\int_\gamma f;
$$

3. for any piecewise regular path $\alpha \colon [p, q] \to \Omega$ with $\alpha(p) = \gamma(b)$, we have

$$
\int_{\gamma + \alpha} f = \int_\gamma f + \int_\alpha f.
$$

Proof

For the second property, we note that

$$
(-\gamma)'(t) = -\gamma'(a + b - t),
$$

and thus,

$$\int_{-\gamma} f = \int_a^b f\big((-\gamma)(t)\big)(-\gamma)'(t)\,dt$$

$$= \int_a^b -f\big(\gamma(a+b-t)\big)\gamma'(a+b-t)\,dt.$$

Making the change of variables $a + b - t = s$, we finally obtain

$$\int_{-\gamma} f = \int_b^a f(\gamma(s))\gamma'(s)\,ds$$

$$= -\int_a^b f(\gamma(s))\gamma'(s)\,ds$$

$$= -\int_\gamma f.$$

The remaining properties follow immediately from the definitions.   □

We also describe two additional properties. For the first one we need the notion of equivalent paths.

Definition 2.40

Two paths $\gamma_1\colon [a_1, b_1] \to \mathbb{C}$ and $\gamma_2\colon [a_2, b_2] \to \mathbb{C}$ are said to be *equivalent* if there exists a differentiable function $\phi\colon [a_2, b_2] \to [a_1, b_1]$ with $\phi' > 0$, $\phi(a_2) = a_1$, and $\phi(b_2) = b_1$, such that $\gamma_2 = \gamma_1 \circ \phi$.

We can now formulate the following result.

Proposition 2.41

If $f\colon \Omega \to \mathbb{C}$ is a continuous function, and γ_1 and γ_2 are equivalent piecewise regular paths in Ω, then

$$\int_{\gamma_1} f = \int_{\gamma_2} f.$$

Proof

We have

$$\int_{\gamma_2} f = \int_{a_2}^{b_2} f(\gamma_2(t))\gamma_2'(t)\, dt$$

$$= \int_{a_2}^{b_2} f((\gamma_1 \circ \phi)(t))\gamma_1'(\phi(t))\phi'(t)\, dt.$$

Making the change of variables $s = \phi(t)$, we obtain

$$\int_{\gamma_2} f = \int_{a_1}^{b_1} f(\gamma_1(s))\gamma_1'(s)\, ds = \int_{\gamma_1} f,$$

which yields the desired identity. $\square$

Finally, we obtain an upper bound for the modulus of the integral.

Proposition 2.42

If $f\colon \Omega \to \mathbb{C}$ is a continuous function and $\gamma\colon [a,b] \to \Omega$ is a piecewise regular path, then

$$\left| \int_\gamma f \right| \le \int_a^b \left| f(\gamma(t))\gamma'(t) \right| dt$$

$$\le L_\gamma \sup\{ |f(\gamma(t))| : t \in [a,b] \}.$$

Proof

Writing $\int_\gamma f = re^{i\theta}$, we obtain

$$\left| \int_\gamma f \right| = r = \int_\gamma e^{-i\theta} f$$

$$= \int_a^b e^{-i\theta} f(\gamma(t))\gamma'(t)\, dt$$

$$= \int_a^b \mathrm{Re}\big[e^{-i\theta} f(\gamma(t))\gamma'(t) \big]\, dt + i \int_a^b \mathrm{Im}\big[e^{-i\theta} f(\gamma(t))\gamma'(t) \big]\, dt.$$

Since $\left|\int_\gamma f\right|$ is a real number, it follows from (1.6) that

$$\left|\int_\gamma f\right| = \int_a^b \operatorname{Re}\left[e^{-i\theta} f(\gamma(t))\gamma'(t)\right] dt$$

$$\leq \int_a^b \left|e^{-i\theta} f(\gamma(t))\gamma'(t)\right| dt.$$

Moreover, since $\left|e^{-i\theta}\right| = 1$, we obtain

$$\left|\int_\gamma f\right| \leq \int_a^b \left|f(\gamma(t))\gamma'(t)\right| dt$$

$$\leq \int_a^b \left|\gamma'(t)\right| dt \cdot \sup\left\{\left|f(\gamma(t))\right| : t \in [a,b]\right\}$$

$$= L_\gamma \sup\left\{\left|f(\gamma(t))\right| : t \in [a,b]\right\}.$$

This yields the desired inequalities. □

Example 2.43

Let us consider the integral

$$\int_\gamma z(z-1)\,dz$$

along the path $\gamma\colon [0,\pi] \to \mathbb{C}$ given by $\gamma(t) = 2e^{it}$. We have

$$L_\gamma = \int_0^\pi \left|2ie^{it}\right| dt = 2\pi.$$

By Proposition 2.42, since $|\gamma(t)| = 2$ for every $t \in [0,\pi]$, we obtain

$$\left|\int_\gamma f\right| \leq L_\gamma \sup\left\{|z(z-1)| : z \in \gamma([0,\pi])\right\}$$

$$\leq 2\pi \sup\left\{\left|z^2\right| + |z| : z \in \gamma([0,\pi])\right\}$$

$$= 2\pi(4+2) = 12\pi.$$

On the other hand,

$$\int_\gamma f = \int_0^\pi \left[\gamma(t)^2 - \gamma(t)\right]\gamma'(t)\,dt$$

$$= \left(\frac{\gamma(t)^3}{3} - \frac{\gamma(t)^2}{2}\right)\Bigg|_{t=0}^{t=\pi} = -\frac{16}{3}.$$

2.5 Primitives

The concept of primitive is useful for the computation of integrals. Let us consider a function $f\colon \Omega \to \mathbb{C}$ in an open set $\Omega \subset \mathbb{C}$.

Definition 2.44

A function $F\colon \Omega \to \mathbb{C}$ is said to be a *primitive* of f in the set Ω if F is holomorphic in Ω and $F' = f$ in Ω.

We first show that in connected open sets all primitives differ by a constant.

Proposition 2.45

If F and G are primitives of f in some connected open set $\Omega \subset \mathbb{C}$, then $F - G$ is constant in Ω.

Proof

We have

$$(F - G)' = F' - G' = f - f = 0$$

in Ω. Hence, it follows from Proposition 2.19 that $F - G$ is constant in Ω. $\square$

Primitives can be used to compute integrals as follows.

Proposition 2.46

If F is a primitive of a continuous function $f\colon \Omega \to \mathbb{C}$ in an open set $\Omega \subset \mathbb{C}$ and $\gamma\colon [a,b] \to \Omega$ is a piecewise regular path, then

$$\int_\gamma f = F(\gamma(b)) - F(\gamma(a)).$$

Proof

For $j = 1,\ldots,n$, let $[a_j, b_j]$, with $b_1 = a_2$, $b_2 = a_3, \ldots$, $b_{n-1} = a_n$, be the subintervals of $[a,b]$ where γ is regular. We note that the function

$$t \mapsto f(\gamma(t))\gamma'(t)$$

is continuous in each interval $[a_j, b_j]$. Therefore,

$$\int_\gamma f = \sum_{j=1}^n \int_{\gamma_j} f = \sum_{j=1}^n \int_{a_j}^{b_j} f(\gamma(t))\gamma'(t)\,dt$$

$$= \sum_{j=1}^n \int_{a_j}^{b_j} F'(\gamma(t))\gamma'(t)\,dt = \sum_{j=1}^n \int_{a_j}^{b_j} (F \circ \gamma)'(t)\,dt$$

$$= \sum_{j=1}^n \left[F(\gamma(b_j)) - F(\gamma(a_j)) \right] = F(\gamma(b)) - F(\gamma(a)).$$

This yields the desired identity. $\square$

Example 2.47

We consider the integral $\int_\gamma (z^3 + 1)\,dz$ along the path $\gamma\colon [0, \pi] \to \mathbb{C}$ given by $\gamma(t) = e^{it}$. Since

$$\left(\frac{z^4}{4} + z \right)' = z^3 + 1,$$

the function $F(z) = z^4/4 + z$ is a primitive of $z^3 + 1$ in $\mathbb{C}$. Therefore,

$$\int_\gamma (z^3 + 1)\,dz = F(\gamma(\pi)) - F(\gamma(0))$$

$$= \left(\frac{1}{4} - 1 \right) - \left(\frac{1}{4} + 1 \right) = -2.$$

We also consider paths with the same initial and final points.

Definition 2.48

A path $\gamma\colon [a, b] \to \mathbb{C}$ is said to be *closed* if $\gamma(a) = \gamma(b)$ (see Figure 2.6).

The following property is an immediate consequence of Proposition 2.46.

Proposition 2.49

If $f\colon \Omega \to \mathbb{C}$ is a continuous function having a primitive in the open set $\Omega \subset \mathbb{C}$ and $\gamma\colon [a, b] \to \Omega$ is a closed piecewise regular path, then

$$\int_\gamma f = 0.$$

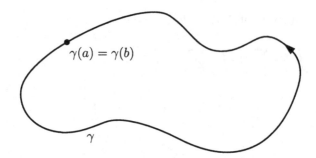

Figure 2.6 A closed path

Now we show that any holomorphic function has primitives. We recall that a set $\Omega \subset \mathbb{C}$ is said to be *convex* if

$$tz + (1 - t)w \in \Omega$$

for every $z, w \in \Omega$ and $t \in [0, 1]$.

Theorem 2.50

If $f \colon \Omega \to \mathbb{C}$ is a holomorphic function in a convex open set $\Omega \subset \mathbb{C}$, then f has a primitive in Ω.

More generally, we have the following result.

Theorem 2.51

If $f \colon \Omega \to \mathbb{C}$ is a continuous function in a convex open set $\Omega \subset \mathbb{C}$ and there exists $p \in \Omega$ such that f is holomorphic in $\Omega \setminus \{p\}$, then f has a primitive in Ω.

Proof

Take $a \in \Omega$. For each $z \in \Omega$, we consider the path $\gamma_z \colon [0, 1] \to \Omega$ given by

$$\gamma_z(t) = a + t(z - a) \tag{2.10}$$

(we recall that Ω is convex). We also consider the function $F \colon \Omega \to \mathbb{C}$ defined by

$$F(z) = \int_{\gamma_z} f. \tag{2.11}$$

Lemma 2.52

We have

$$F(z+h) - F(z) = \int_{\alpha} f, \qquad (2.12)$$

where the path $\alpha \colon [0,1] \to \mathbb{C}$ is given by $\alpha(t) = z + th$.

Proof of the lemma

Let Δ be the triangle whose boundary $\partial \Delta$ is the image of the closed path $\gamma_z + \alpha + (-\gamma_{z+h})$. We note that identity (2.12) is equivalent to

$$\int_{\partial \Delta} f = \int_{\gamma_z} f + \int_{\alpha} f - \int_{\gamma_{z+h}} f$$

$$= F(z) + \int_{\alpha} f - F(z+h) = 0. \qquad (2.13)$$

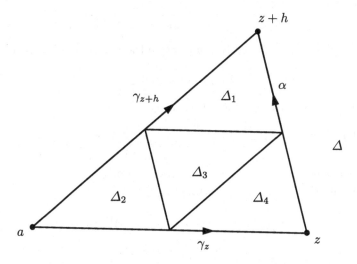

Figure 2.7 Triangles Δ_1, Δ_2, Δ_3 and Δ_4

We first assume that $p \notin \Delta$. We divide the triangle Δ into 4 triangles, say Δ_1, Δ_2, Δ_3 and Δ_4, by adding line segments between the midpoints of the sides of Δ (see Figure 2.7). Then

$$c := \int_{\partial \Delta} f = \sum_{i=1}^{4} \int_{\partial \Delta_i} f,$$

in view of the fact that the integrals along common sides of the triangles Δ_i cancel out, since they have opposite signs. We note that there exists i such that

$$\left| \int_{\partial \Delta_i} f \right| \geq \frac{|c|}{4},$$

since otherwise we would have

$$\left| \sum_{i=1}^{4} \int_{\partial \Delta_i} f \right| < \sum_{i=1}^{4} \frac{|c|}{4} = |c|.$$

One can repeat the argument with this triangle Δ_i in order to obtain a sequence of triangles $\Delta(n) \subset \Delta(n-1)$ such that $\Delta(n)$ is one of the 4 triangles obtained from dividing $\Delta(n-1)$, and

$$\left| \int_{\partial \Delta(n)} f \right| \geq \frac{|c|}{4^n}. \tag{2.14}$$

On the other hand, since f is holomorphic in Δ, for each point $z_0 \in \Delta$, given $\varepsilon > 0$ we have

$$\left| f(z) - f(z_0) - f'(z_0)(z - z_0) \right| < \varepsilon |z - z_0|$$

whenever $|z - z_0|$ is sufficiently small. Since the perimeter of $\Delta(n)$ is

$$L_{\partial \Delta(n)} = 2^{-n} L_{\partial \Delta},$$

where $L_{\partial \Delta}$ is the perimeter of $\partial \Delta$, we obtain

$$\left| \int_{\partial \Delta(n)} \left[f(z) - f(z_0) - f'(z_0)(z - z_0) \right] dz \right| \leq \varepsilon L_{\partial \Delta(n)}^2 = \varepsilon 4^{-n} L_{\partial \Delta}^2 \tag{2.15}$$

for any sufficiently large n. Moreover, since the function $-f(z_0) - f'(z_0)(z - z_0)$ has the primitive $-f(z_0)z - f'(z_0)(z - z_0)^2/2$, we have

$$\int_{\partial \Delta_n} \left[-f(z_0) - f'(z_0)(z - z_0) \right] dz = 0,$$

and it follows from (2.14) and (2.15) that

$$|c| \leq 4^n \left| \int_{\partial \Delta(n)} f \right| \leq \varepsilon L_{\partial \Delta}^2.$$

Letting $\varepsilon \to 0$ we conclude that

$$c = \int_{\partial \Delta} f = 0,$$

which establishes (2.13).

Now we assume that $p \in \Delta$. We note that it is sufficient to consider the case when p is a vertex. Otherwise, being p_1, p_2, p_3 the vertices of Δ, one can consider the three triangles determined by p_i, p_j, p with $i \neq j$. When p belongs to a side of Δ, one of these triangles reduces to a line segment (see Figure 2.8).

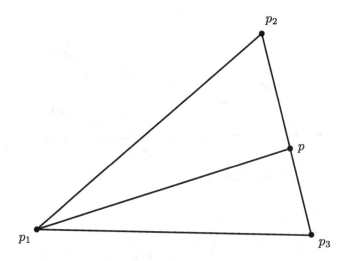

Figure 2.8 Case when p belongs to a side of Δ

When $p = p_3$ is a vertex of Δ, it is sufficient to consider triangles determined by points q_1 and q_2 in the sides containing p (see Figure 2.9). Indeed, by the previous argument, the triangles Δ_1 and Δ_2 respectively with vertices p_1, p_2, q_1 and p_1, q_1, q_2 have zero integral, that is,

$$\int_{\partial \Delta_1} f = \int_{\partial \Delta_2} f = 0.$$

Now let Δ' be the triangle determined by q_1, q_2 and p. Letting $q_1 \to p$ and $q_2 \to p$, we conclude that

$$\left| \int_{\partial \Delta'} f \right| \leq L_{\partial \Delta'} \sup \{ |f(z)| : z \in \Delta' \} \to 0,$$

since $L_{\partial \Delta'} \to 0$. This completes the proof of the lemma. $\square$

We are now ready to show that F is a primitive of f. It follows from

$$\int_\alpha f(z) \, d\zeta = \int_0^1 f(z) h \, dt = f(z) h$$

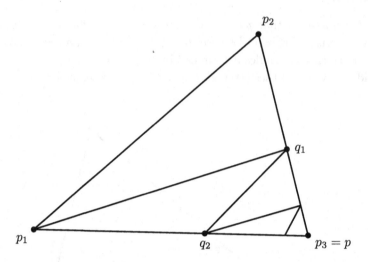

Figure 2.9 Case when p is a vertex of Δ

together with Lemma 2.52 that

$$\frac{F(z+h)-F(z)}{h}-f(z)=\frac{1}{h}\int_{\alpha}\left[f(\zeta)-f(z)\right]d\zeta.$$

Since f is continuous, given $\varepsilon>0$, we have

$$\left|f(\zeta)-f(z)\right|<\varepsilon$$

whenever $|\zeta-z|$ is sufficiently small. Therefore,

$$\left|\frac{F(z+h)-F(z)}{h}-f(z)\right|\leq\frac{1}{|h|}\left|\int_{\alpha}\left[f(\zeta)-f(z)\right]d\zeta\right|$$

$$\leq\frac{\varepsilon L_{\alpha}}{|h|}=\varepsilon$$

whenever $|h|$ is sufficiently small (since $|\zeta-z|\leq|h|$). Letting $\varepsilon\to 0$ we thus obtain $F'(z)=f(z)$, and F is a primitive of f in Ω. $\square$

Example 2.53

For the path γ_z in (2.10) we have $\gamma_z'(t)=z-a$, and by (2.11) a primitive of f is given by

$$F(z)=\int_0^1 f\bigl(a+t(z-a)\bigr)(z-a)\,dt. \tag{2.16}$$

In particular, when $0 \in \Omega$, taking $a = 0$ we obtain

$$F(z) = z \int_0^1 f(tz) \, dt. \tag{2.17}$$

Example 2.54

We have

$$\lim_{z \to 0} \frac{\sin z}{z} = \lim_{z \to 0} \frac{\sin z - \sin 0}{z - 0}$$

$$= (\sin z)' \big|_{z=0}$$

$$= \cos 0 = 1.$$

Hence, the function

$$f(z) = \begin{cases} (\sin z)/z & \text{if } z \neq 0, \\ 1 & \text{if } z = 0 \end{cases}$$

is continuous in $\mathbb{C}$ and holomorphic in $\mathbb{C} \setminus \{0\}$. It thus follows from Theorem 2.51 that f has a primitive in $\mathbb{C}$. Moreover, by (2.17), a primitive is given by

$$F(z) = z \int_0^1 \frac{\sin(tz)}{tz} \, dt = \int_0^1 \frac{\sin(tz)}{t} \, dt.$$

The following result is an immediate consequence of Theorem 2.51 and Proposition 2.49.

Theorem 2.55 (Cauchy's theorem)

If $f \colon \Omega \to \mathbb{C}$ is a continuous function in a convex open set $\Omega \subset \mathbb{C}$ and there exists $p \in \Omega$ such that f is holomorphic in $\Omega \setminus \{p\}$, then

$$\int_\gamma f = 0$$

for any closed piecewise regular path γ in Ω.

2.6   Index of a Closed Path

Now we introduce the notion of the index of a closed path.

Definition 2.56

Given a closed piecewise regular path $\gamma\colon [a,b]\to\mathbb{C}$, we define the *index* of a point $z\in\mathbb{C}\setminus\gamma([a,b])$ with respect to γ by

$$\operatorname{Ind}_\gamma(z)=\frac{1}{2\pi i}\int_\gamma\frac{dw}{w-z}.$$

Example 2.57

Let $\gamma\colon [0,2\pi]\to\mathbb{C}$ be the path given by $\gamma(t)=a+re^{it}$. Then

$$\operatorname{Ind}_\gamma(a)=\frac{1}{2\pi i}\int_0^{2\pi}\frac{rie^{it}}{re^{it}}\,dt$$

$$=\frac{1}{2\pi i}\int_0^{2\pi}i\,dt=1.$$

The following result specifies the values that the index can take.

Theorem 2.58

Let $\gamma\colon [a,b]\to\mathbb{C}$ be a closed piecewise regular path and let $\Omega=\mathbb{C}\setminus\gamma([a,b])$. Then:

1. $\operatorname{Ind}_\gamma(z)\in\mathbb{Z}$ for each $z\in\Omega$;
2. the function $z\mapsto\operatorname{Ind}_\gamma(z)$ is constant in each connected component of Ω;
3. $\operatorname{Ind}_\gamma(z)=0$ for each z in the unbounded connected component of Ω.

Proof

We define a function $\phi\colon [a,b]\to\mathbb{C}$ by

$$\phi(s)=\exp\left(\int_a^s\frac{\gamma'(t)}{\gamma(t)-z}\,dt\right).$$

We have

$$\phi'(s)=\phi(s)\frac{\gamma'(s)}{\gamma(s)-z}$$

in each subinterval $[a_j,b_j]$ of $[a,b]$ where γ is regular. Then

$$\left(\frac{\phi(s)}{\gamma(s)-z}\right)'=\frac{\phi'(s)(\gamma(s)-z)-\gamma'(s)\phi(s)}{(\gamma(s)-z)^2}=0,$$

and for each j there exists $c_j \in \mathbb{C}$ such that

$$\frac{\phi(s)}{\gamma(s) - z} = c_j$$

for every $s \in [a_j, b_j]$. But since γ and ϕ are continuous functions, we conclude that there exists $c \in \mathbb{C}$ such that

$$\frac{\phi(s)}{\gamma(s) - z} = c$$

for every $s \in [a, b]$. In particular,

$$\frac{\phi(s)}{\gamma(s) - z} = \frac{\phi(a)}{\gamma(a) - z} = \frac{1}{\gamma(a) - z},$$

that is,

$$\phi(s) = \frac{\gamma(s) - z}{\gamma(a) - z}.$$

Letting $s = b$, since γ is a closed path, we obtain

$$\phi(b) = \frac{\gamma(b) - z}{\gamma(a) - z} = 1,$$

that is,

$$\phi(b) = \exp\left(\int_a^b \frac{\gamma'(t)}{\gamma(t) - z} \, dt\right)$$

$$= \exp\left(2\pi i \operatorname{Ind}_\gamma(z)\right) = 1. \tag{2.18}$$

We note that

$$e^{2\pi i \alpha} = 1 \quad \Leftrightarrow \quad \alpha \in \mathbb{Z},$$

since $e^{2\pi i \alpha} = \cos(2\pi\alpha) + i\sin(2\pi\alpha)$. It then follows from (2.18) that $\operatorname{Ind}_\gamma(z) \in \mathbb{Z}$.

For the second property, we first note that

$$\left|\operatorname{Ind}_\gamma(z) - \operatorname{Ind}_\gamma(w)\right| = \left|\frac{1}{2\pi i} \int_\gamma \left(\frac{1}{\zeta - z} - \frac{1}{\zeta - w}\right) d\zeta\right|$$

$$= \frac{1}{2\pi} \left|\int_\gamma \frac{z - w}{(\zeta - z)(\zeta - w)} \, d\zeta\right|$$

$$\leq \frac{L_\gamma}{2\pi} \sup\left\{\frac{|z - w|}{|(\gamma(t) - z)(\gamma(t) - w)|} : t \in [a, b]\right\}. \tag{2.19}$$

For w sufficiently close to z, we have

$$\left|\gamma(t) - w\right| \geq \left|\gamma(t) - z\right| - |z - w|,$$

and thus,

$$\frac{1}{\left|(\gamma(t) - z)(\gamma(t) - w)\right|} \leq \frac{1}{|\gamma(t) - z|(|\gamma(t) - z| - |z - w|)}$$
$$\leq \frac{1}{A(A - |z - w|)}$$

for every $z \in \mathbb{C} \setminus \gamma([a, b])$, where

$$A = \inf\left\{|\gamma(t) - z| : t \in [a, b]\right\} > 0.$$

Hence, it follows from (2.19) that

$$\left|\mathrm{Ind}_\gamma(z) - \mathrm{Ind}_\gamma(w)\right| \leq \frac{L_\gamma}{2\pi} \frac{|z - w|}{A(A - |z - w|)},$$

and letting $w \to z$ we obtain

$$\lim_{w \to z} \mathrm{Ind}_\gamma(w) = \mathrm{Ind}_\gamma(z). \tag{2.20}$$

Since the index takes only integer values, it follows from the continuity in (2.20) that the function $z \mapsto \mathrm{Ind}_\gamma(z)$ is constant in each connected component of Ω (we note that since Ω is open, each connected component of Ω is an open set).

For the last property, we note that

$$\left|\mathrm{Ind}_\gamma(z)\right| = \left|\frac{1}{2\pi i} \int_a^b \frac{\gamma'(t)}{\gamma(t) - z} \, dt\right|$$
$$\leq \frac{1}{2\pi} L_\gamma \sup\left\{\frac{|\gamma'(t)|}{|\gamma(t) - z|} : t \in [a, b]\right\}$$
$$\leq \frac{1}{2\pi} L_\gamma \frac{\sup\{\gamma'(t) : t \in [a, b]\}}{|z| - \sup\{\gamma(t) : t \in [a, b]\}}, \tag{2.21}$$

since

$$\left|\gamma(t) - z\right| \geq |z| - \left|\gamma(t)\right|$$

whenever $|z|$ is sufficiently large. In particular, it follows from (2.21) that $|\mathrm{Ind}_\gamma(z)| < 1$ for any sufficiently large $|z|$. Since the index takes only integer values, we obtain $\mathrm{Ind}_\gamma(z) = 0$. It follows again from the continuity in (2.20) that the index is zero in the unbounded connected component of Ω. $\square$

Example 2.59

For each $n \in \mathbb{N}$, let $\gamma \colon [0, 2\pi n] \to \mathbb{C}$ be the path given by $\gamma(t) = a + re^{it}$, looping n times around the point a in the positive direction. Then

$$\mathrm{Ind}_\gamma(a) = \frac{1}{2\pi i} \int_0^{2\pi n} \frac{\gamma'(t)}{\gamma(t) - a} \, dt$$

$$= \frac{1}{2\pi i} \int_0^{2\pi n} \frac{rie^{it}}{re^{it}} \, dt = n.$$

It follows from Theorem 2.58 that

$$\mathrm{Ind}_\gamma(z) = \begin{cases} n & \text{if } |z - a| < r, \\ 0 & \text{if } |z - a| > r. \end{cases}$$

2.7 Cauchy's Integral Formula

Now we establish Cauchy's integral formula for a holomorphic function. In particular, it guarantees that any holomorphic function is uniquely determined by its values along closed paths.

Theorem 2.60

If $f \colon \Omega \to \mathbb{C}$ is a holomorphic function in a convex open set $\Omega \subset \mathbb{C}$ and $\gamma \colon [a, b] \to \Omega$ is a closed piecewise regular path, then

$$f(z)\,\mathrm{Ind}_\gamma(z) = \frac{1}{2\pi i} \int_\gamma \frac{f(w)}{w - z} \, dw \tag{2.22}$$

for every $z \in \Omega \setminus \gamma([a, b])$.

Proof

Let us consider the function $g \colon \Omega \to \mathbb{C}$ defined by

$$g(w) = \begin{cases} (f(w) - f(z))/(w - z) & \text{if } w \in \Omega \setminus \{z\}, \\ f'(z) & \text{if } w = z. \end{cases}$$

Clearly, g is continuous in Ω and holomorphic in $\Omega \setminus \{z\}$. It then follows from Theorem 2.55 that

$$
\begin{aligned}
0 &= \int_\gamma g \\
&= \int_\gamma \frac{f(w) - f(z)}{w - z} \, dw \\
&= \int_\gamma \frac{f(w)}{w - z} \, dw - f(z) \int_\gamma \frac{dw}{w - z} \\
&= \int_\gamma \frac{f(w)}{w - z} \, dw - f(z) 2\pi i \operatorname{Ind}_\gamma(z).
\end{aligned}
$$

This yields the desired identity.   □

Example 2.61

Let $f \colon \mathbb{C} \to \mathbb{C}$ be a holomorphic function in $\mathbb{C}$ and let $\gamma \colon [0, 2\pi] \to \mathbb{C}$ be the path given by $\gamma(t) = z + re^{it}$. Then $\operatorname{Ind}_\gamma(z) = 1$, and by Theorem 2.60 we have

$$
\begin{aligned}
f(z) &= \frac{1}{2\pi i} \int_\gamma \frac{f(w)}{w - z} \, dw \\
&= \frac{1}{2\pi i} \int_0^{2\pi} \frac{f(z + re^{it})}{re^{it}} rie^{it} \, dt \\
&= \frac{1}{2\pi} \int_0^{2\pi} f(z + re^{it}) \, dt.
\end{aligned}
$$

2.8 Integrals and Homotopy of Paths

In this section we show that the integral of a holomorphic function does not change with homotopies of the path. We first recall the notion of homotopy.

Definition 2.62

Two closed paths $\gamma_1, \gamma_2 \colon [a, b] \to \Omega$ are said to be *homotopic* in Ω if there exists a continuous function $H \colon [a, b] \times [0, 1] \to \Omega$ such that (see Figure 2.10):
  1. $H(t, 0) = \gamma_1(t)$ and $H(t, 1) = \gamma_2(t)$ for every $t \in [a, b]$;
  2. $H(a, s) = H(b, s)$ for every $s \in [0, 1]$.
Then the function H is called a *homotopy* between γ_1 and γ_2.

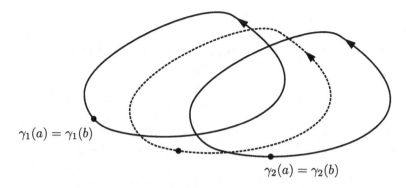

$$\gamma_1(a) = \gamma_1(b)$$

$$\gamma_2(a) = \gamma_2(b)$$

Figure 2.10 Homotopy of paths

We then have the following result.

Theorem 2.63

If $f \colon \Omega \to \mathbb{C}$ is a holomorphic function in an open set $\Omega \subset \mathbb{C}$, and γ_1 and γ_2 are closed piecewise regular paths that are homotopic in Ω, then

$$\int_{\gamma_1} f = \int_{\gamma_2} f. \tag{2.23}$$

Proof

Let H be a homotopy between the paths γ_1 and γ_2. We note that H is uniformly continuous (since it is defined in a compact set). Hence, there exists $n \in \mathbb{N}$ such that

$$\big|H(t,s) - H(t',s')\big| < r$$

for every $(t,s),(t',s') \in [a,b] \times [0,1]$ with

$$|t - t'| < \frac{2(b-a)}{n} \quad \text{and} \quad |s - s'| < \frac{2}{n}. \tag{2.24}$$

Now we consider the points

$$p_{j,k} = H\left(a + \frac{j}{n}(b-a), \frac{k}{n}\right), \quad j,k = 0,\dots,n,$$

and the closed polygons $P_{j,k}$ defined by the points

$$p_{j,k}, \quad p_{j+1,k}, \quad p_{j+1,k+1} \quad \text{and} \quad p_{j,k+1},$$

in this order. It follows from (2.24) that these four points are contained in the ball $B_r(p_{j,k})$ of radius r centered at $p_{j,k}$, and since any ball is a convex set, we also have $P_{j,k} \subset B_r(p_{j,k})$. It then follows from Theorem 2.55 that

$$\int_{\partial P_{j,k}} f = 0, \tag{2.25}$$

where $\partial P_{j,k}$ is the path along the boundary of $P_{j,k}$.

Now we consider the closed polygons Q_k defined by the points

$$p_{0,k}, \quad p_{1,k}, \quad \cdots, \quad p_{n-1,k} \quad \text{and} \quad p_{n,k},$$

in this order, as well as the paths $\alpha_j \colon [j/n, (j+1)/n] \to \mathbb{C}$ and $\beta_j \colon [0,1] \to \mathbb{C}$ given respectively by $\alpha_j(t) = \gamma_1(t)$ and

$$\beta_j(t) = p_{j+1,0} + t(p_{j,0} - p_{j+1,0}).$$

Since $\alpha_j + \beta_j$ is a closed path in the ball $B_r(p_{j,0})$, it follows again from Theorem 2.55 that

$$\int_{\gamma_j} f = -\int_{\beta_j} f = \int_{-\beta_j} f$$

for $j = 0, \ldots, n-1$. Therefore,

$$\int_{\gamma_1} f = \sum_{j=0}^{n-1} \int_{\alpha_j} f = \sum_{j=0}^{n-1} \int_{-\beta_j} f = \int_{\partial Q_0} f. \tag{2.26}$$

One can show in a similar manner that

$$\int_{\gamma_2} f = \int_{\partial Q_n} f. \tag{2.27}$$

On the other hand, it follows from (2.25) that

$$\sum_{j=0}^{n-1} \int_{\partial P_{j,k}} f = 0. \tag{2.28}$$

We note that the path $\partial P_{j,k}$ includes the line segment from $p_{j+1,k}$ to $p_{j+1,k+1}$, in this direction, while $\partial P_{j+1,k}$ includes the same segment but in the opposite direction, and thus the corresponding terms cancel out in the sum in (2.28). Moreover, $\partial P_{0,k}$ includes the line segment from $p_{0,k+1}$ to $p_{0,k}$, in this direction, while $\partial P_{n-1,k}$ includes the same segment but in the opposite direction. In fact, since each path $t \mapsto H(t,s)$ is closed, we have $p_{n,k+1} = p_{0,k+1}$ and $p_{n,k} = p_{0,k}$.

Therefore,

$$0 = \sum_{j=0}^{n-1} \int_{\partial P_{j,k}} f = \int_{\partial Q_k} f - \int_{\partial Q_{k+1}} f,$$

that is,

$$\int_{\partial Q_{k+1}} f = \int_{\partial Q_k} f,$$

for $k = 0, 1, \ldots, n - 1$. Identity (2.23) now follows readily from (2.26) and (2.27). $\square$

The following result is an immediate consequence of Theorem 2.63.

Theorem 2.64

If $f \colon \Omega \to \mathbb{C}$ is a holomorphic function in an open set $\Omega \subset \mathbb{C}$, and γ is a closed piecewise regular path that is homotopic to a constant path in Ω, then

$$\int_\gamma f = 0.$$

We also show that the index does not change with homotopies of the path.

Proposition 2.65

Let γ_1 and γ_2 be closed piecewise regular paths that are homotopic in Ω. Then for each $z \in \mathbb{C} \setminus \Omega$, we have

$$\mathrm{Ind}_{\gamma_1}(z) = \mathrm{Ind}_{\gamma_2}(z). \tag{2.29}$$

Proof

Let us take $z \in \mathbb{C} \setminus \Omega$. We note that the function

$$f(w) = \frac{1}{2\pi i} \cdot \frac{1}{w - z}$$

is holomorphic in $\mathbb{C} \setminus \{z\}$, and thus in particular in Ω. Since

$$\int_{\gamma_j} f = 2\pi i \, \mathrm{Ind}_{\gamma_j}(z)$$

for $j = 1, 2$, identity (2.29) follows readily from Theorem 2.63. $\square$

2.9 Harmonic Conjugate Functions

In this section we discuss the concept of harmonic conjugate functions. We recall that a function $u\colon \Omega \to \mathbb{C}$ with second derivatives in some open set $\Omega \subset \mathbb{C}$ is said to be *harmonic* in Ω if $\Delta u = 0$, where the Laplacian Δu is defined by

$$\Delta u = \frac{\partial^2 u}{\partial x^2} + \frac{\partial^2 u}{\partial y^2}.$$

Definition 2.66

Two harmonic functions $u, v\colon \Omega \to \mathbb{C}$ in the open set $\Omega \subset \mathbb{C}$ are said to be *harmonic conjugate functions* in Ω if u and v satisfy the Cauchy–Riemann equations in Ω.

If the function $f = u + iv$ is holomorphic in an open set $\Omega \subset \mathbb{C}$, then the Cauchy–Riemann equations are satisfied, and

$$\Delta u = \Delta v = 0 \quad \text{in } \Omega$$

(see Problem 2.24). In fact, one can show that $\Delta u = \Delta v = 0$ even without assuming a priori that u and v are of class C^2 (see Exercise 4.36). Therefore, the real and imaginary parts of a holomorphic function are harmonic conjugate functions.

We show that any harmonic function of class C^2 in a simply connected open set has a harmonic conjugate. We first recall the notions of a path connected set and a simply connected set.

Definition 2.67

A set $\Omega \subset \mathbb{C}$ is said to be *path connected* if for each $z, w \in \Omega$ there exists a path $\gamma\colon [a, b] \to \Omega$ with $\gamma(a) = z$ and $\gamma(b) = w$.

In particular, a path connected set is necessarily connected.

Definition 2.68

A set $\Omega \subset \mathbb{C}$ is said to be *simply connected* if it is path connected and any closed path $\gamma\colon [a, b] \to \Omega$ is homotopic to a constant path in Ω.

We then have the following result.

Proposition 2.69

Let $u\colon \Omega \to \mathbb{C}$ be a function of class C^2 in a simply connected open set $\Omega \subset \mathbb{C}$. If $\Delta u = 0$, then there exists a function $v\colon \Omega \to \mathbb{C}$ of class C^2 with $\Delta v = 0$ such that u and v are harmonic conjugate functions. Moreover, the function v is unique up to a constant.

Proof

Since Ω is simply connected and u is of class C^2, it follows from Green's theorem that if α is a closed path in Ω without intersections, then

$$
\int_\alpha -\frac{\partial u}{\partial y}\,dx + \frac{\partial u}{\partial x}\,dy = \int_U \left(\frac{\partial}{\partial x}\left(\frac{\partial u}{\partial x}\right) - \frac{\partial}{\partial y}\left(-\frac{\partial u}{\partial x}\right) \right) dx\,dy
$$

$$
= \int_U \Delta u\,dx\,dy = 0,
\tag{2.30}
$$

where U is the open set whose boundary is the image of α. This shows that given $p \in \Omega$, one can define a function $v\colon \Omega \to \mathbb{C}$ by the line integral

$$
v(x,y) = \int_\gamma -\frac{\partial u}{\partial y}\,dx + \frac{\partial u}{\partial x}\,dy,
\tag{2.31}
$$

where $\gamma\colon [a,b] \to \Omega$ is any path between p and (x,y). Now we show that the Cauchy–Riemann equations are satisfied. It follows from (2.30) that

$$
\frac{\partial v}{\partial x}(x,y) = \lim_{h\to 0} \frac{v(x+h,y)-v(x,y)}{h}
$$

$$
= \lim_{h\to 0} \frac{1}{h} \int_{\gamma_h} -\frac{\partial u}{\partial y}\,dx + \frac{\partial u}{\partial x}\,dy,
$$

where the path $\gamma_h\colon [0,1] \to \mathbb{R}^2$ is given by

$$
\gamma_h(t) = (x+th,y).
$$

Since

$$
\int_{\gamma_h} -\frac{\partial u}{\partial y}\,dx + \frac{\partial u}{\partial x}\,dy = \int_0^1 -\frac{\partial u}{\partial x}(x+th,y)h\,dt,
$$

and the function $-\partial u/\partial x$ is continuous, we obtain

$$
\frac{\partial v}{\partial x}(x,y) = \lim_{h\to 0} \int_0^1 -\frac{\partial u}{\partial x}(x+th,y)\,dt = -\frac{\partial u}{\partial x}(x,y).
$$

One can show in a similar manner that

$$\frac{\partial v}{\partial y}(x,y) = \frac{\partial u}{\partial x}(x,y),$$

and hence, the Cauchy–Riemann equations are satisfied in Ω. Moreover, v is of class C^2 and thus $\Delta v = 0$.

It remains to show that v is unique up to a constant. By Theorem 2.23, the function $f = u + iv$ is holomorphic in Ω. If w is another function of class C^2 with $\Delta w = 0$ such that $f = u + iw$ is holomorphic in Ω, then

$$u + iv - (u + iw) = i(v - w)$$

is also holomorphic in Ω. Since Ω is connected (because it is simply connected) and $i(v - w)$ has constant real part, it follows from Example 2.20 that $v - w$ is constant.   □

The following result can be established in a similar manner.

Proposition 2.70

Let $v\colon \Omega \to \mathbb{C}$ be a function of class C^2 in a simply connected open set $\Omega \subset \mathbb{C}$. If $\Delta v = 0$, then there exists a function $u\colon \Omega \to \mathbb{C}$ of class C^2 with $\Delta u = 0$ such that u and v are harmonic conjugate functions. Moreover, the function u is unique up to a constant.

Proof

Since Ω is simply connected and v is of class C^2, given $p \in \Omega$, one can define a function $u\colon \Omega \to \mathbb{C}$ by the line integral

$$u(x,y) = \int_\gamma \frac{\partial v}{\partial y}\, dx - \frac{\partial v}{\partial x}\, dy,$$

where $\gamma\colon [a,b] \to \Omega$ is any path between p and (x,y). We can now proceed in a similar manner to that in the proof of Proposition 2.69 to show that the Cauchy–Riemann equations are satisfied.   □

We also give some examples.

Example 2.71

We consider the function $f = u + iv$ with real part $u(x, y) = x^2 - xy - y^2$ as in Example 2.15. Since u is of class C^2 and

$$\Delta u = \frac{\partial^2 u}{\partial x^2} + \frac{\partial^2 u}{\partial y^2} = 2 - 2 = 0$$

in the simply connected open set $\Omega = \mathbb{C}$, by Proposition 2.69 there exists a function v of class C^2 such that $f = u + iv$ is holomorphic in $\mathbb{C}$. By (2.31), one can take

$$v(x, y) = \int_\gamma -\frac{\partial u}{\partial y} \, dx + \frac{\partial u}{\partial x} \, dy$$

$$= \int_\gamma (x + 2y) \, dx + (2x - y) \, dy,$$

with the path $\gamma \colon [0, 1] \to \mathbb{C}$ given by $\gamma(t) = (tx, ty)$. We then obtain

$$v(x, y) = \int_0^1 \left[(tx + 2ty)x + (2tx - ty)y \right] dt$$

$$= \left(\frac{1}{2} t^2 x^2 + t^2 yx + t^2 xy - \frac{1}{2} t^2 y^2 \right) \Big|_{t=0}^{t=1}$$

$$= \frac{x^2}{2} + 2xy - \frac{y^2}{2}.$$

Example 2.72

Now we consider the function $u(x, y) = x^2 + y^2$ as in Example 2.16. Since u is of class C^2 and $\Delta u = 4 \neq 0$, the function u is not the real part of any holomorphic function in an open set $\Omega \subset \mathbb{C}$.

Example 2.73

Let us consider the function $u(x, y) = ax^2 + by$, with $a, b \in \mathbb{R}$. Since u is of class C^2 and $\Delta u = 2a$, in order that u is the real part of a holomorphic function in some open set we must have $a = 0$. Moreover, it follows from Proposition 2.69 that if $a = 0$, then there exists a function v of class C^2 in $\mathbb{R}^2$ such that

$$f(x + iy) = u(x, y) + iv(x, y) = by + iv(x, y)$$

is holomorphic in $\mathbb{C}$. One can use the Cauchy–Riemann equations to determine v. Indeed, it follows from the equation $\partial u / \partial x = \partial v / \partial y$ that $\partial v / \partial y = 0$, and hence v does not depend on y. Moreover,

$$\frac{\partial v}{\partial x} = -\frac{\partial u}{\partial y} = -b,$$

and thus $v(x,y) = -bx + c$ for some constant $c \in \mathbb{R}$.

2.10 Solved Problems and Exercises

Problem 2.1

Verify that the function $f(z) = z^2 - z$ is continuous in $\mathbb{C}$.

Solution

Writing

$$f(x + iy) = u(x,y) + iv(x,y), \tag{2.32}$$

with $x, y \in \mathbb{R}$, we obtain

$$u(x,y) = x^2 - y^2 - x \quad \text{and} \quad v(x,y) = 2xy - y.$$

Since u and v are continuous in $\mathbb{R}^2$, the function f is continuous in $\mathbb{C}$.

Problem 2.2

Use the Cauchy–Riemann equations to show that the function $f(z) = e^z + z$ is holomorphic in $\mathbb{C}$.

Solution

One can write the function f in the form (2.32), with

$$u(x,y) = e^x \cos y + x \quad \text{and} \quad v(x,y) = e^x \sin y + y.$$

The Cauchy–Riemann equations

$$\frac{\partial u}{\partial x} = \frac{\partial v}{\partial y} \quad \text{and} \quad \frac{\partial u}{\partial y} = -\frac{\partial v}{\partial x} \tag{2.33}$$

take the form

$$e^x \cos y + 1 = e^x \cos y + 1 \quad \text{and} \quad -e^x \sin y = -e^x \sin y,$$

and thus they are satisfied in $\mathbb{R}^2$. Since u and v are functions of class C^1 in the open set $\mathbb{R}^2$, it follows from Theorem 2.23 that f is holomorphic in $\mathbb{C}$.

Problem 2.3

Show that

$$\left(z^n\right)' = nz^{n-1} \tag{2.34}$$

for every $n \in \mathbb{N}$ and $z \in \mathbb{C}$ (with the convention that $0^0 = 1$).

Solution

Let $f_n(z) = z^n$. For $n = 1$ we have

$$f_1'(z_0) = \lim_{z \to z_0} \frac{z - z_0}{z - z_0} = 1,$$

which establishes (2.34). For $n > 1$, it follows from the identity

$$z^n - z_0^n = (z - z_0) \sum_{k=0}^{n-1} z^k z_0^{n-1-k}$$

that

$$f_n'(z_0) = \lim_{z \to z_0} \frac{z^n - z_0^n}{z - z_0} = \lim_{z \to z_0} \sum_{k=0}^{n-1} z^k z_0^{n-1-k} = nz_0^{n-1}.$$

Problem 2.4

Use the definition of derivative to verify that $|z|$ is not differentiable at $z = 0$.

Solution

Writing $z = |z|e^{i\theta}$, we obtain

$$\frac{|z| - |0|}{z - 0} = \frac{|z|}{z} = \frac{|z|}{|z|e^{i\theta}} = e^{-i\theta}.$$

Since $e^{-i\theta}$ depends on θ, one cannot take the limit when $z \to 0$, and hence f is not differentiable at the origin.

Problem 2.5

Find all points $z \in \mathbb{C}$ at which the function $|z|$ is differentiable.

Solution

We have $|x + iy| = u(x, y) + iv(x, y)$, where

$$u(x, y) = \sqrt{x^2 + y^2} \quad \text{and} \quad v(x, y) = 0.$$

The Cauchy–Riemann equations in (2.33) are thus

$$\frac{x}{\sqrt{x^2 + y^2}} = 0 \quad \text{and} \quad \frac{y}{\sqrt{x^2 + y^2}} = 0.$$

We note that these have no solutions ($x = y = 0$ is not a solution, since one cannot divide by zero). Hence, by Theorem 2.14, the function $|z|$ has no points of differentiability.

Problem 2.6

Find all points of differentiability of the function $f(x + iy) = xy + iy$.

Solution

We write the function f in the form (2.32), with $u(x, y) = xy$ and $v(x, y) = y$. The Cauchy–Riemann equations in (2.33) are thus $y = 1$ and $x = 0$. Hence, by Theorem 2.14, the function f is not differentiable at any point of $\mathbb{C} \setminus \{i\}$. Since the set $\{i\}$ is not open, in order to determine whether f is differentiable at i we must use the definition of derivative, that is, we have to verify whether

$$\frac{f(x + iy) - f(i)}{x + iy - i} = \frac{xy + iy - i}{x + iy - i}$$

has a limit when $x + iy \to i$. Since

$$\frac{xy + iy - i}{x + iy - i} = \frac{x(y - 1) + x + i(y - 1)}{x + i(y - 1)}$$

$$= \frac{x(y - 1)}{x + i(y - 1)} + 1$$

and

$$\left| \frac{x(y - 1)}{x + i(y - 1)} \right| = \frac{|x| \cdot |y - 1|}{|x + i(y - 1)|} \le |x| \to 0$$

when $x + iy \to i$, we conclude that f is differentiable at i, with $f'(i) = 1$.

Problem 2.7

Find all constants $a, b \in \mathbb{R}$ such that the function

$$f(x + iy) = ax^2 + 2xy + by^2 + i(y^2 - x^2)$$

is holomorphic in $\mathbb{C}$.

Solution

We first note that taking $z = x + iy$, we have $z^2 = x^2 - y^2 + i2xy$, and thus,

$$-iz^2 = 2xy + i(y^2 - x^2).$$

Therefore,

$$f(z) = ax^2 + by^2 - iz^2.$$

Since the function $-iz^2$ is holomorphic in $\mathbb{C}$, it is sufficient to find constants $a, b \in \mathbb{R}$ such that the function

$$f(z) + iz^2 = ax^2 + by^2 + i0$$

is holomorphic in $\mathbb{C}$. By Theorem 2.23, this happens if and only if the Cauchy–Riemann equations in (2.33) are satisfied in $\mathbb{R}^2$, that is, if and only if

$$2ax = 0 \quad \text{and} \quad 2by = 0$$

for every $x, y \in \mathbb{R}$. Therefore, $a = b = 0$.

Problem 2.8

Find whether there exists $a \in \mathbb{R}$ such that the function

$$f(x + iy) = ax^2 + 2xy + i(x^2 - y^2 - 2xy)$$

is holomorphic in $\mathbb{C}$.

Solution

By Theorem 2.23, the function f is holomorphic in $\mathbb{C}$ if and only if the Cauchy–Riemann equations are satisfied in $\mathbb{R}^2$. In this case they take the form

$$2ax + 2y = -2y - 2x \quad \text{and} \quad 2x = -(2x - 2y),$$

or equivalently

$$(a+1)x = -2y \quad \text{and} \quad 2x = y. \tag{2.35}$$

We then obtain $(a+1)x = -4x$, and thus $a = -5$. Hence, the equations in (2.35) reduce to the identity $2x = y$, which does not hold for every x and y. Therefore, there exists no $a \in \mathbb{R}$ such that the function f is holomorphic in $\mathbb{C}$.

Problem 2.9

Let $f = u + iv$ be a holomorphic function in $\mathbb{C}$ with real part

$$u(x,y) = 2x^2 - 3xy - 2y^2.$$

Compute explicitly $f(z)$ and $f'(z)$.

Solution

Since f is holomorphic in $\mathbb{C}$, the Cauchy–Riemann equations in (2.33) are satisfied in $\mathbb{R}^2$. It follows from

$$\frac{\partial u}{\partial x} = 4x - 3y$$

and the first equation in (2.33) that

$$\frac{\partial v}{\partial y} = 4x - 3y.$$

Therefore,

$$v(x,y) = 4xy - \frac{3y^2}{2} + C(x)$$

for some differentiable function C. We then obtain

$$\frac{\partial u}{\partial y} = -3x - 4y \quad \text{and} \quad -\frac{\partial v}{\partial x} = -4y - C'(x),$$

and it follows from the second equation in (2.33) that $C'(x) = 3x$. Therefore,

$$C(x) = \frac{3x^2}{2} + c \quad \text{for some } c \in \mathbb{R},$$

and

$$v(x,y) = 4xy - \frac{3y^2}{2} + \frac{3x^2}{2} + c.$$

Hence,

$$f(x+iy) = (2x^2 - 3xy - 2y^2) + i\left(4xy - \frac{3y^2}{2} + \frac{3x^2}{2} + c\right)$$

$$= 2(x^2 - y^2 + 2ixy) + \frac{3}{2}i(x^2 - y^2 + 2ixy) + ic$$

$$= 2z^2 + \frac{3}{2}iz^2 + ic = \left(2 + \frac{3}{2}i\right)z^2 + ic,$$

and thus $f'(z) = (4 + 3i)z$.

Problem 2.10

Find whether there exists a holomorphic function in $\mathbb{C}$ with real part $x^2 - y^2 + y$.

Solution

We note that the function $u(x,y) = x^2 - y^2 + y$ is of class C^2 in the simply connected open set $\mathbb{R}^2$. Since $\Delta u = 0$, by Proposition 2.69 there exists a harmonic conjugate function, that is, a function v such that $f = u + iv$ is holomorphic in $\mathbb{C}$. In other words, there exists a holomorphic function in $\mathbb{C}$ with real part u.

Problem 2.11

Find whether there exists a holomorphic function f in $\mathbb{C}$ with real part $x - y + 1$, and if so determine such a function.

Solution

In order to show that there exists such a function f it is sufficient to observe that $u(x,u) = x - y + 1$ is of class C^2 in the simply connected open set $\mathbb{R}^2$ and that $\Delta u = 0$. Indeed, by Proposition 2.69, this implies that u has a harmonic conjugate function.

Now we determine a holomorphic function

$$f(x+iy) = u(x,y) + iv(x,y)$$

with $u(x,y) = x - y + 1$. The Cauchy–Riemann equations must be satisfied in $\mathbb{R}^2$. It follows from $\partial u/\partial x = 1$ and the first equation in (2.33) that $\partial v/\partial y = 1$. Hence,

$$v(x,y) = y + C(x)$$

for some differentiable function C. We then obtain

$$\frac{\partial u}{\partial y} = -1 \quad \text{and} \quad -\frac{\partial v}{\partial x} = -C'(x),$$

and hence $C'(x) = 1$. Therefore, $C(x) = x + c$ for some constant $c \in \mathbb{R}$, and

$$v(x, y) = y + x + c.$$

We conclude that

$$\begin{aligned}
f(x + iy) &= (x - y + 1) + i(y + x + c) \\
&= (x + iy) + i(x + iy) + 1 + ic \\
&= (1 + i)z + 1 + ic.
\end{aligned}$$

Problem 2.12

Find all values of $a, b \in \mathbb{R}$ for which the function $u(x, y) = ax^2 + xy + by^2$ is the real part of a holomorphic function in $\mathbb{C}$, and determine explicitly all such functions.

Solution

We write $f(x + iy) = u(x, y) + iv(x, y)$. In order that f is holomorphic in $\mathbb{C}$ the Cauchy–Riemann equations must be satisfied in $\mathbb{R}^2$. It follows from

$$\frac{\partial u}{\partial x} = 2ax + y$$

and the first equation in (2.33) that

$$\frac{\partial v}{\partial y} = 2ax + y.$$

Hence,

$$v(x, y) = 2axy + \frac{y^2}{2} + C(x)$$

for some differentiable function C. We obtain

$$\frac{\partial u}{\partial y} = x + 2by \quad \text{and} \quad -\frac{\partial v}{\partial x} = -2ay - C'(x),$$

and thus, $b = -a$ and $C'(x) = -x$. Therefore, $C(x) = -x^2/2 + c$ for some constant $c \in \mathbb{R}$, and

$$v(x, y) = 2axy + \frac{y^2}{2} - \frac{x^2}{2} + c.$$

We conclude that

$$f(x+iy) = \left(ax^2 + xy - ay^2\right) + i\left(2axy + \frac{y^2}{2} - \frac{x^2}{2} + c\right)$$

$$= a\left(x^2 - y^2 + 2ixy\right) - \frac{i}{2}\left(x^2 - y^2 + 2ixy\right) + ic$$

$$= \left(a - \frac{i}{2}\right)z^2 + ic,$$

with $a, c \in \mathbb{R}$.

Problem 2.13

Show that if $f, g \colon \Omega \to \mathbb{C}$ are holomorphic functions in an open set $\Omega \subset \mathbb{C}$, then

$$(f + g)' = f' + g' \quad \text{and} \quad (fg)' = f'g + fg'.$$

Solution

Since f and g are holomorphic in Ω, the derivatives

$$f'(z_0) = \lim_{z \to z_0} \frac{f(z) - f(z_0)}{z - z_0} \quad \text{and} \quad g'(z_0) = \lim_{z \to z_0} \frac{g(z) - g(z_0)}{z - z_0}$$

are well defined for each $z_0 \in \Omega$. Therefore,

$$(f + g)'(z_0) = \lim_{z \to z_0} \frac{f(z) + g(z) - f(z_0) - g(z_0)}{z - z_0}$$

$$= \lim_{z \to z_0} \frac{f(z) - f(z_0)}{z - z_0} + \lim_{z \to z_0} \frac{g(z) - g(z_0)}{z - z_0}$$

$$= f'(z_0) + g'(z_0)$$

and

$$(fg)'(z_0) = \lim_{z \to z_0} \frac{f(z)g(z) - f(z_0)g(z_0)}{z - z_0}$$

$$= \lim_{z \to z_0} \frac{(f(z) - f(z_0))g(z_0) + f(z)(g(z) - g(z_0))}{z - z_0}$$

$$= \lim_{z \to z_0} \frac{(f(z) - f(z_0))g(z_0)}{z - z_0} + \lim_{z \to z_0} \frac{f(z)(g(z) - g(z_0))}{z - z_0}$$

$$= \lim_{z \to z_0} \frac{f(z) - f(z_0)}{z - z_0} g(z_0) + \lim_{z \to z_0} f(z) \cdot \lim_{z \to z_0} \frac{g(z) - g(z_0)}{z - z_0}$$

$$= f'(z_0)g(z_0) + f(z_0)g'(z_0).$$

Problem 2.14

Show that if f and g are holomorphic functions in $\mathbb{C}$ with $f(z_0) = g(z_0) = 0$ and $g'(z_0) \neq 0$, then

$$\lim_{z \to z_0} \frac{f(z)}{g(z)} = \frac{f'(z_0)}{g'(z_0)}.$$

Solution

We have

$$\lim_{z \to z_0} \frac{f(z)}{g(z)} = \lim_{z \to z_0} \frac{f(z) - f(z_0)}{g(z) - g(z_0)}$$

$$= \lim_{z \to z_0} \frac{(f(z) - f(z_0))/(z - z_0)}{(g(z) - g(z_0))/(z - z_0)}$$

$$= \frac{f'(z_0)}{g'(z_0)}.$$

Problem 2.15

Compute the length of the path $\gamma \colon [0,1] \to \mathbb{C}$ given by $\gamma(t) = e^{it} \cos t$ (see Figure 2.11).

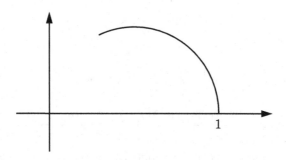

Figure 2.11 Curve defined by the path γ in Problem 2.15

Solution

The length of a piecewise regular path $\gamma\colon [a,b] \to \mathbb{C}$ is given by (2.9). We then have

$$L_\gamma = \int_0^1 \left| ie^{it}\cos t - e^{it}\sin t \right| dt$$

$$= \int_0^1 \left| e^{it} \right| \cdot \left| i\cos t - \sin t \right| dt$$

$$= \int_0^1 1 \, dt = 1.$$

Problem 2.16

Compute the integral

$$\int_\gamma \left(z^2 - \bar{z} \right) dz$$

along the path $\gamma\colon [0,1] \to \mathbb{C}$ given by $\gamma(t) = e^{it}$.

Solution

We have

$$\int_\gamma \left(z^2 - \bar{z} \right) dz = \int_0^1 \left(\gamma(t)^2 - \overline{\gamma(t)} \right) \gamma'(t)\, dt,$$

and hence,

$$\int_\gamma \left(z^2 - \bar{z} \right) dz = \int_0^1 \left(e^{2it} - e^{-it} \right) ie^{it}\, dt$$

$$= i \int_0^1 \left(e^{3it} - 1 \right) dt$$

$$= \left(\frac{1}{3}e^{3it} - it \right) \Big|_{t=0}^{t=1} = \frac{1}{3}e^{3i} - i - \frac{1}{3}.$$

Problem 2.17

For each $n \in \mathbb{Z}$, compute the integral

$$\int_\gamma \cos(nz)\, dz$$

along the path $\gamma\colon [0,1] \to \mathbb{C}$ given by $\gamma(t) = e^{\pi it}$.

Solution

Let $f_n(z) = \cos(nz)$. If $n = 0$, then $F_0(z) = z$ is a primitive of $f_0(z) = 1$, and thus,

$$
\int_\gamma f_0(z)\,dz = \int_\gamma 1\,dz = F_0(\gamma(t))\Big|_{t=0}^{t=1}
$$
$$
= e^{\pi i t}\Big|_{t=0}^{t=1} = e^{\pi i} - 1 = -2.
$$

If $n \neq 0$, then $F_n(z) = \sin(nz)/n$ is a primitive of f_n, and thus,

$$
\int_\gamma f_n(z)\,dz = F_n(\gamma(t))\Big|_{t=0}^{t=1} = \frac{1}{n}\sin(n\gamma(t))\Big|_{t=0}^{t=1}
$$
$$
= \frac{1}{n}\sin(n e^{\pi i}) - \frac{1}{n}\sin(n e^0)
$$
$$
= \frac{1}{n}\sin(-n) - \frac{1}{n}\sin n = -\frac{2}{n}\sin n.
$$

Problem 2.18

Compute the integral $\int_\gamma z\,dz$, where $\gamma\colon [a,b] \to \mathbb{C}$ is a path looping once along the boundary of the square defined by the condition $|x| + |y| \leq 3$ (see Figure 2.12), in the positive direction.

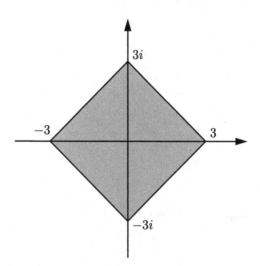

Figure 2.12 Square defined by the condition $|x| + |y| \leq 3$

Solution

We note that the function $f(z) = z$ is holomorphic in $\mathbb{C}$, and that the boundary of the square defined by the condition $|x| + |y| \le 3$ is homotopic to the circle of radius 3 centered at 0. It thus follows from Theorem 2.63 that

$$\int_\gamma z \, dz = \int_\alpha z \, dz,$$

where the path $\alpha\colon [0,1] \to \mathbb{C}$ is given by $\alpha(t) = 3e^{2\pi i t}$. Hence,

$$\int_\gamma z \, dz = \int_0^1 3e^{2\pi i t} 6\pi i e^{2\pi i t} \, dt$$

$$= 18\pi i \int_0^1 e^{4\pi i t} \, dt$$

$$= \frac{9}{2}\left(e^{4\pi i} - 1\right) = 0.$$

Problem 2.19

For each $n \in \mathbb{N}$, show that

$$\int_0^{2\pi} (2\cos t)^{2n} \, dt = 2\pi \binom{2n}{n}. \tag{2.36}$$

Solution

Let us consider the integral

$$I = \int_\gamma \frac{1}{z}\left(z + \frac{1}{z}\right)^{2n} dz,$$

where the path $\gamma\colon [0, 2\pi] \to \mathbb{C}$ is given by $\gamma(t) = e^{it}$. We have

$$I = \int_\gamma \frac{1}{z} \sum_{k=0}^{2n} \binom{2n}{k} z^k \left(z^{-1}\right)^{2n-k} dz$$

$$= \sum_{k=0}^n \binom{2n}{k} \int_\gamma z^{2k-2n-1} \, dz. \tag{2.37}$$

Since

$$\int_\gamma z^p \, dz = \begin{cases} 2\pi i & \text{if } p = -1, \\ 0 & \text{if } p \in \mathbb{Z} \setminus \{-1\}, \end{cases}$$

the only nonzero term in (2.37) occurs when $2k - 2n - 1 = -1$, that is, $k = n$, and we obtain

$$I = \binom{2n}{n} \int_\gamma z^{-1} \, dz = 2\pi i \binom{2n}{n}. \tag{2.38}$$

On the other hand,

$$I = \int_0^{2\pi} e^{-it} \left(e^{it} + e^{-it} \right)^{2n} i e^{it} \, dt = i \int_0^{2\pi} (2\cos t)^{2n} \, dt. \tag{2.39}$$

Comparing (2.38) and (2.39), we obtain identity (2.36).

Problem 2.20

For the path $\gamma \colon [0, \pi] \to \mathbb{C}$ given by $\gamma(t) = e^{it}$, show that

$$\left| \int_\gamma \frac{e^z}{z} \, dz \right| \le \pi e. \tag{2.40}$$

Solution

The length of γ is given by

$$L_\gamma = \int_0^\pi \left| \gamma'(t) \right| dt = \int_0^\pi \left| i e^{it} \right| dt = \pi,$$

and thus, by Proposition 2.42,

$$\left| \int_\gamma \frac{e^z}{z} \, dz \right| \le L_\gamma \sup \left\{ \left| \frac{e^z}{z} \right| : z \in \gamma([0, \pi]) \right\}$$

$$= \pi \sup \left\{ \frac{\left| e^{\gamma(t)} \right|}{\left| \gamma(t) \right|} : t \in [0, \pi] \right\}. \tag{2.41}$$

Since $|\gamma(t)| = 1$ and

$$\left| e^{\gamma(t)} \right| = \left| e^{\cos t + i \sin t} \right| = \left| e^{\cos t} e^{i \sin t} \right| = e^{\cos t} \le e$$

for each $t \in [0, \pi]$, inequality (2.40) follows readily from (2.41).

Problem 2.21

Find all functions $u \colon \mathbb{R}^2 \to \mathbb{R}$ of class C^1 such that

$$f(x + iy) = u(x, y) + iu(x, y) \tag{2.42}$$

is a holomorphic function in $\mathbb{C}$.

Solution

By Theorem 2.23, in order that f is holomorphic in $\mathbb{C}$, the Cauchy–Riemann equations in (2.33) must be satisfied in $\mathbb{R}^2$ with $u = v$, that is,

$$\frac{\partial u}{\partial x} = \frac{\partial u}{\partial y} \quad \text{and} \quad \frac{\partial u}{\partial y} = -\frac{\partial u}{\partial x}.$$

In particular, we have

$$\frac{\partial u}{\partial x} = -\frac{\partial u}{\partial x} \quad \text{and} \quad \frac{\partial u}{\partial y} = -\frac{\partial u}{\partial y},$$

and thus,

$$\frac{\partial u}{\partial x} = \frac{\partial u}{\partial y} = 0. \tag{2.43}$$

Since the open set $\mathbb{R}^2$ is connected, it follows from (2.43) that u is constant. Therefore, the holomorphic functions in $\mathbb{C}$ of the form (2.42) are the constant functions $a + ia$, with $a \in \mathbb{R}$.

Problem 2.22

Show that if f and $\overline{f}$ are holomorphic functions in $\mathbb{C}$, then f is constant in $\mathbb{C}$.

Solution

Writing the function f in the form (2.32), we obtain

$$\overline{f(x + iy)} = \overline{u(x, y) + iv(x, y)} = u(x, y) - iv(x, y).$$

Since f and $\overline{f}$ are holomorphic in $\mathbb{C}$, in addition to the Cauchy–Riemann equations in (2.33) for the function f, the Cauchy–Riemann equations for $\overline{f} = u - iv$ are also satisfied, that is,

$$\frac{\partial u}{\partial x} = -\frac{\partial v}{\partial y} \quad \text{and} \quad \frac{\partial u}{\partial y} = \frac{\partial v}{\partial x}. \tag{2.44}$$

It follows from (2.33) and (2.44) that

$$\frac{\partial v}{\partial y} = -\frac{\partial v}{\partial y} \quad \text{and} \quad -\frac{\partial v}{\partial x} = \frac{\partial v}{\partial x},$$

and hence,

$$\frac{\partial v}{\partial x} = \frac{\partial v}{\partial y} = 0. \tag{2.45}$$

Since the open set $\mathbb{R}^2$ is connected, it follows from (2.45) that v is constant. It then follows from Example 2.20 that f is constant.

Problem 2.23

For the function $u\colon \mathbb{R}^2 \to \mathbb{R}$ given by $u(x,y) = e^x \sin y$:
1. find a function v such that $f(x + iy) = u(x,y) + iv(x,y)$ is holomorphic in $\mathbb{C}$ and $f(0) = -i$;
2. compute the integral $\int_\gamma (f(z)/z)\, dz$, where γ is the circle of radius 4 centered at the origin, looping three times in the negative direction.

Solution

1. In order that f is holomorphic in $\mathbb{C}$, the Cauchy–Riemann equations must be satisfied in $\mathbb{R}^2$, and thus,

$$e^x \sin y = \frac{\partial v}{\partial y} \quad \text{and} \quad e^x \cos y = -\frac{\partial v}{\partial x}. \tag{2.46}$$

It follows from the first equation that

$$v(x,y) = -e^x \cos y + C(x)$$

for some differentiable function C. Thus, it follows from the second equation in (2.46) that $e^x \cos y = e^x \cos y - C'(x)$, and hence, $C(x) = c$ for some constant $c \in \mathbb{R}$. We then obtain

$$f(x + iy) = e^x \sin y + i\big(-e^x \cos y + c\big) = -ie^z + ic,$$

and it follows from $f(0) = -i$ that $c = 0$. Hence, $f(z) = -ie^z$.

2. By Cauchy's integral formula in (2.22), since $\text{Ind}_\gamma(0) = -3$, we obtain

$$\int_\gamma \frac{f(z)}{z}\, dz = 2\pi i f(0)\,\text{Ind}_\gamma(0) = 2\pi i \cdot (-i) \cdot (-3) = -6\pi.$$

Problem 2.24

Let $f = u + iv$ be a holomorphic function in an open set $\Omega \subset \mathbb{C}$. Show that if u and v are of class C^2, then

$$\Delta u = \Delta v = 0 \quad \text{in } \Omega.$$

Solution

Since f is holomorphic in Ω, the Cauchy–Riemann equations are satisfied in Ω. Taking derivatives in these equations with respect to x and y we obtain respectively

$$\frac{\partial^2 u}{\partial x^2} = \frac{\partial^2 v}{\partial x \partial y} \quad \text{and} \quad \frac{\partial^2 u}{\partial x \partial y} = -\frac{\partial^2 v}{\partial x^2}, \tag{2.47}$$

and

$$\frac{\partial^2 u}{\partial y \partial x} = \frac{\partial^2 v}{\partial y^2} \quad \text{and} \quad \frac{\partial^2 u}{\partial y^2} = -\frac{\partial^2 v}{\partial y \partial x}. \tag{2.48}$$

On the other hand, since u and v are of class C^2, we have

$$\frac{\partial^2 u}{\partial x \partial y} = \frac{\partial^2 u}{\partial y \partial x} \quad \text{and} \quad \frac{\partial^2 v}{\partial x \partial y} = \frac{\partial^2 v}{\partial y \partial x}.$$

Thus, combining the first equation in (2.47) with the second in (2.48), we obtain

$$\Delta u = \frac{\partial^2 u}{\partial x^2} + \frac{\partial^2 u}{\partial y^2} = 0.$$

Analogously, combining the second equation in (2.47) with the first in (2.48), we obtain

$$\Delta v = \frac{\partial^2 v}{\partial x^2} + \frac{\partial^2 v}{\partial y^2} = 0.$$

Problem 2.25

Let $f = u + iv$ be a holomorphic function in an open set $\Omega \subset \mathbb{C}$. Show that if u and v are of class C^2, then $\Delta(uv) = 0$ in Ω.

Solution

We obtain

$$\Delta(uv) = \frac{\partial^2(uv)}{\partial x^2} + \frac{\partial^2(uv)}{\partial y^2}$$

$$= \frac{\partial^2 u}{\partial x^2} v + 2\frac{\partial u}{\partial x}\frac{\partial v}{\partial x} + u\frac{\partial^2 v}{\partial x^2} + \frac{\partial^2 u}{\partial y^2} v + 2\frac{\partial u}{\partial y}\frac{\partial v}{\partial y} + u\frac{\partial^2 v}{\partial y^2}$$

$$= (\Delta u)v + u\Delta v + 2\frac{\partial u}{\partial x}\frac{\partial v}{\partial x} + 2\frac{\partial u}{\partial y}\frac{\partial v}{\partial y}. \tag{2.49}$$

On the other hand, by Problem 2.24, we have $\Delta u = \Delta v = 0$ in Ω. Together with the Cauchy–Riemann equations, this implies that

$$\Delta(uv) = 2\frac{\partial u}{\partial x}\frac{\partial v}{\partial x} + 2\frac{\partial u}{\partial y}\frac{\partial v}{\partial y}$$

$$= 2\frac{\partial v}{\partial y}\left(-\frac{\partial u}{\partial y}\right) + 2\frac{\partial u}{\partial y}\frac{\partial v}{\partial y} = 0.$$

Problem 2.26

Let $f = u + iv$ be a holomorphic function in an open set $\Omega \subset \mathbb{C}$. Show that if u and v are of class C^2, then $\Delta(u^2 + v^2) \geq 0$ in Ω.

Solution

By Problem 2.24, we have $\Delta u = \Delta v = 0$ in Ω. Setting $u = v$ in (2.49), we then obtain

$$\frac{1}{2}\Delta(u^2 + v^2) = u\Delta u + \left(\frac{\partial u}{\partial x}\right)^2 + \left(\frac{\partial u}{\partial y}\right)^2 + v\Delta v + \left(\frac{\partial v}{\partial x}\right)^2 + \left(\frac{\partial v}{\partial y}\right)^2$$

$$= \left(\frac{\partial u}{\partial x}\right)^2 + \left(\frac{\partial u}{\partial y}\right)^2 + \left(\frac{\partial v}{\partial x}\right)^2 + \left(\frac{\partial v}{\partial y}\right)^2 \geq 0.$$

Problem 2.27

Let f be a holomorphic function in some open set $\Omega \subset \mathbb{C}$ such that

$$|f(z) - 1| < 1 \quad \text{for } z \in \Omega. \tag{2.50}$$

Show that

$$\int_\gamma \frac{f'(z)}{f(z)}\,dz = 0$$

for any closed piecewise regular path γ in Ω.

Solution

It follows from (2.50) that f never vanishes in Ω. Therefore, the function $g\colon \Omega \to \mathbb{C}$ given by $g(z) = \log f(z)$ is well defined. It also follows from (2.50) that the image of f does not intersect the half-line $\mathbb{R}_0^- \subset \mathbb{C}$, and thus g is holomorphic in Ω. We then have

$$g'(z) = \frac{f'(z)}{f(z)},$$

and g is a primitive of f'/f. Since the path γ is closed, it follows from Proposition 2.49 that

$$\int_\gamma \frac{f'(z)}{f(z)}\, dz = \int_\gamma g'(z)\, dz = 0.$$

Problem 2.28

Show that

$$\frac{1}{2\pi} \int_0^{2\pi} \frac{R^2 - r^2}{R^2 - 2Rr\cos\theta + r^2}\, d\theta = 1, \quad 0 < r < R.$$

Solution

We have

$$\frac{R + re^{i\theta}}{R - re^{i\theta}} = \frac{(R + re^{i\theta})(R - re^{-i\theta})}{(R - re^{i\theta})(R - re^{-i\theta})}$$

$$= \frac{R^2 - r^2 + 2irR\sin\theta}{R^2 - 2rR\cos\theta + r^2}.$$

Therefore,

$$\frac{1}{2\pi} \int_0^{2\pi} \frac{R^2 - r^2}{R^2 - 2rR\cos\theta + r^2}\, d\theta = \frac{1}{2\pi} \int_0^{2\pi} \mathrm{Re}\left(\frac{R + re^{i\theta}}{R - re^{i\theta}}\right) d\theta$$

$$= \mathrm{Re}\left(\frac{1}{2\pi} \int_0^{2\pi} \frac{R + re^{i\theta}}{R - re^{i\theta}}\, d\theta\right)$$

$$= \mathrm{Re}\left(\frac{1}{2\pi i} \int_\gamma \frac{R + z}{z(R - z)}\, dz\right),$$

where the path $\gamma\colon [0, 2\pi] \to \mathbb{C}$ is given by $\gamma(\theta) = re^{i\theta}$. Moreover,

$$\frac{1}{2\pi i} \int_\gamma \frac{R+z}{z(R-z)}\, dz = \frac{1}{2\pi i} \int_\gamma \left(\frac{1}{z} + \frac{2}{R-z} \right) dz$$

$$= \frac{1}{2\pi i} \int_\gamma \frac{1}{z}\, dz + \frac{1}{2\pi i} \int_\gamma \frac{2}{R-z}\, dz$$

$$= 1 + \frac{1}{2\pi i} \int_\gamma \frac{2}{R-z}\, dz.$$

On the other hand, since the function $f(z) = 2/(R - z)$ is holomorphic for $|z| < R$, it follows from Cauchy's theorem (Theorem 2.55) that

$$\int_\gamma \frac{2}{R-z}\, dz = 0,$$

and hence,

$$\frac{1}{2\pi} \int_0^{2\pi} \frac{R^2 - r^2}{R^2 - 2rR\cos\theta + r^2}\, d\theta = \mathrm{Re}\left(1 + \frac{1}{2\pi i} \int_\gamma \frac{2}{R-z}\, dz \right) = 1.$$

Problem 2.29

Verify that the function $f(z) = (z + 1)\log z$ is continuous at $z = -1$.

Solution

Since $f(-1) = 0$, in order to verify that f is continuous at $z = -1$, one must show that

$$\lim_{z \to -1} f(z) = 0. \qquad (2.51)$$

We first observe that since

$$\log z = \log |z| + i\arg z,$$

with $\arg z \in (-\pi, \pi]$, we have

$$|\log z| = \sqrt{(\log |z|)^2 + (\arg z)^2} \le \sqrt{(\log |z|)^2 + \pi^2}.$$

Hence,

$$|\log z| \leq \sqrt{1 + \pi^2}$$

for $|z| < e$, and thus,

$$|f(z)| = |z + 1| \cdot |\log z| \leq |z + 1| \sqrt{1 + \pi^2} \to 0$$

when $z \to -1$ (we note that when we let $z \to -1$, one can always assume that $|z| < e$, since $|-1| < e$). This shows that (2.51) holds, and the function f is continuous at $z = -1$.

Problem 2.30

Find all continuous functions $f \colon \mathbb{C} \to \mathbb{C}$ such that $f(z)^2 = 1$ for $z \in \mathbb{C}$.

Solution

It follows from $f(z)^2 = 1$ that $f(z) = 1$ or $f(z) = -1$, for each $z \in \mathbb{C}$. We show that f takes only one of these values. Otherwise, there would exist $z_1, z_2 \in \mathbb{C}$ with $f(z_1) = 1$ and $f(z_2) = -1$, but by the continuity of f there would also exist a point z in the line segment between z_1 and z_2 with $f(z) \neq 1$ and $f(z) \neq -1$. But this contradicts the fact that f can only take the values 1 and -1. Therefore, either $f = 1$ or $f = -1$.

Problem 2.31

Compute the integral

$$\int_0^\infty \frac{\sin(t^2)}{t} \, dt.$$

Solution

Given $r, R > 0$, with $r < R$, we consider the path $\gamma = \gamma_1 + \gamma_2 + \gamma_3 + \gamma_4$, where

$$\gamma_1 \colon [r, R] \to \mathbb{C} \quad \text{is given by } \gamma_1(t) = t,$$
$$\gamma_2 \colon [0, \pi/2] \to \mathbb{C} \quad \text{is given by } \gamma_2(t) = Re^{it},$$
$$\gamma_3 \colon [r, R] \to \mathbb{C} \quad \text{is given by } \gamma_3(t) = i(r + R - t),$$
$$\gamma_4 \colon [0, \pi/2] \to \mathbb{C} \quad \text{is given by } \gamma_4(t) = e^{i(\pi/2 - t)}$$

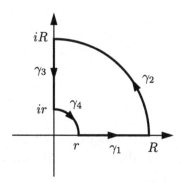

Figure 2.13   Path $\gamma = \gamma_1 + \gamma_2 + \gamma_3 + \gamma_4$

(see Figure 2.13). We also consider the function $f(z) = e^{iz^2}/z$. It follows from Cauchy's theorem (Theorem 2.55) that

$$
0 = \int_{\gamma_1} f + \int_{\gamma_2} f + \int_{\gamma_3} f + \int_{\gamma_4} f
$$

$$
= \int_r^R \frac{e^{it^2}}{t}\, dt + i \int_0^{\pi/2} e^{i(Re^{it})^2}\, dt
$$

$$
+ \int_r^R \frac{e^{-i(r+R-t)^2}}{t}\, dt + i \int_0^{\pi/2} e^{i[re^{i(\pi/2-t)}]^2}\, dt
$$

$$
= \int_r^R \frac{e^{it^2}}{t}\, dt + i \int_0^{\pi/2} e^{i(Re^{it})^2}\, dt
$$

$$
- \int_r^R \frac{e^{-it^2}}{t}\, dt - i \int_0^{\pi/2} e^{i(re^{it})^2}\, dt. \tag{2.52}
$$

On the other hand,

$$
\left| i \int_0^{\pi/2} e^{i(Re^{it})^2}\, dt \right| \le \int_0^{\pi/2} \left| e^{i(Re^{it})^2} \right| dt
$$

$$
= \int_0^{\pi/2} e^{-R^2 \sin(2t)}\, dt
$$

$$
= 2 \int_0^{\pi/4} e^{-R^2 \sin(2t)}\, dt.
$$

Now we consider the function

$$
h(t) = \sin(2t) - 4t/\pi.
$$

Since

$$h''(t) = -4\sin(2t) < 0 \quad \text{for } t \in (0, \pi/4),$$

the derivative

$$h'(t) = 2\cos(2t) - 4/\pi$$

is strictly decreasing in $[0, \pi/4]$. Hence, since $h'(0) > 0$ and $h'(\pi/4) < 0$, there exists a unique $s \in (0, \pi/4)$ such that h is increasing in $[0, s]$ and decreasing in $[s, \pi/4]$. Since $h(0) = h(\pi/4) = 0$, we conclude that $h(t) \geq 0$ for $t \in [0, \pi/4]$. Therefore,

$$\left| i \int_0^{\pi/2} e^{i(Re^{it})^2} \, dt \right| \leq 2 \int_0^{\pi/4} e^{-R^2 \sin(2t)} \, dt$$

$$\leq 2 \int_0^{\pi/4} e^{-4R^2 t/\pi} \, dt$$

$$= \frac{\pi}{2R^2} \left(1 - e^{-R^2}\right) \to 0$$

when $R \to +\infty$. It then follows from (2.52) that

$$0 = \int_r^R \frac{e^{it^2}}{t} \, dt + i \int_0^{\pi/2} e^{i(Re^{it})^2} \, dt - \int_r^R \frac{e^{-it^2}}{t} \, dt - i \int_0^{\pi/2} e^{i(re^{it})^2} \, dt$$

$$\to \int_r^\infty \frac{e^{it^2} - e^{-it^2}}{t} \, dt - i \int_0^{\pi/2} e^{i(re^{it})^2} \, dt$$

$$= 2i \int_r^\infty \frac{\sin(t^2)}{t} \, dt - i \int_0^{\pi/2} e^{i(re^{it})^2} \, dt$$

when $R \to +\infty$, and thus,

$$\int_r^\infty \frac{\sin(t^2)}{t} \, dt = \frac{1}{2} \int_0^{\pi/2} e^{i(re^{it})^2} \, dt. \tag{2.53}$$

Since the function e^{iz^2} is continuous, given $\varepsilon > 0$, there exists $r > 0$ such that $|e^{iz^2} - 1| < \varepsilon$ for every $z \in \mathbb{C}$ with $|z| \leq r$. Therefore,

$$\left| \int_0^{\pi/2} e^{i(re^{it})^2} \, dt - \frac{\pi}{2} \right| = \left| \int_0^{\pi/2} \left(e^{i(re^{it})^2} - 1 \right) dt \right|$$

$$\leq \int_0^{\pi/2} \left| e^{i(re^{it})^2} - 1 \right| dt \leq \frac{\varepsilon\pi}{2},$$

and it follows from (2.53) that

$$\left| \int_r^\infty \frac{\sin(t^2)}{t} \, dt - \frac{\pi}{4} \right| = \frac{1}{2} \left| \int_0^{\pi/2} e^{i(re^{it})^2} \, dt - \frac{\pi}{2} \right|$$

$$\leq \frac{1}{2} \int_0^{\pi/2} \left| e^{i(re^{it})^2} - 1 \right| dt \leq \frac{\varepsilon \pi}{4}$$

for any sufficiently small r. Letting $r \to 0$ and then $\varepsilon \to 0$, we conclude that

$$\int_0^\infty \frac{\sin(t^2)}{t} \, dt = \frac{\pi}{4}.$$

EXERCISES

2.1. Compute the limit, if it exists:
   (a) $\lim\limits_{z \to 0} \dfrac{z}{\bar{z}}$;
   (b) $\lim\limits_{z \to i}(\operatorname{Im} z - \operatorname{Re} z)$;
   (c) $\lim\limits_{z \to 3} z^z$.

2.2. Verify that the functions $\operatorname{Re} z$, $\operatorname{Im} z$ and $|z|$ are continuous in $\mathbb{C}$.

2.3. Find whether the function $f(z) = z + \cos z$ is continuous in $\mathbb{C}$.

2.4. Determine the set of points $z \in \mathbb{C}$ where the function is continuous:
   (a) $x|z|$;
   (b) $\begin{cases} z^3/|z|^2 & \text{if } z \neq 0, \\ 0 & \text{if } z = 0; \end{cases}$
   (c) $(z+1)\log z$.

2.5. Verify that the function $f(z) = (1 - \log z)\log z$ is not continuous.

2.6. Determine the set of points $z \in \mathbb{C}$ where the function is differentiable:
   (a) $\operatorname{Re} z \cdot \operatorname{Im} z$;
   (b) $\operatorname{Re} z + \operatorname{Im} z$;
   (c) $z^2 - |z|^2$;
   (d) $|z|(z-1)$.

2.7. Determine the set of points $z \in \mathbb{C}$ where the function is differentiable:
   (a) $e^x \cos y - i e^x \sin y$;
   (b) $x^2 y + ixy$;
   (c) $x(y-1) + ix^2(y-1)$.

2.8. Compute $(\log \log z)'$ and indicate its domain.

2.9. Let f be a holomorphic function in $\mathbb{C}$ with real part $xy - x^2 + y^2 - 1$ such that $f(0) = -1$. Find $f(z)$ explicitly and compute the second derivative $f''(z)$.

2.10. Find the constants $a, b \in \mathbb{R}$ for which the function u is the real part of a holomorphic function in $\mathbb{C}$:

(a) $u(x, y) = ax + by$;

(b) $u(x, y) = ax^2 - bxy$;

(c) $u(x, y) = ax^2 - by^2 + xy$;

(d) $u(x, y) = ax^2 + 3xy - by^4$;

(e) $u(x, y) = ax^2 + \cos x \cos y + by^2$.

2.11. For the values of $a, b \in \mathbb{R}$ obtained in Exercise 2.10, find a holomorphic function f in $\mathbb{C}$ with real part u.

2.12. Find whether there exists $a \in \mathbb{R}$ such that the function

$$f(x + iy) = ax^2 + 2xy + i(x^2 - y^2 - 2xy)$$

is holomorphic in $\mathbb{C}$.

2.13. Find all constants $a, b \in \mathbb{R}$ such that the function

$$f(x + iy) = ax^2 + 2xy + by^2 + i(y^2 - x^2)$$

is holomorphic in $\mathbb{C}$.

2.14. For each $a, b, c \in \mathbb{C}$, compute the integral

$$\int_\gamma (az^2 + bz + c)\, dz,$$

where the path $\gamma \colon [0, 1] \to \mathbb{C}$ is given by $\gamma(t) = it$.

2.15. Compute the integral $\int_\gamma (3z^2 + 3)\, dz$ along a path $\gamma \colon [a, b] \to \mathbb{C}$ with $\gamma(a) = 3$ and $\gamma(b) = 2 + i$.

2.16. Compute the integral:

(a) $\int_\gamma z \bar{z}^2\, dz$ along the path $\gamma \colon [0, 1] \to \mathbb{C}$ given by $\gamma(t) = 2e^{it}$;

(b) $\int_\gamma (e^z / z)\, dz$ along the path $\gamma \colon [0, 2] \to \mathbb{C}$ given by $\gamma(t) = e^{2\pi i t}$.

2.17. Find a primitive of the function $y + e^x \cos y - i(x - e^x \sin y)$ in $\mathbb{C}$.

2.18. Identify each statement as true or false.

(a) The function $f \colon \mathbb{C} \to \mathbb{C}$ given by $f(z) = (|z|^2 - 2)\bar{z}$ is differentiable at $z = 0$.

(b) There exists a closed regular path γ such that $\int_\gamma \sin z\, dz \neq 0$.

(c) $\int_\gamma e^z\, dz = 0$ for some closed path γ whose image is the boundary of a square.

(d) There exists a holomorphic function f in $\mathbb{C} \setminus \{0\}$ such that $f'(z) = 1/z$ in $\mathbb{C} \setminus \{0\}$.

(e) If f is holomorphic in $\mathbb{C}$ and has real part $4xy + 2e^x \sin y$, then $f(z) = -2i(e^z + z^2)$.

(f) The largest open ball centered at the origin where the function $z^2 + z$ is one-to-one has radius 1.

(g) For the path $\gamma\colon [0,2\pi] \to \mathbb{C}$ given by $\gamma(t) = 2e^{it}$, we have

$$\left| \int_\gamma \frac{\cos z}{z}\, dz \right| < 2\pi e.$$

2.19. Compute the derivative

$$\frac{d}{ds} \int_\gamma \left(s^2 z + sz^2 \right) dz$$

for each $s \in \mathbb{R}$, where the path $\gamma\colon [0,1] \to \mathbb{C}$ is given by $\gamma(t) = e^{\pi it}$.

2.20. Compute the derivative

$$\frac{d}{ds} \int_\gamma \frac{e^{s(z+1)}}{z}\, dz$$

for each $s \in \mathbb{R}$, where the path $\gamma\colon [0,1] \to \mathbb{C}$ is given by $\gamma(t) = e^{2\pi it}$.

2.21. Compute the index $\mathrm{Ind}_\gamma(-1)$ for the path $\gamma\colon [0,2] \to \mathbb{C}$ given by

$$\gamma(t) = \left[1 + t(2 - t) \right] e^{2\pi it}$$

(see Figure 2.14).

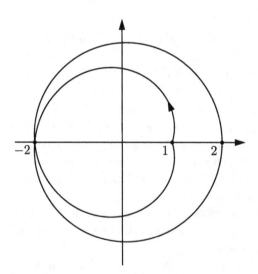

Figure 2.14 Path γ in Exercise 2.21

2.22. Find all functions $u\colon \mathbb{R}^2 \to \mathbb{R}$ of class C^1 such that $f = u + iu^2$ is a holomorphic function in $\mathbb{C}$.

2.23. Find all holomorphic functions in $\mathbb{C}$ whose real part is twice the imaginary part.

2.24. For a function

$$f\left(re^{i\theta}\right) = a(r,\theta) + ib(r,\theta),$$

show that the Cauchy–Riemann equations are equivalent to

$$\frac{\partial a}{\partial r} = \frac{1}{r}\frac{\partial b}{\partial \theta} \quad \text{and} \quad \frac{\partial b}{\partial r} = -\frac{1}{r}\frac{\partial a}{\partial \theta}.$$

2.25. Show that the function $\log z$ is holomorphic in the open set $\mathbb{C} \setminus \mathbb{R}_0^-$. Hint: use Exercise 2.24.

2.26. Find all points where the function is differentiable:
   (a) $\log(z-1)$;
   (b) $(z-1)\log(z-1)$.

2.27. Show that

$$\left| \int_\gamma \frac{e^z}{z}\, dz \right| < \pi e$$

for the path $\gamma\colon [0,\pi] \to \mathbb{C}$ given by $\gamma(t) = e^{it}$.

2.28. For a path $\gamma\colon [0,1] \to \mathbb{C}$ satisfying $|\gamma(t)| < 1$ for every $t \in [0,1]$, show that

$$\sum_{n=1}^{\infty} \int_\gamma n z^{n-1}\, dz = \int_\gamma \frac{dz}{(1-z)^2}.$$

2.29. Given a function $f(x+iy) = u(x,y) + iv(x,y)$, since

$$x = \frac{z + \bar{z}}{2} \quad \text{and} \quad y = \frac{z - \bar{z}}{2i},$$

one can define

$$g(z,\bar{z}) = f\left(\frac{z+\bar{z}}{2} + i\frac{z-\bar{z}}{2i}\right) = f(x+iy). \qquad (2.54)$$

Show that f satisfies the Cauchy–Riemann equations at (x_0, y_0) if and only if

$$\frac{\partial g}{\partial \bar{z}}(x_0 + iy_0, x_0 - iy_0) = 0.$$

2.30. For the function g in (2.54), show that if

$$\frac{\partial g}{\partial z} = \frac{\partial g}{\partial \bar{z}} = 0$$

in some connected open set $\Omega \subset \mathbb{C}$, then f is constant in Ω.

2.31. Show that the integral $\int_\gamma \overline{f(z)} f'(z)\, dz$ is purely imaginary for any closed piecewise regular path γ, and any function f of class C^1 in an open set containing the image of γ.

2.32. Show that if $f \colon \mathbb{C} \to \mathbb{C}$ is a bounded continuous function, then

$$\lim_{r \to \infty} \int_{\gamma_r} \frac{f(z)}{z^2}\, dz = 0 \quad \text{and} \quad \lim_{r \to 0} \int_{\gamma_r} \frac{f(z)}{z}\, dz = 2\pi i f(0),$$

where the path $\gamma_r \colon [0, 2\pi] \to \mathbb{C}$ is given by $\gamma_r(t) = r e^{it}$.

2.33. Let $f \colon \mathbb{C} \to \mathbb{C}$ be a holomorphic function in $\mathbb{C}$.

(a) For the path $\gamma \colon [0, 2\pi] \to \mathbb{C}$ given by $\gamma(t) = z + r e^{it}$, show that

$$|f(z)| \le \frac{1}{2\pi} \int_0^{2\pi} |f(z + r e^{it})|\, dt.$$

(b) Show that if the function f has a maximum in the closed ball $\{w \in \mathbb{C} : |w - z| \le r\}$, then it occurs at the boundary.

3
Sequences and Series

This chapter gives an introduction to the study of sequences and series, both of complex and real numbers. We note that the convergence of sequences and series of complex numbers can always be reduced to the convergence of sequences and series of real numbers. We also consider the uniform convergence of functions, and we show that in the presence of uniform convergence both limits and series commute with the integral.

3.1 Sequences

Let $(z_n)_n = (z_n)_{n \in \mathbb{N}}$ be a sequence of complex numbers. We first introduce the notion of a convergent sequence.

Definition 3.1

A sequence $(z_n)_n$ is said to be *convergent* if there exists $z \in \mathbb{C}$ such that $|z_n - z| \to 0$ when $n \to \infty$. In this case, the number z is called the *limit* of the sequence $(z_n)_n$, and we write

$$\lim_{n \to \infty} z_n = z.$$

Otherwise, the sequence $(z_n)_n$ is said to be *divergent*.

L. Barreira, C. Valls, *Complex Analysis and Differential Equations*,
Springer Undergraduate Mathematics Series,
DOI 10.1007/978-1-4471-4008-5_3, © Springer-Verlag London 2012

We also introduce the notion of a Cauchy sequence.

Definition 3.2

We say that $(z_n)_n$ is a *Cauchy sequence* if given $\varepsilon > 0$, there exists $p \in \mathbb{N}$ such that

$$|z_n - z_m| < \varepsilon \quad \text{for every } n, m > p.$$

Now we show that the two notions are equivalent.

Proposition 3.3

A sequence $(z_n)_n$ of complex numbers is convergent if and only if it is a Cauchy sequence.

Proof

If the sequence $(z_n)_n$ is convergent and has limit z, then given $\varepsilon > 0$, there exists $p \in \mathbb{N}$ such that $|z_n - z| < \varepsilon$ for every $n > p$. Therefore,

$$|z_n - z_m| \le |z_n - z| + |z_m - z| < 2\varepsilon$$

for every $n, m > p$, and $(z_n)_n$ is a Cauchy sequence.

Now we assume that $(z_n)_n$ is a Cauchy sequence. Let us write $z_n = x_n + iy_n$, with $x_n, y_n \in \mathbb{R}$ for each $n \in \mathbb{N}$. Since

$$|z_n - z_m|^2 = (x_n - x_m)^2 + (y_n - y_m)^2,$$

we obtain

$$|x_n - x_m| \le |z_n - z_m| \quad \text{and} \quad |y_n - y_m| \le |z_n - z_m|,$$

and hence, the sequences $(x_n)_n$ and $(y_n)_n$ of real numbers are also Cauchy sequences. Therefore, they are convergent (we recall that in $\mathbb{R}$ the Cauchy sequences are exactly the convergent sequences). Now let

$$x = \lim_{n \to \infty} x_n \quad \text{and} \quad y = \lim_{n \to \infty} y_n.$$

Taking $z = x + iy$, we obtain

$$|z_n - z|^2 = (x_n - x)^2 + (y_n - y)^2 \to 0$$

when $n \to \infty$. This shows that the sequence $(z_n)_n$ is convergent and that its limit is z. $\square$

We also show that a sequence is convergent if and only if its real and imaginary parts are convergent.

Proposition 3.4

If $z_n = x_n + iy_n$, with $x_n, y_n \in \mathbb{R}$ for each $n \in \mathbb{N}$, then $(z_n)_n$ is convergent if and only if $(x_n)_n$ and $(y_n)_n$ are convergent. In this case, we have

$$\lim_{n\to\infty} z_n = \lim_{n\to\infty} x_n + i \lim_{n\to\infty} y_n. \tag{3.1}$$

Proof

Given $z = x + iy$, we have

$$|z_n - z|^2 = (x_n - x)^2 + (y_n - y)^2.$$

Therefore, $z_n \to z$ if and only if

$$x_n \to x \quad \text{and} \quad y_n \to y.$$

This yields the desired property. $\qquad\square$

Example 3.5

For the sequence

$$z_n = \frac{1 + in}{n + 1},$$

we have

$$\lim_{n\to\infty} z_n = \lim_{n\to\infty} \left(\frac{1}{n+1} + i\frac{n}{n+1} \right)$$

$$= \lim_{n\to\infty} \frac{1}{n+1} + i \lim_{n\to\infty} \frac{n}{n+1} = i.$$

3.2 Series of Complex Numbers

Now we consider series $\sum_{n=0}^{\infty} z_n$ of complex numbers $z_n \in \mathbb{C}$.

Definition 3.6

A series $\sum_{n=0}^{\infty} z_n$ of complex numbers is said to be *convergent* if the sequence of partial sums $\left(\sum_{n=1}^{m} z_n\right)_m$ is convergent, in which case the limit of the sequence is called the *sum* of the series. Otherwise, the series is said to be *divergent*.

Example 3.7

Let us consider the series $\sum_{n=0}^{\infty} z^n$. We have

$$\sum_{n=0}^{m} z^n = \frac{1 - z^{m+1}}{1 - z},$$

since it is the sum of the geometric progression z^n. If $|z| < 1$, then $|z^{m+1}| = |z|^{m+1} \to 0$ when $m \to \infty$, and thus,

$$\sum_{n=0}^{m} z^n \to \frac{1}{1 - z}$$

when $m \to \infty$. Hence, the series $\sum_{n=0}^{\infty} z^n$ is convergent for $|z| < 1$, and

$$\sum_{n=0}^{\infty} z^n = \frac{1}{1 - z}.$$

Sometimes, the following property allows to establish in a simple manner the divergence of a series.

Proposition 3.8

If the series $\sum_{n=0}^{\infty} z_n$ is convergent, then $z_n \to 0$ when $n \to \infty$.

Proof

We have

$$z_m = \sum_{n=0}^{m} z_n - \sum_{n=0}^{m-1} z_n. \tag{3.2}$$

Since both $\sum_{n=0}^{m} z_n$ and $\sum_{n=0}^{m-1} z_n$ converge to $\sum_{n=0}^{\infty} z_n$ when $m \to \infty$, it follows from (3.2) that $z_m \to 0$ when $m \to \infty$. $\square$

Example 3.9

By Proposition 3.8, the series $\sum_{n=0}^{\infty} z^n$ is divergent for $|z| \geq 1$, since in this case

$$\left| z^n \right| = |z|^n \geq 1,$$

and thus the sequence z^n does not converge to zero.

Now we introduce the notion of an absolutely convergent series.

Definition 3.10

A series $\sum_{n=0}^{\infty} z_n$ is said to be *absolutely convergent* if the series of the moduli $\sum_{n=0}^{\infty} |z_n|$ is convergent.

We show that absolutely convergent series are convergent.

Proposition 3.11

If the series $\sum_{n=0}^{\infty} z_n$ is absolutely convergent, then it is also convergent, and

$$\left| \sum_{n=0}^{\infty} z_n \right| \leq \sum_{n=0}^{\infty} |z_n|.$$

Proof

For $p \geq q$, we have

$$\left| \sum_{n=0}^{p} z_n - \sum_{n=0}^{q} z_n \right| = \left| \sum_{n=q+1}^{p} z_n \right| \leq \sum_{n=q+1}^{p} |z_n|$$

$$= \sum_{n=0}^{p} |z_n| - \sum_{n=0}^{q} |z_n|$$

$$= \left| \sum_{n=0}^{p} |z_n| - \sum_{n=0}^{q} |z_n| \right|. \tag{3.3}$$

By Proposition 3.3, since the sequence $\sum_{n=0}^{p} |z_n|$ is convergent it is a Cauchy sequence. Thus, it follows from (3.3) that $\sum_{n=0}^{p} z_n$ is also a Cauchy sequence, and again by Proposition 3.3 it is convergent. This shows that the series $\sum_{n=0}^{\infty} z_n$ is convergent.

Moreover,

$$\left| \sum_{n=0}^{p} z_n \right| \le \sum_{n=0}^{p} |z_n|$$

for each $p \in \mathbb{N}$, and hence,

$$\left| \sum_{n=0}^{\infty} z_n \right| = \left| \lim_{p\to\infty} \sum_{n=0}^{p} z_n \right|$$

$$= \lim_{p\to\infty} \left| \sum_{n=0}^{p} z_n \right|$$

$$\le \lim_{p\to\infty} \sum_{n=0}^{p} |z_n| = \sum_{n=0}^{\infty} |z_n|.$$

This completes the proof of the proposition.   □

Example 3.12

Let us consider the series $\sum_{n=0}^{\infty} (-1/2)^n$. We have

$$\sum_{n=0}^{\infty} |(-1/2)^n| = \sum_{n=0}^{\infty} 2^{-n} = \lim_{m\to\infty} \sum_{n=0}^{m} 2^{-n}$$

$$= \lim_{m\to\infty} \frac{1 - 2^{-(m+1)}}{1 - 2^{-1}} = \frac{1}{1 - 2^{-1}} = 2.$$

Therefore, $\sum_{n=0}^{\infty} (-1/2)^n$ is absolutely convergent, and by Proposition 3.11 it is also convergent. In fact, we have

$$\sum_{n=0}^{\infty} (-1/2)^n = \lim_{m\to\infty} \sum_{n=0}^{m} (-1/2)^n$$

$$= \lim_{m\to\infty} \frac{1 - (-1/2)^m}{1 - (-1/2)} = \frac{1}{3/2} = \frac{2}{3}.$$

We also show that a series is convergent if and only if the corresponding series of real and imaginary parts are convergent. We consider a series $\sum_{n=0}^{\infty} z_n$ of complex numbers, and for each $n \ge 0$ we write $z_n = x_n + iy_n$, with $x_n, y_n \in \mathbb{R}$.

Proposition 3.13

The series $\sum_{n=0}^{\infty} z_n$ is convergent if and only if the series $\sum_{n=0}^{\infty} x_n$ and $\sum_{n=0}^{\infty} y_n$ are convergent. In this case, we have

$$\sum_{n=0}^{\infty} z_n = \sum_{n=0}^{\infty} x_n + i \sum_{n=0}^{\infty} y_n. \tag{3.4}$$

Proof

We have

$$\sum_{n=0}^{m} z_n = \sum_{n=0}^{m} x_n + i \sum_{n=0}^{m} y_n.$$

Therefore, by Proposition 3.4, the series $\sum_{n=0}^{\infty} z_n$ is convergent if and only if the series $\sum_{n=0}^{\infty} x_n$ and $\sum_{n=0}^{\infty} y_n$ are convergent. Identity (3.4) follows readily from (3.1). $\qquad\square$

Proposition 3.13 tells us that in order to study the convergence of series of complex numbers it is sufficient to consider series of real numbers.

3.3 Series of Real Numbers

In this section we consider series $\sum_{n=0}^{\infty} x_n$ of real numbers $x_n \in \mathbb{R}$.

Example 3.14

Let us consider the series

$$\sum_{n=1}^{\infty} \frac{1}{n(n+1)}.$$

Since

$$\frac{1}{n(n+1)} = \frac{1}{n} - \frac{1}{n+1},$$

we have

$$\sum_{n=1}^{m} \frac{1}{n(n+1)} = 1 - \frac{1}{m+1} \to 1$$

when $m \to \infty$. Therefore,

$$\sum_{n=1}^{\infty} \frac{1}{n(n+1)} = 1.$$

More generally, we have the following result.

Proposition 3.15

If $x_n = a_n - a_{n+1}$, where $(a_n)_n$ is a convergent sequence, then

$$\sum_{n=0}^{\infty}(a_n - a_{n+1}) = a_0 - \lim_{n\to\infty} a_n. \tag{3.5}$$

Proof

We have

$$\sum_{n=0}^{m}(a_n - a_{n+1}) = (a_0 - a_1) + (a_1 - a_2) + \cdots + (a_m - a_{m+1})$$

$$= a_0 - a_{m+1}.$$

Letting $m \to \infty$ yields the desired result.   □

Series such as that in (3.5) are usually called telescopic series.

Example 3.16

We show that the series

$$\sum_{n=1}^{\infty} \frac{1}{n^\alpha}$$

is convergent for $\alpha > 1$ and divergent for $\alpha \leq 1$. For $\alpha > 1$ it is sufficient to note that

$$\sum_{n=2}^{\infty} \frac{1}{n^\alpha} \leq \int_1^\infty \frac{dx}{x^\alpha} = \left.\frac{x^{-\alpha+1}}{-\alpha+1}\right|_{x=1}^{x=\infty} = \frac{1}{\alpha-1}.$$

For $\alpha \leq 1$ it is sufficient to consider $\alpha = 1$ since for any $\alpha \leq 1$ and $n \in \mathbb{N}$, we have

$$\frac{1}{n^\alpha} \geq \frac{1}{n}.$$

For $\alpha = 1$ we obtain

$$\sum_{n=1}^{\infty} \frac{1}{n} \geq \int_{1}^{\infty} \frac{dx}{x} = \log x \Big|_{x=1}^{x=\infty} = \infty.$$

Now we consider series of nonnegative numbers and we describe some tests for convergence and divergence.

Proposition 3.17

The series $\sum_{n=0}^{\infty} x_n$, with $x_n \geq 0$, is convergent if and only if the sequence of partial sums $\left(\sum_{n=1}^{m} z_n \right)_m$ is bounded from above.

Proof

It is sufficient to note that a nondecreasing sequence is convergent if and only if it is bounded from above. □

Proposition 3.17 has the following immediate consequence.

Proposition 3.18

Let us assume that $0 \leq x_n \leq y_n$ for every $n \geq 0$.
1. If $\sum_{n=0}^{\infty} y_n$ is convergent, then $\sum_{n=0}^{\infty} x_n$ is convergent.
2. If $\sum_{n=0}^{\infty} x_n$ is divergent, then $\sum_{n=0}^{\infty} y_n$ is divergent.

As a corollary of Proposition 3.18 we obtain the following result.

Proposition 3.19

If $x_n, y_n \geq 0$ for every $n \geq 0$, and $x_n/y_n \to \alpha > 0$ when $n \to \infty$, then $\sum_{n=0}^{\infty} x_n$ is convergent if and only if $\sum_{n=0}^{\infty} y_n$ is convergent.

Proof

Since $x_n/y_n \to \alpha > 0$ when $n \to \infty$, there exists $p \in \mathbb{N}$ such that

$$\frac{\alpha}{2} < \frac{x_n}{y_n} < 2\alpha$$

for every $n > p$. Hence,

$$x_n < 2\alpha y_n \quad \text{and} \quad y_n < \frac{2}{\alpha} x_n$$

for every $n > p$. The desired property now follows readily from Proposition 3.18. □

We also illustrate how to use this convergence test.

Example 3.20

We consider the series

$$\sum_{n=1}^{\infty} \frac{n^2}{n^7 + 2}. \tag{3.6}$$

Since

$$\frac{n^2}{n^7 + 2} : \frac{1}{n^5} = \frac{n^7}{n^7 + 2} \to 1$$

when $n \to \infty$, it follows from Proposition 3.19 that the series in (3.6) is convergent if and only if the series $\sum_{n=1}^{\infty} 1/n^5$ is convergent. But we know from Example 3.16 that the latter is convergent, and thus the series in (3.6) is also convergent.

The following is another convergence test.

Proposition 3.21

For a series $\sum_{n=0}^{\infty} x_n$, with $x_n > 0$:
1. if there exists $\alpha < 1$ such that $x_{n+1}/x_n \le \alpha$ for any sufficiently large n, then $\sum_{n=0}^{\infty} x_n$ is convergent;
2. if $x_{n+1}/x_n \ge 1$ for any sufficiently large n, then $\sum_{n=0}^{\infty} x_n$ is divergent.

Proof

Under the assumption in the first property, there exists $n_0 \in \mathbb{N}$ such that

$$x_{n+1} \le \alpha x_n \quad \text{for } n \ge n_0.$$

By induction, we obtain

$$x_n \le \alpha^{n-n_0} x_{n_0} \quad \text{for } n \ge n_0.$$

Therefore,

$$\sum_{n=n_0}^{m} x_n \le \sum_{n=n_0}^{m} \alpha^{n-n_0} x_{n_0} = \frac{1 - \alpha^{m-n_0+1}}{1 - \alpha} x_{n_0} \to \frac{x_{n_0}}{1 - \alpha} \qquad (3.7)$$

when $m \to \infty$, and the series $\sum_{n=0}^{\infty} x_n$ is convergent.

Under the assumption in the second property, we have $x_{n+1} \ge x_n > 0$ for any sufficiently large n. This implies that the sequence x_n cannot converge to zero, and thus the series $\sum_{n=0}^{\infty} x_n$ is divergent. $\qquad \square$

As a corollary of Proposition 3.21 we obtain the following result.

Proposition 3.22 (Ratio test)

For the series $\sum_{n=0}^{\infty} x_n$, with $x_n > 0$, let us assume that $x_{n+1}/x_n \to \alpha$ when $n \to \infty$.

1. If $\alpha < 1$, then $\sum_{n=0}^{\infty} x_n$ is convergent.
2. If $\alpha > 1$, then $\sum_{n=0}^{\infty} x_n$ is divergent.

Example 3.23

We consider the series $\sum_{n=1}^{\infty} n/5^n$. We have

$$\frac{n+1}{5^{n+1}} : \frac{n}{5^n} = \frac{n+1}{n} \cdot \frac{5^n}{5^{n+1}} = \frac{n+1}{5n} \to \frac{1}{5} < 1$$

when $n \to \infty$. It follows from Proposition 3.22 that the series is convergent.

Example 3.24

Now let us consider the series

$$\sum_{n=1}^{\infty} \frac{e^n}{n!}.$$

We have

$$\frac{e^{n+1}}{(n+1)!} : \frac{e^n}{n!} = \frac{e^{n+1} n!}{e^n (n+1)!} = \frac{e}{n+1} \to 0 < 1$$

when $n \to \infty$. It follows from Proposition 3.22 that the series is convergent.

The following is still another convergence test.

Proposition 3.25

For the series $\sum_{n=0}^{\infty} x_n$, with $x_n \geq 0$:
1. if there exists $\alpha < 1$ such that $\sqrt[n]{x_n} < \alpha$ for any sufficiently large n, then $\sum_{n=0}^{\infty} x_n$ is convergent;
2. if $\sqrt[n]{x_n} \geq 1$ for infinitely many values of n, then $\sum_{n=0}^{\infty} x_n$ is divergent.

Proof

Under the assumption in the first property, there exists an integer $n_0 \in \mathbb{N}$ such that $x_n < \alpha^n$ for every $n \geq n_0$. Proceeding as in (3.7), we conclude that the series $\sum_{n=0}^{\infty} x_n$ is convergent.

Under the assumption in the second property, we have $x_n \geq 1$ for infinitely many values of n. Hence, the sequence x_n cannot converge to zero, and the series $\sum_{n=0}^{\infty} x_n$ is divergent. □

In order to formulate a corollary of Proposition 3.25, we recall that the *upper limit* of a sequence $(x_n)_n$ of real numbers is defined by

$$\limsup_{n \to \infty} x_n = \lim_{n \to \infty} \sup\{x_m : m \geq n\}.$$

Proposition 3.26 (Root test)

For the series $\sum_{n=0}^{\infty} x_n$, with $x_n \geq 0$:
1. if $\limsup_{n \to \infty} \sqrt[n]{x_n} < 1$, then $\sum_{n=0}^{\infty} x_n$ is convergent;
2. if $\limsup_{n \to \infty} \sqrt[n]{x_n} > 1$, then $\sum_{n=0}^{\infty} x_n$ is divergent.

Proof

For the first property, since

$$\beta := \limsup_{n \to \infty} \sqrt[n]{x_n} < 1,$$

given $\varepsilon > 0$ with $\beta + \varepsilon < 1$, there exists $p \in \mathbb{N}$ such that

$$\sqrt[n]{x_n} < \beta + \varepsilon < 1 \quad \text{for every } n > p.$$

Now we can apply the first property in Proposition 3.25 to conclude that $\sum_{n=0}^{\infty} x_n$ is convergent.

For the second property, we note that $\sqrt[n]{x_n} > 1$ for infinitely many values of n. Thus, we can apply the second property in Proposition 3.25 to conclude that $\sum_{n=0}^{\infty} x_n$ is divergent. □

Example 3.27

Let us consider the series

$$\sum_{n=1}^{\infty} \frac{3+n}{n^n}. \tag{3.8}$$

We have

$$\sqrt[n]{\frac{3+n}{n^n}} = \frac{\sqrt[n]{3+n}}{n} \to 0 < 1$$

when $n \to \infty$, since $\sqrt[n]{3+n} \to 1$. Indeed,

$$\log \sqrt[n]{3+n} = \frac{\log(3+n)}{n} \to 0,$$

and thus $\sqrt[n]{3+n} \to 1$. It follows from Proposition 3.26 that the series in (3.8) is convergent.

Example 3.28

We note that for a series $\sum_{n=0}^{\infty} x_n$, with $x_n \geq 0$, if

$$\limsup_{n \to \infty} \sqrt[n]{x_n} = 1, \tag{3.9}$$

then the series can be convergent or divergent. For example, (3.9) holds when $x_n = 1/n$ and the series $\sum_{n=1}^{\infty} 1/n$ is divergent. On the other hand, (3.9) also holds when $x_n = 1/n^2$ and the series $\sum_{n=1}^{\infty} 1/n^2$ is convergent (see Example 3.16).

Finally, we consider series whose terms are alternately positive and negative.

Proposition 3.29 (Leibniz's test)

If $x_n \searrow 0$ when $n \to \infty$, then the series $\sum_{n=0}^{\infty} (-1)^n x_n$ is convergent.

Proof

Let us consider the sequence

$$S_m = \sum_{n=1}^{m} (-1)^n x_n.$$

Since $x_n \searrow 0$ when $n \to \infty$, we have

$$S_{2m+2} - S_{2m} = x_{2m+2} - x_{2m+1} < 0$$

and

$$S_{2m+3} - S_{2m+1} = -x_{2m+3} + x_{2m+2} > 0.$$

Hence, the sequences $(S_{2m})_m$ and $(S_{2m+1})_m$ are, respectively, decreasing and increasing. Moreover,

$$S_{2m+1} = S_{2m} - x_{2m+1} < S_{2m} \tag{3.10}$$

for every $m \in \mathbb{N}$. Therefore,

$$S_1 < \lim_{m \to \infty} S_{2m+1} \leq \lim_{m \to \infty} S_{2m} < S_2.$$

The first and third inequalities follow from the monotonicity of the sequences $(S_{2m})_m$ and $(S_{2m+1})_m$. Since $x_n \searrow 0$ when $n \to \infty$, it also follows from (3.10) that

$$\lim_{m \to \infty} S_{2m} - \lim_{m \to \infty} S_{2m+1} = \lim_{m \to \infty} x_{2m+1} = 0.$$

Since all sublimits of the sequence $(S_m)_m$ are equal, we conclude that the series $\sum_{n=0}^{\infty} (-1)^n x_n$ is convergent. $\qquad\square$

3.4 Uniform Convergence

In this section we study the notion of uniform convergence. Let $(f_n)_n$ be a sequence of functions $f_n \colon \Omega \to \mathbb{C}$ in a set $\Omega \subset \mathbb{C}$.

Definition 3.30

We say that the sequence $(f_n)_n$ is *uniformly convergent* on Ω if:
1. the limit

$$f(z) = \lim_{n \to \infty} f_n(z)$$

   exists for every $z \in \Omega$;
2. given $\varepsilon > 0$, there exists $p \in \mathbb{N}$ such that

$$|f_n(z) - f(z)| < \varepsilon \tag{3.11}$$

   for every $n \geq p$ and $z \in \Omega$.

Example 3.31

Now we consider the functions $f_n(z) = (z + n)/n$. Clearly, $f_n(z) \to 1$ when $n \to \infty$, for every $z \in \mathbb{C}$. On the other hand,

$$\left| f_n(z) - 1 \right| = |z/n| < \varepsilon$$

for $|z| < \varepsilon n$, but not for $z \in \mathbb{C}$ (for any n). Therefore, the convergence is not uniform on $\mathbb{C}$.

Example 3.32

Let us consider the functions $f_n(z) = \sum_{k=0}^{n} z^k$. By Example 3.7, we have

$$\lim_{n \to \infty} f_n(z) = \frac{1}{1 - z}$$

for $|z| < 1$. We show that the convergence is uniform on

$$\Omega = \left\{ z \in \mathbb{C} : |z| < r \right\}$$

for every $r < 1$. Indeed,

$$\left| \frac{1}{1 - z} - f_n(z) \right| = \left| \sum_{k=n+1}^{\infty} z^k \right|$$

$$\leq \sum_{k=n+1}^{\infty} r^k = \lim_{m \to \infty} \sum_{k=n+1}^{m} r^k$$

$$= \lim_{m \to \infty} \frac{r^{n+1}(1 - r^{m-n})}{1 - r} = \frac{r^{n+1}}{1 - r} \to 0$$

when $n \to \infty$.

The limit of a uniformly convergent sequence of continuous functions is also a continuous function.

Theorem 3.33

Let $(f_n)_n$ be a sequence of continuous functions converging uniformly on Ω. Then the limit function f is continuous in Ω.

Proof

Given $\varepsilon > 0$, there exists $p \in \mathbb{N}$ such that (3.11) holds for every $n \geq p$ and $z \in \Omega$. Therefore,

$$\left|f(z) - f(z_0)\right| \leq \left|f(z) - f_p(z)\right| + \left|f_p(z) - f_p(z_0)\right| + \left|f_p(z_0) - f(z_0)\right|$$
$$< 2\varepsilon + \left|f_p(z) - f_p(z_0)\right|$$

for every $z, z_0 \in \Omega$. Since f_p is continuous, we obtain

$$\left|f(z) - f(z_0)\right| < 3\varepsilon$$

whenever $|z - z_0|$ is sufficiently small. Since ε is arbitrary, we conclude that

$$\lim_{z \to z_0} f(z) = f(z_0)$$

for every $z_0 \in \Omega$, and the function f is continuous in Ω. $\square$

The following statement shows that in the presence of uniform convergence the limit commutes with the integral.

Proposition 3.34

Let $(f_n)_n$ be a sequence of continuous functions converging uniformly on Ω to a function f. If $\gamma\colon [a, b] \to \Omega$ is a piecewise regular path, then

$$\lim_{n \to \infty} \int_\gamma f_n = \int_\gamma f.$$

Proof

We have

$$\left|\int_\gamma f_n - \int_\gamma f\right| = \left|\int_\gamma (f_n - f)\right|$$
$$\leq L_\gamma \sup\left\{\left|f_n(\gamma(t)) - f(\gamma(t))\right| : t \in [a, b]\right\}$$
$$\leq L_\gamma \sup\left\{\left|f_n(z) - f(z)\right| : z \in \Omega\right\} \to 0$$

when $n \to \infty$. This yields the desired identity. $\square$

Now we consider the notion of uniform convergence for a series of functions.

Definition 3.35

We say that the series of functions $\sum_{n=1}^{\infty} f_n$ is *uniformly convergent* on Ω if the sequence of partial sums $\sum_{n=1}^{m} f_n$ converges uniformly on Ω when $m \to \infty$.

The following is a test for the uniform convergence of series of functions.

Theorem 3.36 (Weierstrass' test)

Given functions $f_n \colon \Omega \to \mathbb{C}$ for $n \in \mathbb{N}$, if there exist constants $a_n > 0$ such that $\sum_{n=1}^{\infty} a_n$ is convergent and

$$|f_n(z)| \le a_n \quad \text{for } n \in \mathbb{N}, \ z \in \Omega,$$

then the series $\sum_{n=1}^{\infty} f_n$ is uniformly convergent on Ω.

Proof

We first observe that

$$\left| \sum_{n=1}^{p} f_n(z) - \sum_{n=1}^{q} f_n(z) \right| \le \sum_{n=q+1}^{p} |f_n(z)| \le \sum_{n=q+1}^{\infty} a_n \tag{3.12}$$

for each $z \in \Omega$ and $p > q$. Since $\sum_{n=1}^{\infty} a_n$ is convergent, it follows from (3.12) that $\sum_{n=1}^{p} f_n(z)$ is a Cauchy sequence, and thus the limit

$$f(z) = \sum_{n=1}^{\infty} f_n(z)$$

exists for each $z \in \Omega$. Letting $p \to \infty$ in (3.12), we obtain

$$\left| f(z) - \sum_{n=1}^{q} f_n(z) \right| \le \sum_{n=q+1}^{\infty} a_n.$$

Hence, given $\varepsilon > 0$, there exists $q \in \mathbb{N}$ such that

$$\left| f(z) - \sum_{n=1}^{m} f_n(z) \right| \le \sum_{n=m+1}^{\infty} a_n \le \sum_{n=q+1}^{\infty} a_n < \varepsilon$$

for every $m \ge p$. This shows that the convergence is uniform. $\qquad\square$

Example 3.37

Let us consider the functions $f_n \colon \Omega \to \mathbb{C}$ in the set $\Omega = \{z \in \mathbb{C} : |z| < 1\}$, given by

$$f_n(z) = z^n/n^2.$$

We have $|f_n(z)| \le 1/n^2$ for $z \in \Omega$, and by Example 3.16 the series $\sum_{n=1}^{\infty} 1/n^2$ is convergent. It then follows from Theorem 3.36 that the series $\sum_{n=1}^{\infty} f_n$ is uniformly convergent on Ω.

We also show that in the presence of uniform convergence a series commutes with the integral.

Proposition 3.38

If $f = \sum_{n=1}^{\infty} f_n$ is a series of continuous functions converging uniformly on Ω, then f is continuous in Ω. Moreover,

$$\int_{\gamma} f = \sum_{n=1}^{\infty} \int_{\gamma} f_n$$

for any piecewise regular path γ in Ω.

Proof

We consider the sequence of continuous functions

$$g_n = \sum_{k=1}^{n} f_k.$$

Since g_n converges uniformly to f on Ω, it follows from Theorem 3.33 that f is continuous in Ω. Moreover, by Proposition 3.34, we have

$$\lim_{n\to\infty} \int_{\gamma} g_n = \int_{\gamma} f. \qquad (3.13)$$

On the other hand,

$$\int_{\gamma} g_n = \sum_{k=1}^{n} \int_{\gamma} f_k,$$

and thus,

$$\lim_{n\to\infty} \int_{\gamma} g_n = \sum_{k=1}^{\infty} \int_{\gamma} f_k. \qquad (3.14)$$

Comparing (3.13) and (3.14), we obtain the desired identity. $\qquad \square$

We conclude this section with a result concerning the differentiation of a series term by term.

Proposition 3.39

Given functions $f_n \colon \Omega \to \mathbb{C}$ of class C^1 in an open set $\Omega \subset \mathbb{C}$, if the series $\sum_{n=1}^{\infty} f_n$ is convergent in Ω and the series $\sum_{n=1}^{\infty} f_n'$ is uniformly convergent on Ω, then

$$\left(\sum_{n=1}^{\infty} f_n \right)' = \sum_{n=1}^{\infty} f_n' \quad \text{in } \Omega.$$

Proof

Let

$$f = \sum_{n=1}^{\infty} f_n \quad \text{and} \quad g = \sum_{n=1}^{\infty} f_n'.$$

By Proposition 3.38, the function g is continuous. Now we take $z \in \Omega$ and $h \in \mathbb{C}$ such that the line segment between z and $z + h$ is contained in Ω. We also consider the path $\gamma_h \colon [0,1] \to \mathbb{C}$ given by $\gamma_h(t) = z + th$. Again by Proposition 3.38, we have

$$\int_{\gamma_h} g = \sum_{n=1}^{\infty} \int_{\gamma_h} f_n'$$

$$= \sum_{n=1}^{\infty} \left[f_n(z+h) - f_n(z) \right] = f(z+h) - f(z). \tag{3.15}$$

We want to show that

$$\lim_{h \to 0} \frac{1}{h} \int_{\gamma_h} g = g(z). \tag{3.16}$$

It follows from

$$g(z) - \frac{1}{h} \int_{\gamma_h} g = \frac{1}{h} \int_{\gamma_h} \left[g(z) - g(w) \right] dw$$

that

$$\left| g(z) - \frac{1}{h} \int_{\gamma_h} g \right| \leq \frac{L_{\gamma_h}}{|h|} \sup \left\{ \left| g(z) - g(z+th) \right| : t \in [0,1] \right\}$$

$$= \sup \left\{ \left| g(z) - g(z+th) \right| : t \in [0,1] \right\}.$$

Since g is continuous, we obtain (3.16). It then follows from (3.15) that $f'(z) = g(z)$ for every $z \in \Omega$. This establishes the identity in the theorem. ☐

3.5 Solved Problems and Exercises

Problem 3.1

Verify that the sequence $z_n = (1/2 - i/3)^n$ is convergent.

Solution

Since

$$\left| \frac{1}{2} - \frac{i}{3} \right| = \sqrt{\frac{1}{4} + \frac{1}{9}} = \frac{\sqrt{13}}{6} < 1,$$

we have

$$|z_n| = \left| \frac{1}{2} - \frac{i}{3} \right|^n = \left(\frac{\sqrt{13}}{6} \right)^n \to 0$$

when $n \to \infty$. Hence, the sequence z_n converges to 0.

Problem 3.2

Verify that if the sequence z_n is convergent, then $|z_n|$ is also convergent.

Solution

Let

$$z = \lim_{n \to \infty} z_n.$$

We have

$$\left| |z_n| - |z| \right| \le |z_n - z| \to 0$$

when $n \to \infty$, and thus, the sequence $|z_n|$ converges to $|z|$.

Problem 3.3

Verify that the series $\sum_{n=1}^{\infty} \cos(1/n)$ is divergent.

Solution

In order that a series $\sum_{n=1}^{\infty} z_n$, with $z_n \in \mathbb{C}$, is convergent it is necessary that $z_n \to 0$ when $n \to \infty$ (see Proposition 3.8). In the present case, we have

$$z_n = \cos(1/n) \to 1 \quad \text{when } n \to \infty,$$

and hence, the series $\sum_{n=1}^{\infty} \cos(1/n)$ is divergent.

Problem 3.4

Find whether the series $\sum_{n=1}^{\infty} (-1)^n / n^2$ is absolutely convergent.

Solution

We have

$$\sum_{n=1}^{\infty} \left| \frac{(-1)^n}{n^2} \right| = \sum_{n=1}^{\infty} \frac{1}{n^2},$$

and it follows from Example 3.16 that the series $\sum_{n=1}^{\infty} 1/n^2$ is convergent.

Alternatively, note that the function $1/x^2$ is decreasing, and hence (see Figure 3.1)

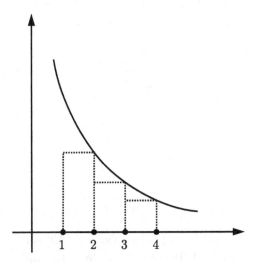

Figure 3.1 Graph of the function $1/x^2$

$$\sum_{n=2}^{\infty} \frac{1}{n^2} \le \int_{1}^{+\infty} \frac{1}{x^2} \, dx = \lim_{a \to +\infty} \int_{1}^{a} \frac{1}{x^2} \, dx$$

$$= \lim_{a \to +\infty} -\frac{1}{x}\bigg|_{x=1}^{x=a} = \lim_{a \to +\infty} \left(1 - \frac{1}{a}\right) = 1 < +\infty. \qquad (3.17)$$

Thus, the series $\sum_{n=1}^{\infty} (-1)^n / n^2$ is absolutely convergent.

Problem 3.5

Find whether the series

$$\sum_{n=1}^{\infty} \frac{n^2 \cos(1/n)}{n^4 + 1} \qquad (3.18)$$

is convergent.

Solution

Since

$$\left| \frac{n^2 \cos(1/n)}{n^4 + 1} \right| < \frac{n^2}{n^4 + 1} < \frac{1}{n^2},$$

it follows from Propositions 3.11 and 3.18 together with (3.17) that the series in (3.18) is convergent.

Problem 3.6

Show that the series

$$\sum_{n=1}^{\infty} \frac{n + 3}{4 + n^3} \qquad (3.19)$$

is convergent.

Solution

Since

$$\frac{n + 3}{4 + n^3} : \frac{1}{n^2} = \frac{n^3 + 3n^2}{4 + n^3} = \frac{1 + 3/n}{1 + 4/n^3} \to 1$$

when $n \to \infty$, by Proposition 3.19 the series in (3.19) is convergent if and only if $\sum_{n=1}^{\infty} 1/n^2$ is convergent. It then follows from Example 3.16 or from (3.17) that the series in (3.19) is convergent.

Problem 3.7

Show that the series

$$\sum_{n=1}^{\infty} \frac{\cos(n\pi)}{n^3 - n + 1} \tag{3.20}$$

is absolutely convergent.

Solution

We have $\cos(n\pi) = (-1)^n$. Since

$$\left| \frac{(-1)^n}{n^3 - n + 1} \right| : \frac{1}{n^3} = \frac{n^3}{n^3 - n + 1} \to 1$$

when $n \to \infty$, by Proposition 3.19 the series in (3.20) is absolutely convergent if and only if $\sum_{n=1}^{\infty} 1/n^3$ is convergent. By Example 3.16, the series $\sum_{n=1}^{\infty} 1/n^3$ is convergent.

Alternatively, since the function $1/x^3$ is decreasing, proceeding as in (3.17) we obtain

$$\sum_{n=2}^{\infty} \frac{1}{n^3} \le \int_1^{+\infty} \frac{1}{x^3}\, dx = \frac{1}{2} < +\infty.$$

Thus, the series in (3.20) is absolutely convergent.

Problem 3.8

Find whether the series $\sum_{n=2}^{\infty} 1/(n \log n)$ is convergent.

Solution

For $x > 1$ we have

$$\left(\frac{1}{x \log x} \right)' = -\frac{\log x + 1}{(x \log x)^2} < 0,$$

and thus, the function $1/(x \log x)$ is decreasing. Therefore,

$$\sum_{n=2}^{\infty} \frac{1}{n \log n} \ge \int_2^{+\infty} \frac{1}{x \log x}\, dx$$

$$= \lim_{a \to +\infty} \log \log x \Big|_{x=2}^{x=a} = +\infty$$

(see Figure 3.2), and the series $\sum_{n=2}^{\infty} 1/(n \log n)$ is divergent.

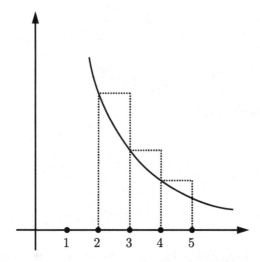

Figure 3.2 Graph of the function $1/(x \log x)$

Problem 3.9

Use the root test in Proposition 3.26 to show that the series $\sum_{n=1}^{\infty}(4^n+1)/n^n$ is convergent.

Solution

We have

$$\sqrt[n]{\frac{4^n+1}{n^n}} = \frac{\sqrt[n]{4^n+1}}{n} \to 0 < 1 \tag{3.21}$$

when $n \to \infty$, since $\sqrt[n]{4^n+1} \to 4$ when $n \to \infty$. Indeed,

$$\log \sqrt[n]{4^n+1} = \frac{\log(4^n+1)}{n}$$

$$= \frac{n\log 4 + \log(1+1/4^n)}{n} \to \log 4$$

when $n \to \infty$, and thus $\sqrt[n]{4^n+1} \to 4$ when $n \to \infty$. By the root test, it follows from (3.21) that the series $\sum_{n=1}^{\infty}(4^n+1)/n^n$ is convergent.

Problem 3.10

Verify that if $z_n = x_n + iy_n$, with $x_n, y_n \in \mathbb{R}$, is a convergent sequence, then the sequence $x_n y_n$ is also convergent.

Solution

Since z_n is convergent, by Proposition 3.4 the sequences x_n and y_n are also convergent. Let

$$x = \lim_{n \to \infty} x_n \quad \text{and} \quad y = \lim_{n \to \infty} y_n$$

be their limits. Moreover, any convergent sequence of real numbers is bounded, and hence there exists $M > 0$ such that

$$|x_n| \leq M \quad \text{and} \quad |y_n| \leq M$$

for every $n \in \mathbb{N}$. We then obtain

$$|x_n y_n - xy| = |(x_n - x)y_n + x(y_n - y)|$$
$$\leq M|x_n - x| + |x| \cdot |y_n - y| \to 0$$

when $n \to \infty$, and thus, the sequence $x_n y_n$ converges to xy.

Problem 3.11

Verify that the series

$$\sum_{n=1}^{\infty} \left(1 - \frac{4 - (-1)^n}{n}\right)^{n^2} \tag{3.22}$$

is convergent.

Solution

Since

$$4 - (-1)^n = \begin{cases} 3 & \text{if } n \text{ is even,} \\ 5 & \text{if } n \text{ is odd,} \end{cases}$$

we have

$$1 - \frac{4 - (-1)^n}{n} \leq 1 - \frac{3}{n}.$$

It then follows from

$$\lim_{n \to \infty} \left(1 + \frac{x}{n}\right)^n = e^x \tag{3.23}$$

that

$$\sqrt[n]{\left(1 - \frac{4 - (-1)^n}{n}\right)^{n^2}} \leq \sqrt[n]{\left(1 - \frac{3}{n}\right)^{n^2}}$$

$$= \left(1 - \frac{3}{n}\right)^n \to e^{-3} < 1$$

when $n \to \infty$. By the root test (see Proposition 3.26), the series in (3.22) is convergent.

Problem 3.12

Given $x_n, y_n \in \mathbb{R}$ for $n \in \mathbb{N}$, find whether the identity

$$\sum_{n=1}^{\infty} x_n \sum_{n=1}^{\infty} y_n = \sum_{n=1}^{\infty} x_n y_n \tag{3.24}$$

is always satisfied.

Solution

For example, if $x_n = y_n = 1/n$, then

$$\sum_{n=1}^{\infty} x_n y_n = \sum_{n=1}^{\infty} \frac{1}{n^2} < +\infty,$$

but

$$\sum_{n=1}^{\infty} x_n = \sum_{n=1}^{\infty} y_n = \sum_{n=1}^{\infty} \frac{1}{n} = +\infty.$$

This shows that identity (3.24) is not always satisfied.

Problem 3.13

Find the sums of the series

$$\sum_{n=1}^{\infty} \frac{1}{n(n+2)} \quad \text{and} \quad \sum_{n=1}^{\infty} \frac{1}{n(n+3)}.$$

Solution

Since

$$\frac{1}{n(n+2)} = \frac{1}{2n} - \frac{1}{2(n+2)},$$

we have

$$\sum_{n=1}^{m} \frac{1}{n(n+2)} = \frac{1}{2} + \frac{1}{2 \cdot 2} - \frac{1}{2(m+1)} - \frac{1}{2(m+2)} \to \frac{3}{4}$$

when $m \to \infty$. Hence,

$$\sum_{n=1}^{\infty} \frac{1}{n(n+2)} = \frac{3}{4}.$$

For the second series, since

$$\frac{1}{n(n+3)} = \frac{1}{3n} - \frac{1}{3(n+3)},$$

we have

$$\sum_{n=1}^{m} \frac{1}{n(n+3)} = \frac{1}{3} + \frac{1}{3 \cdot 2} + \frac{1}{3 \cdot 3} - \frac{1}{3(m+1)} - \frac{1}{3(m+2)} - \frac{1}{3(m+3)}$$

$$\to \frac{11}{18}$$

when $m \to \infty$. Hence,

$$\sum_{n=1}^{\infty} \frac{1}{n(n+3)} = \frac{11}{18}.$$

Problem 3.14

Knowing that

$$\sum_{n=0}^{\infty} \frac{1}{n!} = e,$$

compute the sum of the series

$$\sum_{n=1}^{\infty} \frac{n^2 + 1}{n!}.$$

Solution

We have

$$n^2 + 1 = n(n + 1 - 1) + 1 = n(n - 1) + n + 1,$$

and hence,

$$\frac{n^2 + 1}{n!} = \frac{1}{(n - 2)!} + \frac{1}{(n - 1)!} + \frac{1}{n!}$$

for $n \geq 2$. Therefore,

$$\sum_{n=1}^{\infty} \frac{n^2 + 1}{n!} = 2 + \sum_{n=2}^{\infty} \frac{n^2 + 1}{2}$$

$$= 2 + \sum_{n=2}^{\infty} \left(\frac{1}{(n - 2)!} + \frac{1}{(n - 1)!} + \frac{1}{n!} \right)$$

$$= 2 + e + e - 1 + e - 2 = 3e - 1.$$

Problem 3.15

Compute

$$\sum_{n=1}^{\infty} \frac{4n + 3}{4^n}.$$

Solution

Let us consider the sequence

$$S_n = \sum_{m=1}^{n} \frac{4m + 3}{4^m}.$$

We have

$$\frac{3}{4} S_n = S_n - \frac{1}{4} S_n$$

$$= \sum_{m=1}^{n} \frac{4m + 3}{4^m} - \sum_{m=1}^{n} \frac{4m + 3}{4^{m+1}}$$

$$= \sum_{m=1}^{n} \frac{4m + 3}{4^m} - \sum_{m=2}^{n+1} \frac{4m - 1}{4^m}$$

$$= \frac{7}{4} + \sum_{m=2}^{n} \frac{4m+3}{4^m} - \sum_{m=2}^{n} \frac{4m-1}{4^m} - \frac{4n+3}{4^{n+1}}$$

$$= \frac{7}{4} + \sum_{m=2}^{n} \frac{1}{4^{m-1}} - \frac{4n+3}{4^{n+1}}$$

$$= \frac{7}{4} + \frac{1}{4} \cdot \frac{1 - 1/4^{n-1}}{1 - 1/4} - \frac{4n+3}{4^{n+1}}.$$

Hence,

$$\frac{3}{4} S_n \to \frac{7}{4} + \frac{1}{3}$$

when $n \to \infty$, and thus,

$$\sum_{n=1}^{\infty} \frac{4n+3}{4^n} = \lim_{n\to\infty} S_n = \frac{4}{3}\left(\frac{7}{4} + \frac{1}{3}\right) = \frac{25}{9}.$$

Problem 3.16

Find the sums of the series

$$\sum_{n=0}^{\infty} a^n \cos(bn) \quad \text{and} \quad \sum_{n=0}^{\infty} a^n \sin(bn) \qquad (3.25)$$

for each $a \in (-1, 1)$.

Solution

We first note that

$$\left| a^n \cos(bn) \right| \le |a|^n \quad \text{and} \quad \left| a^n \sin(bn) \right| \le |a|^n. \qquad (3.26)$$

Since the series $\sum_{n=0}^{\infty} |a|^n$ is convergent for $a \in (-1, 1)$ (compare with Example 3.7), it follows from (3.26) that the two series in (3.25) are also convergent. Now we define

$$S_n = \sum_{m=0}^{n} a^m \cos(bm) \quad \text{and} \quad T_n = \sum_{m=0}^{n} a^m \sin(bm).$$

We have

$$S_n + iT_n = \sum_{m=0}^{n} a^m \big[\cos(bm) + i\sin(bm)\big]$$

$$= \sum_{m=0}^{n} a^m e^{ibm} = \sum_{m=0}^{n} \left(ae^{ib}\right)^m$$

$$= \frac{1 - \left(ae^{ib}\right)^{n+1}}{1 - ae^{ib}} \to \frac{1}{1 - ae^{ib}} \qquad (3.27)$$

when $n \to \infty$. Moreover,

$$\frac{1}{1 - ae^{ib}} = \frac{1}{(1 - a\cos b) - ia\sin b}$$

$$= \frac{(1 - a\cos b) + ia\sin b}{(1 - a\cos b)^2 + a^2\sin^2 b}$$

$$= \frac{1 - a\cos b + ia\sin b}{1 - 2a\cos b + a^2}.$$

Since

$$\sum_{n=0}^{\infty} a^n \cos(bn) = \lim_{n\to\infty} S_n \quad \text{and} \quad \sum_{n=0}^{\infty} a^n \sin(bn) = \lim_{n\to\infty} T_n,$$

it follows from (3.27) that

$$\sum_{n=0}^{\infty} a^n \cos(bn) = \frac{1 - a\cos b}{1 - 2a\cos b + a^2}$$

and

$$\sum_{n=0}^{\infty} a^n \sin(bn) = \frac{a\sin b}{1 - 2a\cos b + a^2}.$$

Problem 3.17

Show that the series

$$\sum_{n=1}^{\infty} \frac{1}{n^z} \qquad (3.28)$$

is absolutely convergent for $\operatorname{Re} z > 1$.

Solution

Writing $z = x + iy$, with $x, y \in \mathbb{R}$, we have

$$\left|n^z\right| = \left|n^{x+iy}\right| = \left|e^{(x+iy)\log n}\right| = e^{x\log n} = n^x,$$

and thus, the series in (3.28) is absolutely convergent if and only if $\sum_{n=1}^{\infty} 1/n^x$ is convergent. Proceeding as in (3.17), we obtain

$$\sum_{n=2}^{\infty} \frac{1}{n^x} \leq \int_1^{+\infty} \frac{1}{s^x}\, ds = \frac{1}{x-1} < +\infty,$$

and thus, the series in (3.28) is absolutely convergent.

Problem 3.18

Verify that the sequence of functions $f_n(z) = z^{n+1}/n^3$ converges uniformly on the set $\Omega = \{z \in \mathbb{C} : |z| < 1\}$.

Solution

For $|z| < 1$, we have

$$\left|f_n(z)\right| \leq \frac{|z|}{n^3} < \frac{1}{n^3} \to 0$$

when $n \to \infty$. Hence,

$$\sup\{\left|f_n(z)\right| : z \in \Omega\} \leq \frac{1}{n^3} \to 0$$

when $n \to \infty$, and thus the sequence f_n converges uniformly to 0 on Ω.

Problem 3.19

Show that the sequence of functions $f_n(z) = e^{-nz}$ converges uniformly on the set $\Omega = \{z \in \mathbb{C} : \operatorname{Re} z > 5\}$.

Solution

Writing $z = x + iy$, with $x, y \in \mathbb{R}$, we have $f_n(z) = e^{-nx - niy}$. For $\operatorname{Re} z = x > 5$, we obtain ·

$$\left|f_n(z)\right| = e^{-nx} < e^{-5n} \to 0$$

when $n \to \infty$. Hence,

$$\sup\{|f_n(z)| : z \in \Omega\} \le e^{-5n} \to 0$$

when $n \to \infty$, and thus the sequence f_n converges uniformly to 0 on Ω.

Problem 3.20

Verify that for each $r < 1$ the sequence of functions $f_n(z) = \sum_{k=0}^{n} z^{2k}$ converges uniformly to $1/(1-z^2)$ on the set $\Omega_r = \{z \in \mathbb{C} : |z| < r\}$.

Solution

For $|z| < 1$, we have

$$f_n(z) = \sum_{k=0}^{n} (z^2)^k = \frac{1 - z^{2(n+1)}}{1 - z^2} \to \frac{1}{1 - z^2}$$

when $n \to \infty$. Moreover, given $r < 1$, for each $z \in \Omega_r$ we have

$$\left| \frac{1}{1-z^2} - f_n(z) \right| = \left| \sum_{k=n+1}^{\infty} z^{2k} \right| \le \sum_{k=n+1}^{\infty} |z^{2k}|$$

$$\le \sum_{k=n+1}^{\infty} r^{2k} = \lim_{m \to \infty} \sum_{k=n+1}^{m} r^{2k}$$

$$= \lim_{m \to \infty} \frac{r^{2(n+1)} (1 - r^{2(m-n)})}{1 - r^2} = \frac{r^{2(n+1)}}{1 - r^2} \to 0$$

when $n \to \infty$. Hence,

$$\sup\left\{ \left| f_n(z) - \frac{1}{1-z^2} \right| : z \in \Omega_r \right\} \le \frac{r^{2(n+1)}}{1 - r^2} \to 0$$

when $n \to \infty$, and thus the sequence f_n converges uniformly to $1/(1-z^2)$ on the set Ω_r, for each $r < 1$.

Problem 3.21

Find whether the sequence of functions f_n in Problem 3.20 converges uniformly on the set $\{z \in \mathbb{C} : |z| \le 1\}$.

Solution

If $|z| = 1$, then $|z^{2k}| = |z|^{2k} = 1$, and thus the sequence f_n does not converge. Hence, it also does not converge uniformly on the set $\{z \in \mathbb{C} : |z| \le 1\}$.

Problem 3.22

Show that if the series $\sum_{n=1}^{\infty} x_n$, with $x_n \ge 0$, is convergent, then the series

$$\sum_{n=1}^{\infty} \log(1 + x_n) \tag{3.29}$$

is also convergent.

Solution

Let us consider the function $f \colon \mathbb{R}_0^+ \to \mathbb{R}$ defined by $f(x) = x - \log(1+x)$. Since

$$f'(x) = 1 - \frac{1}{1+x} > 0$$

for $x > 0$ and $f(0) = 0$, we have $f(x) \ge 0$ for $x \ge 0$. Hence,

$$0 \le \log(1 + x_n) \le x_n \tag{3.30}$$

for each $n \in \mathbb{N}$. Since the series $\sum_{n=1}^{\infty} x_n$ is convergent, it follows from Proposition 3.18 and (3.30) that the series in (3.29) is also convergent.

Problem 3.23

Show that if the series $\sum_{n=1}^{\infty} z_n$, with $z_n \in \mathbb{C} \setminus \{-1\}$, is absolutely convergent, then the series $\sum_{n=1}^{\infty} z_n/(1 + z_n)$ is also absolutely convergent.

Solution

Since the series $\sum_{n=1}^{\infty} z_n$ is convergent, we have $z_n \to 0$ when $n \to \infty$. In particular, there exists $p \in \mathbb{N}$ such that

$$|z_n| \le 1/2 \quad \text{for } n \ge p.$$

Since $|1 + z_n| \ge 1 - |z_n|$, we then obtain

$$\sum_{n=p}^{\infty} \left| \frac{z_n}{1 + z_n} \right| \le \sum_{n=p}^{\infty} \frac{|z_n|}{1 - |z_n|} \le 2 \sum_{n=p}^{\infty} |z_n| < +\infty,$$

and the series $\sum_{n=1}^{\infty} z_n/(1+z_n)$ is absolutely convergent.

Problem 3.24

Show that if $f\colon \mathbb{R} \to \mathbb{R}$ is a differentiable function with $f(0)=0$ and $|f'(x)| \le 1$ for every $x \in \mathbb{R}$, then the series

$$\sum_{n=1}^{\infty} f\left(\frac{x}{n^2}\right) \tag{3.31}$$

is absolutely convergent for each $x \in \mathbb{R}$.

Solution

Since $f(0)=0$, we have

$$f\left(\frac{x}{n^2}\right) = f\left(\frac{x}{n^2}\right) - f(0) = f'(x_n)\frac{x}{n^2}$$

for some point x_n between 0 and x/n^2. Hence,

$$\left| f\left(\frac{x}{n^2}\right) \right| \le |f'(x_n)| \cdot \left| \frac{x}{n^2} \right| \le \frac{|x|}{n^2},$$

and thus, the series in (3.31) is absolutely convergent for each $x \in \mathbb{R}$.

Problem 3.25

Consider the functions $f_n\colon \mathbb{R} \to \mathbb{R}$ given by

$$f_n(x) = \frac{1}{n}e^{-n^2 x^2}$$

for each $n \in \mathbb{N}$. Show that the sequence f_n converges uniformly to zero, and that the sequence f_n' converges to zero but not uniformly.

Solution

Since

$$|f_n(x)| \le \frac{1}{n} \quad \text{for every } n \in \mathbb{N},\ x \in \mathbb{R},$$

the sequence f_n converges uniformly to zero. On the other hand, we have

$$f_n'(x) = -\frac{2nx}{e^{n^2 x^2}},$$

and hence,

$$\lim_{n\to\infty} f_n'(x) = 0 \quad \text{for } x \in \mathbb{R};$$

that is, the sequence of functions f_n' converges to zero at every point. Nevertheless, we also have $|f_n'(\pm 1/n)| = 2/e$, and thus the sequence f_n' does not converge uniformly to zero.

Problem 3.26

Compute the derivative $\left(\sum_{n=1}^{\infty} z^{2n}/n\right)'$ for $|z| < 1$.

Solution

Since $|z^{2n}/n| \le |z|^{2n}$, the series $\sum_{n=1}^{\infty} z^{2n}/n$ is convergent for $|z| < 1$. Moreover, by Problem 3.20, the series of derivatives

$$\sum_{n=1}^{\infty} \left(\frac{z^{2n}}{n}\right)' = 2 \sum_{n=1}^{\infty} z^{2n-1}$$

$$= 2z \sum_{k=0}^{\infty} z^{2k}$$

is uniformly convergent on the set $\{z \in \mathbb{C} : |z| < r\}$, for each $r < 1$. It then follows from Proposition 3.39 and Problem 3.20 that

$$\left(\sum_{n=1}^{\infty} \frac{z^{2n}}{n}\right)' = \sum_{n=1}^{\infty} \left(\frac{z^{2n}}{n}\right)'$$

$$= 2z \sum_{k=0}^{\infty} z^{2k} = \frac{2z}{1-z^2}$$

in each set $\{z \in \mathbb{C} : |z| < r\}$, and thus for $|z| < 1$.

Problem 3.27

Show that if the series $\sum_{n=1}^{\infty} x_n$, with $x_n > 0$, is convergent, then the series

$$\sum_{n=1}^{\infty} \frac{1}{n^a} \sqrt{\frac{x_n}{n}} \tag{3.32}$$

is convergent for each $a > 0$.

Solution

Since

$$0 \le (x - y)^2 = x^2 - 2xy + y^2,$$

we have

$$xy \le \frac{x^2 + y^2}{2}$$

for every $x, y \in \mathbb{R}$. Therefore,

$$\frac{1}{n^p} \sqrt{\frac{x_n}{n}} = \sqrt{x_n} \frac{1}{n^{a+1/2}} \le \frac{1}{2}\left(x_n + \frac{1}{n^{2a+1}} \right),$$

and hence,

$$\sum_{n=1}^{\infty} \frac{1}{n^a} \sqrt{\frac{x_n}{n}} \le \frac{1}{2} \sum_{n=1}^{\infty} x_n + \frac{1}{2} \sum_{n=1}^{\infty} \frac{1}{n^{2a+1}}. \tag{3.33}$$

By hypothesis, the series $\sum_{n=1}^{\infty} x_n$ is convergent. On the other hand, by Example 3.16, since $a > 0$, the last series in (3.33) is also convergent. It then follows from (3.33) that the series in (3.32) is convergent.

Problem 3.28

Use identity (3.23) to show that $e^{x+y} = e^x e^y$ for every $x, y \in \mathbb{R}$.

Solution

It follows from (3.23) that

$$e^x e^y = \lim_{n \to \infty} \left[\left(1 + \frac{x}{n} \right)^n \left(1 + \frac{y}{n} \right)^n \right]$$

$$= \lim_{n \to \infty} \left(1 + \frac{x+y}{n} + \frac{xy}{n^2} \right)^n.$$

On the other hand, given $\varepsilon > 0$, there exists $p \in \mathbb{N}$ such that $|xy|/n < \varepsilon$ for every $n > p$. Therefore,

$$e^x e^y \le \lim_{n \to \infty} \left(1 + \frac{x+y}{n} + \frac{|xy|}{n^2} \right)^n$$

$$\le \lim_{n \to \infty} \left(1 + \frac{x+y+\varepsilon}{n} \right)^n = e^{x+y+\varepsilon},$$

and, analogously,

$$e^x e^y \geq \lim_{n \to \infty} \left(1 + \frac{x+y}{n} - \frac{|xy|}{n^2}\right)^n$$

$$\geq \lim_{n \to \infty} \left(1 + \frac{x+y-\varepsilon}{n}\right)^n = e^{x+y-\varepsilon}.$$

Hence,

$$e^{x+y-\varepsilon} \leq e^x e^y \leq e^{x+y+\varepsilon}.$$

Letting $\varepsilon \to 0$, we conclude that $e^{x+y} = e^x e^y$.

EXERCISES

3.1. Verify that the sequence is convergent:
 (a) $1/(n-i)$;
 (b) $(n+i)/(n-i)$;
 (c) e^{3ni}/n^2;
 (d) $\sin(1/n^2)\cos(1/n^2)$.

3.2. Verify that the sequence e^{in} is divergent.

3.3. Compute the limit of the sequence, if it exists:
 (a) $i^n/(n+1)$;
 (b) $(4+in)/(n+1)$;
 (c) $\cosh(in)/n$;
 (d) $i^{n+1} - i^n$.

3.4. Find whether the series is convergent or divergent:
 (a) $\sum_{n=0}^{\infty} 1/(n^5 + 2)$;
 (b) $\sum_{n=1}^{\infty} 2^{-n}n^2$;
 (c) $\sum_{n=1}^{\infty} (n+1)/3^n$;
 (d) $\sum_{n=1}^{\infty} 1/(n\sqrt[3]{n})$;
 (e) $\sum_{n=0}^{\infty} n/\sqrt{n^6 + n^2 + 1}$;
 (f) $\sum_{n=0}^{\infty} (\sqrt{n+1} - \sqrt{n})$;
 (g) $\sum_{n=0}^{\infty} 6^n/n!$;
 (h) $\sum_{n=0}^{\infty} n3^n/e^n$;
 (i) $\sum_{n=0}^{\infty} n!/(n^3 + 4^n)$.

3.5. Verify that series $\sum_{n=1}^{\infty} n!/6^n$ is divergent.

3.6. Find whether the series is convergent or divergent:
   (a) $\sum_{n=1}^{\infty}(-1)^n/(n^2+\cos n)$;
   (b) $\sum_{n=1}^{\infty}[\cos(n+1)-\cos n]$;
   (c) $\sum_{n=1}^{\infty}\cos(1/n)\sin(1/n)$;
   (d) $\sum_{n=1}^{\infty}\cos[(n+1)/(n-1)]\sin(1/n^2)$.

3.7. Find whether the series is convergent or divergent:
   (a) $\sum_{n=2}^{\infty}n/\log n$;
   (b) $\sum_{n=2}^{\infty}1/(\log n+n\log n)$;
   (c) $\sum_{n=1}^{\infty}(n-\log n)/(n+\log n)^2$;
   (d) $\sum_{n=2}^{\infty}1/\log n!$.

3.8. Verify that the series $\sum_{n=2}^{\infty}1/(n\sqrt{n}\log^3 n)$ is convergent.

3.9. Find whether the series is absolutely convergent:
   (a) $\sum_{n=1}^{\infty}(-1)^n/n$;
   (b) $\sum_{n=1}^{\infty}(-i)^n/n^2$;
   (c) $\sum_{n=1}^{\infty}(3i)^{n+1}/(n^3+1)$.

3.10. Compute the sum of the series:
   (a) $\sum_{n=1}^{\infty}4^{-n}$;
   (b) $\sum_{n=1}^{\infty}5^{-(2n+1)}$;
   (c) $\sum_{n=1}^{\infty}1/(n^2+4n)$.

3.11. Compute the sum of the series:
   (a) $\sum_{n=0}^{\infty}(an^2+bn+c)/(n+1)!$;
   (b) $\sum_{n=0}^{\infty}(an^2+bn+c)/(n+2)!$.

3.12. Find $a\in\mathbb{R}$ such that the series

$$\sum_{n=1}^{\infty}\frac{a^{n+1}}{n+1}x^n$$

is convergent for $x=-5$ and divergent for $x=5$.

3.13. Verify that the series

$$\sum_{n=1}^{\infty}\left(1+\frac{5-2(-1)^n}{n}\right)^{n^2}$$

is divergent.

3.14. Find whether the series

$$\sum_{n=1}^{\infty}\left(1+\frac{(-1)^n-2}{n}\right)^{n^2}$$

is convergent.

3.15. Show that the series

$$\sum_{n=1}^{\infty}\frac{1}{n^a}\sin\left(\frac{(-1)^n}{n^2}\right)$$

is absolutely convergent for each $a > -1$.

3.16. Show that if the series $\sum_{n=1}^{\infty}|z_n|^2$, with $z_n \in \mathbb{C}$, is convergent, then $\sum_{n=1}^{\infty}z_n^3$ is also convergent.

3.17. Find all complex numbers $z \in \mathbb{C}$ for which the series

$$\sum_{n=30}^{\infty}\left(\frac{z}{z+1}\right)^n$$

is absolutely convergent.

3.18. Show that

$$\lim_{n\to\infty}\left(1+\frac{z}{n}\right)^n = e^z.$$

Hint: compute the modulus and the argument of $(1+z/n)^n$.

3.19. Use Exercise 3.18 to show that $e^{z+w} = e^z e^w$.

3.20. Compute

$$\lim_{n\to\infty}\left(1+\frac{z}{n}+\frac{iz}{n^2}\right)^n.$$

3.21. Let $\sum_{n=1}^{\infty}f_n$ be a series of continuous functions in a set $K \subset \mathbb{R}^p$.

(a) Show that if there exist constants $a_n > 0$ such that $\sum_{n=1}^{\infty}a_n$ is convergent and $|f_n(x)| \le a_n$ for every $n \in \mathbb{N}$ and $x \in K$, then the series $\sum_{n=1}^{\infty}f_n$ is uniformly convergent on K.

(b) Show that if the series $\sum_{n=1}^{\infty}f_n$ is uniformly convergent on K, then $\sum_{n=1}^{\infty}f_n$ is continuous in K.

(c) Show that if the functions f_n are of class C^1 in an open set K, with $\sum_{n=1}^{\infty}f_n$ convergent in K and $\sum_{n=1}^{\infty}f_n'$ uniformly convergent in K, then

$$\left(\sum_{n=1}^{\infty}f_n\right)' = \sum_{n=1}^{\infty}f_n' \quad\text{in } K.$$

3.22. Show that the sequence of functions is uniformly convergent:
   (a) $f_n(z) = z^{2n}$ for $|z| < 1/3$;
   (b) $f_n(z) = z^3/(n^2 + z^2)$ for $|z| < 1$.

3.23. Find whether the sequence of functions is uniformly convergent:
   (a) $f_n(z) = z^n$ for $|z| < 1$;
   (b) $f_n(z) = nz^n$ for $|z| < 1/2$.

3.24. Show that the series is uniformly convergent:
   (a) $\sum_{n=1}^{\infty} e^{-inz}$ for $\operatorname{Re} z > 3$;
   (b) $\sum_{n=1}^{\infty} nz^n$ for $|z| < 1/3$.

3.25. Verify that $\left(\sum_{n=1}^{\infty} z^n\right)' = \sum_{n=1}^{\infty} nz^{n-1}$ for $|z| < 1$.

3.26. Compute:
   (a) $\left(\sum_{n=1}^{\infty} z^{2n+1}\right)'$ for $|z| < 1$;
   (b) $\left(\sum_{n=2}^{\infty} z^n/(n-1)\right)'$ for $|z| < 1$.

3.27. Verify that for each $r < 1$ the series $\sum_{n=0}^{\infty} p_n z^n$, with $p_n \in \{-1, 1\}$ for each $n \in \mathbb{N}$, is uniformly convergent on the set $\{z \in \mathbb{C} : |z| < r\}$.

3.28. Find whether the identity is always satisfied:
   (a) $\sum_{n=1}^{\infty} z_n^2 = \left(\sum_{n=1}^{\infty} z_n\right)^2$;
   (b) $\sum_{n=1}^{\infty} |z_n| = \left|\sum_{n=1}^{\infty} z_n\right|$.

3.29. Let $f : [1, +\infty) \to \mathbb{R}^+$ be a decreasing function that is integrable in each bounded interval. Show that the series $\sum_{n=1}^{\infty} f(n)$ is convergent if and only if $\int_1^{\infty} f(x)\,dx < +\infty$.

3.30. Verify that if $x_n, y_n \in \mathbb{R}$ for $n \in \mathbb{N}$, then

$$2 \sum_{n=1}^{\infty} x_n y_n \leq \sum_{n=1}^{\infty} \left(x_n^2 + y_n^2\right)$$

whenever the two series are convergent.

3.31. Verify that if $x_n, y_n \in \mathbb{R}$ for $n \in \mathbb{N}$, then

$$4 \sum_{n=1}^{\infty} x_n y_n \left(x_n^2 + y_n^2\right) \leq \sum_{n=1}^{\infty} \left(x_n^4 + 6x_n^2 y_n^2 + y_n^4\right)$$

whenever the two series are convergent.

3.32. Use Exercises 3.30 and 3.31 to show that if $x_n, y_n \in \mathbb{R}$ for $n \in \mathbb{N}$, then

$$\sum_{n=1}^{\infty} x_n y_n \left(x_n^2 + y_n^2\right) \leq \sum_{n=1}^{\infty} \left(x_n^4 + y_n^4\right)$$

whenever the two series are convergent.

4
Analytic Functions

In this chapter we introduce the notion of an analytic function as a function that can be represented by power series. In particular, we show that the analytic functions are exactly the holomorphic functions. We also study the notion of a singularity with the help of power series with positive and negative powers—the Laurent series. Finally, we show how to compute integrals of a class of functions with singularities—the meromorphic functions—and we describe applications to the computation of improper integrals in the real line.

4.1 Power Series

We consider the series

$$\sum_{n=0}^{\infty} c_n(z-a)^n, \qquad (4.1)$$

where $z, a, c_n \in \mathbb{C}$ for each $n \in \mathbb{N} \cup \{0\}$, with the convention that $0^0 = 1$. We call it a *power series* centered at a.

L. Barreira, C. Valls, *Complex Analysis and Differential Equations*,
Springer Undergraduate Mathematics Series,
DOI 10.1007/978-1-4471-4008-5_4, © Springer-Verlag London 2012

Definition 4.1

The number

$$R = 1/\limsup_{n \to \infty} \sqrt[n]{|c_n|}$$

is called the *radius of convergence* of the power series in (4.1).

We note that R can also take the value $+\infty$.

Example 4.2

The radius of convergence of the power series $\sum_{n=0}^{\infty}(z - i)^n/5^n$ is

$$R = 1/\limsup_{n \to \infty} \sqrt[n]{1/5^n} = 5.$$

Example 4.3

The radius of convergence of the power series $\sum_{n=1}^{\infty}(z - 1)^n/n$ is

$$R = 1/\limsup_{n \to \infty} \sqrt[n]{1/n} = 1,$$

since $\sqrt[n]{1/n} = 1/\sqrt[n]{n} \to 1$ when $n \to \infty$.

Sometimes it is possible to compute the radius of convergence as follows.

Proposition 4.4

We have

$$R = \lim_{n \to \infty} \left| \frac{c_n}{c_{n+1}} \right| \tag{4.2}$$

whenever the limit exists.

Proof

Let ρ be the limit on the right-hand side of (4.2), assuming that it exists. Given $\varepsilon > 0$, there exists $p \in \mathbb{N}$ such that

$$(1/\rho - \varepsilon)|c_n| \le |c_{n+1}| \le (1/\rho + \varepsilon)|c_n|$$

for every $n \geq p$. Hence,

$$(1/\rho - \varepsilon)^{n-p}|c_p| \leq |c_n| \leq (1/\rho + \varepsilon)^{n-p}|c_p|,$$

and thus also

$$(1/\rho - \varepsilon)^{1-p/n} \leq \sqrt[n]{|c_n|} \leq (1/\rho + \varepsilon)^{1-p/n}|c_p|^{1/n},$$

for every $n \geq p$. Letting $n \to \infty$, we obtain

$$1/\rho - \varepsilon \leq \limsup_{n \to \infty} \sqrt[n]{|c_n|} \leq 1/\rho + \varepsilon.$$

Since ε is arbitrary, we conclude that $R = \rho$. $\qquad\square$

Example 4.5

The radius of convergence of the power series $\sum_{n=0}^{\infty} z^n 3^n/(4^n + n)$ is

$$R = \lim_{n \to \infty} \frac{3^n}{4^n + n} \cdot \frac{4^{n+1} + n + 1}{3^{n+1}}$$

$$= \frac{1}{3} \lim_{n \to \infty} \frac{4 + (n+1)/4^n}{1 + n/4^n} = \frac{4}{3}.$$

Example 4.6

The radius of convergence of the power series $\sum_{n=0}^{\infty} z^n/n!$ is

$$R = \lim_{n \to \infty} \frac{(n+1)!}{n!} = \lim_{n \to \infty} (n+1) = +\infty.$$

In order to compute R using the formula

$$R = 1/\limsup_{n \to \infty} \sqrt[n]{1/n!},$$

we observe that

$$n! = n(n-1) \cdots 2 \cdot 1 \geq 2^{n-1}, \quad n \geq 2,$$

$$n! = n(n-1) \cdots 3 \cdot 2 \cdot 1 \geq 3^{n-2}, \quad n \geq 3,$$

$$\cdots$$

$$n! = n(n-1) \cdots k(k-1) \cdots 2 \cdot 1 \geq k^{n-k+1}, \quad n \geq k.$$

Therefore,

$$\sqrt[n]{n!} \geq \sqrt[n]{k^{n-k+1}} \to k$$

when $n \to \infty$. Since k is arbitrary, we conclude that $\sqrt[n]{n!} \to +\infty$ when $n \to \infty$. We finally obtain

$$R = 1/\limsup_{n\to\infty} \left(1/\sqrt[n]{n!}\right) = \lim_{n\to\infty} \sqrt[n]{n!} = +\infty.$$

Example 4.7

The radius of convergence of the power series $\sum_{n=1}^{\infty}[2+(-1)^n](z-4)^n$ is

$$R = 1/\limsup_{n\to\infty} \sqrt[n]{2+(-1)^n} = 1.$$

In this case the limit in (4.2) does not exist, and thus, one cannot use Proposition 4.4 to compute the radius of convergence. Indeed,

$$\frac{c_{n+1}}{c_n} = \frac{2-(-1)^n}{2+(-1)^n} = \begin{cases} 1/3 & \text{if } n \text{ is even,} \\ 3 & \text{if } n \text{ is odd.} \end{cases}$$

The following result discusses the convergence of power series. We denote the open ball of radius $r > 0$ centered at $a \in \mathbb{C}$ by

$$B_r(a) = \left\{ z \in \mathbb{C} : |z - a| < r \right\}$$

(see Figure 4.1), and its closure by

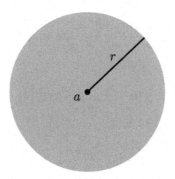

Figure 4.1   Open ball $B_r(a)$

$$\overline{B_r(a)} = \left\{ z \in \mathbb{C} : |z - a| \le r \right\}.$$

We continue to denote by R the radius of convergence.

Theorem 4.8

The power series in (4.1) is:
1. absolutely convergent for $z \in B_R(a)$;
2. divergent for $z \notin \overline{B_R(a)}$;
3. uniformly convergent on $\overline{B_r(a)}$ for each $r < R$, that is, given $r < R$ and $\varepsilon > 0$, there exists $p \in \mathbb{N}$ such that

$$\left| \sum_{n=0}^{\infty} c_n(z-a)^n - \sum_{n=0}^{m} c_n(z-a)^n \right| < \varepsilon$$

for every $m > p$ and $z \in \overline{B_r(a)}$.

Proof

Given $z \in \overline{B_r(a)}$, we have

$$\sum_{n=0}^{\infty} |c_n(z-a)^n| \le \sum_{n=0}^{\infty} |c_n| r^n.$$

On the other hand, given $\varepsilon > 0$, there exists $p \in \mathbb{N}$ such that

$$\sqrt[n]{|c_n|} < \frac{1}{R} + \varepsilon$$

for every $n \ge p$. Therefore,

$$\sum_{n=p}^{\infty} |c_n(z-a)^n| \le \sum_{n=p}^{\infty} \left(\frac{1}{R} + \varepsilon \right)^n r^n. \tag{4.3}$$

If $r < R$, then, taking ε sufficiently small, we have

$$\left(\frac{1}{R} + \varepsilon \right) r < 1,$$

and it follows from (4.3) that the power series in (4.1) is absolutely convergent. This establishes the first property.

For the second property, we note that given $\varepsilon > 0$, there exists a sequence of natural numbers $k_n \nearrow \infty$ when $n \to \infty$ such that

$$\sqrt[k_n]{|c_{k_n}|} > \frac{1}{R} - \varepsilon.$$

Therefore, if $|z - a| > R$, then, taking ε sufficiently small, we have

$$\left| c_{k_n}(z-a)^{k_n} \right|^{1/k_n} > \left(\frac{1}{R} - \varepsilon \right)|z-a| > 1.$$

This implies that the series in (4.1) is divergent.

Finally, proceeding as in the first property, if $\rho = (1/R + \varepsilon)r < 1$, then

$$\sum_{n=m}^{\infty} \left| c_n(z-a)^n \right| \leq \sum_{n=m}^{\infty} \rho^n = \frac{\rho^m}{1 - \rho}$$

for every $m \geq p$. This shows that the convergence is uniform. $\qquad \square$

Example 4.9

Let us consider the power series $\sum_{n=1}^{\infty} z^n/n$. Its radius of convergence is

$$R = 1/\limsup_{n \to \infty} \sqrt[n]{1/n} = 1.$$

Therefore, the series is absolutely convergent for $|z| < 1$, divergent for $|z| > 1$, and uniformly convergent on each ball $\overline{B_r(0)}$ with $r < 1$.

For points with $|z| = 1$ one can have convergence or divergence. For example, for $z = 1$ the series is divergent while for $z = -1$ it is convergent. We are not going to describe any general method to study what happens at the boundary of the domain of convergence of a power series.

Now we introduce the notion of an analytic function.

Definition 4.10

A function $f \colon \Omega \to \mathbb{C}$ is said to be *analytic* in an open set $\Omega \subset \mathbb{C}$ if for each ball $B_r(a) \subset \Omega$ there exists a power series

$$\sum_{n=0}^{\infty} c_n(z-a)^n$$

converging to $f(z)$ for each $z \in B_r(a)$.

Example 4.11

The function $f(z) = 1/(z-1)$ is analytic in $\mathbb{C} \setminus \{1\}$. Indeed, given $a \in \mathbb{C} \setminus \{1\}$, one can write

$$
\begin{aligned}
\frac{1}{z-1} &= \frac{1}{z-a-(1-a)} \\
&= \frac{1}{a-1} \cdot \frac{1}{1-(z-a)/(1-a)} \\
&= \frac{1}{a-1} \sum_{n=0}^{\infty} \left(\frac{z-a}{1-a}\right)^n \\
&= \sum_{n=0}^{\infty} -\frac{(z-a)^n}{(1-a)^{n+1}}
\end{aligned}
$$

whenever $|(z-a)/(1-a)| < 1$, that is, for $|z-a| < |1-a|$.

We show that analytic functions are of class C^{∞}, and thus, in particular, they are holomorphic.

Theorem 4.12

If the function $f \colon \Omega \to \mathbb{C}$ is analytic in Ω, then f is of class C^{∞} in Ω and all its derivatives $f^{(k)}$ are also analytic in Ω, with

$$
f^{(k)}(z) = \sum_{n=k}^{\infty} \frac{n!}{(n-k)!} c_n (z-a)^{n-k} \tag{4.4}
$$

for each $k \in \mathbb{N}$ and $z \in B_r(a) \subset \Omega$. Moreover,

$$
c_k = \frac{f^{(k)}(a)}{k!}, \quad k \in \mathbb{N} \cup \{0\}. \tag{4.5}
$$

Proof

We note that it is sufficient to establish (4.4) for $k = 1$. Indeed, for $k > 1$ identity (4.4) then follows by induction. Moreover, setting $z = a$ in (4.4), we obtain

$$
f^{(k)}(a) = \frac{k!}{(k-k)!} c_k = k! c_k,
$$

which establishes (4.5). Finally, since

$$\limsup_{n\to\infty} \sqrt[n]{\left|\frac{n!}{(n-k)!}c_n\right|} = \limsup_{n\to\infty} \sqrt[n]{n(n-1)\cdots(n-k+1)|c_n|}$$

$$= \limsup_{n\to\infty} \sqrt[n]{|c_n|},$$

the radius of convergence of the power series in (4.4) coincides with the radius of convergence of the power series of f in (4.1).

Let us then establish (4.4) for $k=1$. We consider the function

$$g(z) = \sum_{n=1}^{\infty} nc_n(z-a)^{n-1}.$$

For $w = z + h$, we have

$$\frac{f(w)-f(z)}{h} - g(z) = \sum_{n=1}^{\infty} c_n\left(\frac{(w-a)^n - (z-a)^n}{h} - n(z-a)^{n-1}\right).$$

Note that the first term of the series (for $n=1$) is zero. On the other hand, one can use induction to show that

$$\frac{(w-a)^n - (z-a)^n}{h} - n(z-a)^{n-1} = h\sum_{k=1}^{n-1} k(w-a)^{n-k-1}(z-a)^{k-1} \quad (4.6)$$

for $n \geq 2$. Indeed, for $n=2$ we have

$$\frac{(w-a)^2 - (z-a)^2}{h} - 2(z-a) = \frac{(w-z)(w+z-2a)}{h} - 2(z-a) = h.$$

Moreover, assuming that (4.6) holds for a given n, we obtain

$$\frac{(w-a)^{n+1} - (z-a)^{n+1}}{h} - (n+1)(z-a)^n$$

$$= \frac{(w-a)[(w-a)^n - (z-a)^n] + (z-a)^n(w-z)}{h} - (n+1)(z-a)^n$$

$$= (w-a)\left(\frac{(w-a)^n - (z-a)^n}{h} - n(z-a)^{n-1}\right)$$

$$\quad + (w-a)n(z-a)^{n-1} + (z-a)^n - (n+1)(z-a)^n$$

$$= (w-a)h\sum_{k=1}^{n-1} k(w-a)^{n-k-1}(z-a)^{k-1}$$

$$\quad + n(w-a)(z-a)^{n-1} - n(z-a)^n$$

$$= h \sum_{k=1}^{n-1} k(w-a)^{(n+1)-k-1}(z-a)^{k-1} + n(w-a-z+a)(z-a)^{n-1}$$

$$= h \sum_{k=1}^{n} k(w-a)^{(n+1)-k-1}(z-a)^{k-1},$$

which establishes identity (4.6) with n replaced by $n+1$. Given $z, w \in B_r(a)$, it follows from (4.6) that

$$\left| \frac{f(w)-f(z)}{h} - g(z) \right| \le \sum_{n=2}^{\infty} |c_n| \cdot \left| \frac{(w-a)^n - (z-a)^n}{h} - n(z-a)^{n-1} \right|$$

$$\le \sum_{n=2}^{\infty} |h| \cdot |c_n| \sum_{k=1}^{n-1} k|w-a|^{n-k-1}|z-a|^{k-1}$$

$$\le |h| \sum_{n=2}^{\infty} |c_n| r^{n-2} \sum_{k=1}^{n-1} k$$

$$= |h| \sum_{n=2}^{\infty} |c_n| \frac{n(n-1)}{2} r^{n-2}. \tag{4.7}$$

Since

$$\limsup_{n\to\infty} \sqrt[n]{|c_n| \frac{n(n-1)}{2}} = \limsup_{n\to\infty} \sqrt[n]{|c_n|}$$

and $B_r(a) \subset \Omega$, we have

$$r < 1/\limsup_{n\to\infty} \sqrt[n]{|c_n|}$$

$$= 1/\limsup_{n\to\infty} \sqrt[n]{|c_n| n(n-1)/2}.$$

Hence, by Theorem 4.8 the last series in (4.7) is convergent. Letting $h \to 0$ we finally obtain

$$f'(z) = \lim_{h\to 0} \frac{f(z+h)-f(z)}{h} = g(z).$$

This yields the desired identity. $\square$

In particular, Theorem 4.12 shows that all analytic functions are holomorphic. Now we show that all holomorphic functions are analytic.

Theorem 4.13

If $f\colon \Omega \to \mathbb{C}$ is a holomorphic function in an open set $\Omega \subset \mathbb{C}$, then f is analytic in Ω.

Proof

Given $a \in \Omega$ and $r > 0$ such that $B_r(a) \subset \Omega$, we consider the restriction of f to the ball $B_r(a)$, which is a convex open set. We also consider the path $\gamma_s\colon [0, 2\pi] \to \mathbb{C}$ given by

$$\gamma_s(t) = a + se^{it},$$

for $s < r$. By Theorem 2.60, for each $z \in B_s(a)$ we have

$$f(z) = \frac{1}{2\pi i} \int_{\gamma_s} \frac{f(w)}{w - z}\, dw, \tag{4.8}$$

since $\operatorname{Ind}_{\gamma_s}(z) = 1$. On the other hand, given $z \in B_s(a)$ and $w \in \mathbb{C}$ with $|w - a| = s$, we have

$$\frac{1}{w - z} = \frac{1}{w - a - (z - a)}$$

$$= \frac{1}{w - a} \cdot \frac{1}{1 - (z - a)/(w - a)}$$

$$= \sum_{n=0}^{\infty} \frac{(z - a)^n}{(w - a)^{n+1}},$$

since $|z - a| < |w - a| = s$. It then follows from (4.8) and Proposition 3.38 that

$$f(z) = \sum_{n=0}^{\infty} \left(\frac{1}{2\pi i} \int_{\gamma_s} \frac{f(w)}{(w - a)^{n+1}}\, dw \right)(z - a)^n \tag{4.9}$$

for each $z \in B_s(a)$. By Theorem 2.63, the numbers

$$c_n = \frac{1}{2\pi i} \int_{\gamma_s} \frac{f(w)}{(w - a)^{n+1}}\, dw$$

are independent of s. Therefore, letting $s \to r$ we conclude that identity (4.9) holds for every $z \in B_r(a)$, and thus f is analytic in Ω. $\qquad\square$

We give some examples.

Example 4.14

Using the formula for the coefficients in (4.5), we obtain, for example:

$$e^z = \sum_{n=0}^{\infty} \frac{1}{n!} z^n, \quad z \in \mathbb{C};$$

$$\log(1 - z) = \sum_{n=1}^{\infty} -\frac{1}{n} z^n, \quad |z| < 1;$$

$$\sin z = \sum_{n=0}^{\infty} \frac{(-1)^n}{(2n + 1)!} z^{2n+1}, \quad z \in \mathbb{C}; \tag{4.10}$$

$$\cos z = \sum_{n=0}^{\infty} \frac{(-1)^n}{(2n)!} z^{2n}, \quad z \in \mathbb{C}.$$

In fact, the formulas in (4.10) for e^z, $\sin z$ and $\cos z$ are the usual definitions of these functions (even in $\mathbb{R}$).

Example 4.15

Using the series for the sine function in (4.10), we obtain

$$\frac{\sin z}{z} = \sum_{n=0}^{\infty} \frac{(-1)^n}{(2n + 1)!} z^{2n} \tag{4.11}$$

for $z \neq 0$. Since analytic functions (such as the series on the right-hand side of (4.11)) are differentiable, they are also continuous, and thus,

$$\lim_{z \to 0} \frac{\sin z}{z} = \sum_{n=0}^{\infty} \frac{(-1)^n}{(2n + 1)!} 0^{2n} = 1,$$

since $0^0 = 1$. Analogously, using the series for the cosine function in (4.10), we obtain

$$\frac{\cos z - 1}{z^2} = \sum_{n=1}^{\infty} \frac{(-1)^n}{(2n)!} z^{2n-2}$$

for $z \neq 0$, and hence,

$$\lim_{z \to 0} \frac{\cos z - 1}{z^2} = -\frac{1}{2}.$$

It follows from Theorems 4.12 and 4.13 that holomorphic functions are C^∞. We also have the following corollary.

Theorem 4.16

A function f is holomorphic in an open set $\Omega \subset \mathbb{C}$ if and only if it is analytic in Ω.

Comparing (4.5) with (4.9), we obtain the following result.

Theorem 4.17 (Cauchy's integral formula for the derivatives)

Let f be a holomorphic function in an open set $\Omega \subset \mathbb{C}$. If $B_r(z) \subset \Omega$ and γ is a closed piecewise regular path in $B_r(z) \setminus \{z\}$, then

$$f^{(k)}(z) \operatorname{Ind}_\gamma(z) = \frac{k!}{2\pi i} \int_\gamma \frac{f(w)}{(w-z)^{k+1}} \, dw, \quad k \in \mathbb{N} \cup \{0\}. \qquad (4.12)$$

We give two applications of Cauchy's integral formula for the first derivative.

Theorem 4.18 (Liouville's theorem)

If a holomorphic function $f \colon \mathbb{C} \to \mathbb{C}$ is bounded, then it is constant.

Proof

For each $z \in \mathbb{C}$ and $r > 0$, we consider the path $\gamma \colon [0, 2\pi] \to \mathbb{C}$ given by $\gamma(t) = z + re^{it}$. Then $\operatorname{Ind}_\gamma(z) = 1$ and it follows from (4.12) with $k = 1$ that

$$f'(z) = \frac{1}{2\pi i} \int_\gamma \frac{f(w)}{(w-z)^2} \, dw.$$

Therefore, by Proposition 2.42,

$$\begin{aligned}
|f'(z)| &= \frac{1}{2\pi} \left| \int_\gamma \frac{f(w)}{(w-z)^2} \, dw \right| \\
&\leq \frac{1}{2\pi} L_\gamma \sup \left\{ \frac{|f(\gamma(t))|}{|re^{it}|^2} : t \in [0, 2\pi] \right\} \\
&= \frac{1}{r} \sup \left\{ |f(\gamma(t))| : t \in [0, 2\pi] \right\}. \qquad (4.13)
\end{aligned}$$

On the other hand, since f is bounded, there exists $L > 0$ such that

$$\sup \left\{ |f(\gamma(t))| : t \in [0, 2\pi] \right\} \leq L.$$

It then follows from (4.13) that

$$|f'(z)| \leq \frac{L}{r} \to 0$$

when $r \to \infty$, and $f'(z) = 0$ for every $z \in \mathbb{C}$. By Proposition 2.19, we conclude that f is constant. $\quad\square$

Now we establish the Fundamental theorem of algebra, as a consequence of Theorem 4.18.

Theorem 4.19 (Fundamental theorem of algebra)

Any nonconstant polynomial $P(z)$ with coefficients in $\mathbb{C}$ has zeros in $\mathbb{C}$.

Proof

We proceed by contradiction. If $P(z)$ had no zeros in $\mathbb{C}$, then $f(z) = 1/P(z)$ would be a holomorphic function in $\mathbb{C}$. Now we write

$$P(z) = z^n + a_{n-1}z^{n-1} + \cdots + a_1 z + a_0$$

for some $a_0, a_1, \ldots, a_{n-1} \in \mathbb{C}$ and $n \in \mathbb{N}$. Letting $r = |z|$, we obtain

$$
\begin{aligned}
|P(z)| &\geq |z|^n - |a_{n-1}z^{n-1} + \cdots + a_1 z + a_0| \\
&\geq r^n - \left(|a_{n-1}|r^{n-1} + \cdots + |a_1|r + |a_0|\right) \\
&\geq r^{n-1}\left(r - |a_{n-1}| - \cdots - |a_1| - |a_0|\right) \to +\infty
\end{aligned}
$$

when $r \to \infty$. This shows that the function f is bounded. It then follows from Theorem 4.18 that f is constant, which yields a contradiction (because P is not constant). Hence, P must have zeros in $\mathbb{C}$. $\quad\square$

Finally, as an application of Theorem 4.16, we show that the uniform limit of a sequence of holomorphic functions is still a holomorphic function.

Theorem 4.20

Let $(f_n)_n$ be a sequence of holomorphic functions converging uniformly on an open set $\Omega \subset \mathbb{C}$. Then the limit function f is holomorphic in Ω.

Proof

Given $a \in \Omega$, let us take $r > 0$ such that $B_r(a) \subset \Omega$. Since each function f_n is holomorphic, it follows from Theorem 2.55 that

$$\int_\gamma f_n = 0$$

for any closed piecewise regular path γ in the ball $B_r(a)$. By Proposition 3.34, we obtain

$$\int_\gamma f = \lim_{n \to \infty} \int_\gamma f_n = 0.$$

Proceeding as in the proof of Theorem 2.51, one can show that in the ball $B_r(a)$ the function f has the primitive F in (2.16); that is, f is the derivative of a holomorphic function in $B_r(a)$. Since holomorphic functions are analytic, f is the derivative of an analytic function, and hence it is also analytic. Finally, since analytic functions are holomorphic, we conclude that f is holomorphic. $\square$

4.2 Zeros

We show in this section that the zeros of á nonzero analytic function are isolated.

Theorem 4.21

Let $f \colon \Omega \to \mathbb{C}$ be a nonzero analytic function in a connected open set $\Omega \subset \mathbb{C}$. Then $\{z \in \Omega : f(z) = 0\}$ is a set of isolated points.

Proof

Let us assume that the set

$$A = \{z \in \Omega : f^{(m)}(z) = 0 \text{ for } m \in \mathbb{N} \cup \{0\}\}$$

is nonempty. Given $z \in \overline{A} \cap \Omega$, there exists a sequence $(z_n)_n$ in A converging to z. By Theorem 4.12, the function f is C^∞ in Ω, and thus,

$$f^{(m)}(z) = \lim_{n \to \infty} f^{(m)}(z_n) = 0$$

for every $m \in \mathbb{N} \cup \{0\}$. This shows that $z \in A$, that is,

$$\overline{A} \cap \Omega = A. \tag{4.14}$$

On the other hand, given $a \in A$, let us take $r > 0$ such that $B_r(a) \subset \Omega$. Since f is analytic, we have

$$f(z) = \sum_{n=0}^{\infty} \frac{f^{(n)}(a)}{n!}(z - a)^n = 0, \quad z \in B_r(a).$$

This shows that $B_r(a) \subset A$, and hence the set A is open. Therefore, if $A \neq \Omega$, then A and $\Omega \setminus A$ are nonempty sets with

$$\Omega = A \cup (\Omega \setminus A),$$

such that

$$\overline{A} \cap (\Omega \setminus A) = (\overline{A} \cap \Omega) \setminus (\overline{A} \cap A) = \varnothing$$

(by (4.14)) and

$$A \cap \overline{\Omega \setminus A} = \varnothing$$

(since A is open). But this is impossible, since Ω is connected. Hence, $A = \varnothing$.

Given $a \in \Omega$ with $f(a) = 0$, let us take $r > 0$ such that $B_r(a) \subset \Omega$. Since f is analytic, we have

$$f(z) = \sum_{n=0}^{\infty} c_n(z - a)^n, \quad z \in B_r(a) \tag{4.15}$$

for some constants $c_n \in \mathbb{C}$. Moreover, since $A = \varnothing$, there exists $n \in \mathbb{N} \cup \{0\}$ such that $c_n \neq 0$ in (4.15). Let m be the smallest integer with this property. Then

$$f(z) = \sum_{n=m}^{\infty} c_n(z - a)^n = (z - a)^m g(z)$$

for $z \in B_r(a)$, where

$$g(z) = \sum_{n=m}^{\infty} c_n(z - a)^{n-m}.$$

Since g is analytic, by Theorem 4.12 it is also continuous. Hence, since $g(a) = c_m \neq 0$, there exists $s \leq r$ such that $g(z) \neq 0$ for every $z \in B_s(a)$. Therefore, the function f does not vanish in $B_s(a) \setminus \{a\}$, and the zero a is isolated. $\square$

Example 4.22

The set of zeros of the function $f(z) = \sin(z/\pi)$ is $\mathbb{Z}$.

Example 4.23

Let us show that if $f, g \colon \mathbb{C} \to \mathbb{C}$ are analytic functions and $f = g$ in $\mathbb{R}$, then $f = g$ in $\mathbb{C}$. It is sufficient to note that the analytic function $f - g$ has zeros that are not isolated, and thus $f - g = 0$ in $\mathbb{C}$.

4.3 Laurent Series and Singularities

In this section we consider functions that are not necessarily holomorphic. We first introduce the notion of an isolated singularity.

Definition 4.24

When a function f is holomorphic in $B_r(a) \setminus \{a\}$ for some $r > 0$, but is not holomorphic in $B_r(a)$, the point a is called an *isolated singularity* of f.

We give some examples.

Example 4.25

The function $f(z) = 1/z$ has an isolated singularity at $z = 0$.

Example 4.26

The function $f(z) = (\sin z)/z$ has an isolated singularity at $z = 0$. However, the function

$$g(z) = \begin{cases} (\sin z)/z & \text{if } z \neq 0, \\ 1 & \text{if } z = 0, \end{cases}$$

which can be represented by the power series

$$\frac{\sin z}{z} = \frac{1}{z} \sum_{n=0}^{\infty} \frac{(-1)^n}{(2n+1)!} z^{2n+1} = \sum_{n=0}^{\infty} \frac{(-1)^n}{(2n+1)!} z^{2n} \tag{4.16}$$

for $z \neq 0$, is holomorphic. Indeed, for $z = 0$ the last series in (4.16) takes the value 1 (recall that $z^0 = 1$), and thus it coincides with g.

Example 4.27

The function

$$f(z) = \begin{cases} z & \text{if } z \neq 3, \\ 2 & \text{if } z = 3 \end{cases}$$

has an isolated singularity at $z = 3$.

Example 4.28

The function $f(z) = 1/\sin(1/z)$ is not defined at the points $z = 0$ and $z = 1/(k\pi)$ for $k \in \mathbb{Z} \setminus \{0\}$. Indeed, the denominator $\sin(1/z)$ is not defined at $z = 0$. Moreover, for $z = 1/(k\pi)$, we have $\sin(1/z) = \sin(k\pi) = 0$, and hence the function f is also not defined at these points. On the other hand, $z_k \to 0$ when $k \to \infty$, and thus, $z = 0$ is not an isolated singularity of f.

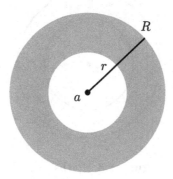

Figure 4.2   Ring $\{z \in \mathbb{C} : r < |z - a| < R\}$

Again, if a is an isolated singularity of f, then there exists $r > 0$ such that f is holomorphic in $B_r(a) \setminus \{a\}$. In this case, and more generally for holomorphic functions in a ring (see Figure 4.2), one can show that f is given by a power series with positive and negative powers.

Theorem 4.29 (Laurent series)

If f is a holomorphic function in the ring

$$A = \{z \in \mathbb{C} : r < |z - a| < R\}, \tag{4.17}$$

then there exist unique constants $c_n \in \mathbb{C}$ for $n \in \mathbb{Z}$ such that

$$f(z) = \sum_{n \in \mathbb{Z}} c_n(z-a)^n, \quad z \in A. \tag{4.18}$$

Before proving the theorem, we note that property (4.18) means that for each $z \in A$ the sequence

$$f_m(z) = \sum_{n=-m}^{m} c_n(z-a)^n$$

converges to $f(z)$ when $m \to \infty$, that is, we have

$$f(z) = \lim_{m \to \infty} f_m(z).$$

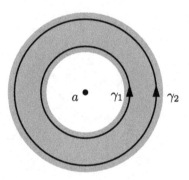

Figure 4.3 Paths γ_1 and γ_2

Proof of Theorem 4.29

Given $\rho_1, \rho_2 \in (r, R)$ with $\rho_1 < \rho_2$, we define paths $\gamma_1, \gamma_2 \colon [0, 2\pi] \to \mathbb{C}$ by

$$\gamma_j(t) = a + \rho_j e^{it}, \quad j = 1, 2$$

(see Figure 4.3). For each $z \in A$ such that $\rho_1 < |z - a| < \rho_2$, we consider the function $g \colon A \to \mathbb{C}$ given by

$$g(w) = \begin{cases} (f(w) - f(z))/(w - z) & \text{if } w \in A \setminus \{z\}, \\ f'(z) & \text{if } w = z. \end{cases}$$

Since f is holomorphic, by Theorem 4.13 the function g is given by a power series in each ring $B_r(a) \setminus \{z\} \subset A \setminus \{z\}$. But since g is continuous in A, it is given by the same power series in the ball $B_r(a)$ (since all power series are continuous functions). By Theorem 4.12, we conclude that g is holomorphic in A. It then follows from Theorem 2.63 that

$$\int_{\gamma_1} g = \int_{\gamma_2} g,$$

that is,

$$\int_{\gamma_1} \frac{f(w) - f(z)}{w - z} \, dw = \int_{\gamma_2} \frac{f(w) - f(z)}{w - z} \, dw.$$

We have $\mathrm{Ind}_{\gamma_1}(z) = 0$ and $\mathrm{Ind}_{\gamma_2}(z) = 1$. Therefore,

$$\int_{\gamma_1} \frac{f(w)}{w - z} \, dw = \int_{\gamma_2} \frac{f(w)}{w - z} \, dw - 2\pi i f(z),$$

and we obtain

$$f(z) = f_1(z) + f_2(z),$$

where

$$f_1(z) = \frac{1}{2\pi i} \int_{\gamma_2} \frac{f(w)}{w - z} \, dw$$

and

$$f_2(z) = -\frac{1}{2\pi i} \int_{\gamma_1} \frac{f(w)}{w - z} \, dw.$$

Now we observe that

$$\frac{1}{w - z} = \sum_{n=0}^{\infty} \frac{(z - a)^n}{(w - a)^{n+1}} \quad \text{for } |w - a| = \rho_2,$$

and

$$\frac{1}{w - z} = -\sum_{n=1}^{\infty} \frac{(w - a)^{n-1}}{(z - a)^n} \quad \text{for } |w - a| = \rho_1.$$

By Theorem 4.8 and Proposition 3.38, one can integrate these series term by term to obtain

$$f_1(z) = \sum_{n=0}^{\infty} c_n (z - a)^n,$$

with

$$c_n = \frac{1}{2\pi i} \int_{\gamma_2} \frac{f(w)}{(w-a)^{n+1}} \, dw, \quad n \in \mathbb{N} \cup \{0\},$$

and

$$f_2(z) = \sum_{n=1}^{\infty} d_n (z-a)^{-n},$$

with

$$d_n = \frac{1}{2\pi i} \int_{\gamma_1} f(w)(w-a)^{n-1} \, dw, \quad n \in \mathbb{N}. \tag{4.19}$$

Taking $c_{-n} = d_n$ for $n \in \mathbb{N} \cup \{0\}$, we conclude that

$$f(z) = f_1(z) + f_2(z) = \sum_{n \in \mathbb{Z}} c_n (z-a)^n.$$

It remains to establish the uniqueness of the constants c_n. Let us assume that

$$f(z) = \sum_{n \in \mathbb{Z}} b_n (z-a)^n, \quad z \in A \tag{4.20}$$

for some other constants $b_n \in \mathbb{C}$ for $n \in \mathbb{Z}$. It follows from (4.18) and (4.20) that

$$\sum_{n=0}^{\infty} (c_n - b_n)(z-a)^n = \sum_{n=1}^{\infty} (b_{-n} - c_{-n})(z-a)^{-n}. \tag{4.21}$$

Now we define a function $h \colon \mathbb{C} \to \mathbb{C}$ by

$$h(z) = \sum_{n=0}^{\infty} (c_n - b_n)(z-a)^n$$

for $|z - a| < R$, and by

$$h(z) = \sum_{n=1}^{\infty} (b_{-n} - c_{-n})(z-a)^{-n}$$

for $|z - a| > r$. It follows from (4.21) that h is well defined, and hence it is holomorphic in $\mathbb{C}$. Moreover, since the second series in (4.21) has no term of order zero, we have $h(z) \to 0$ when $|z| \to \infty$. This shows that h is bounded, and hence, it follows from Theorem 4.18 that h is constant. We thus obtain $h = 0$, since $h(z) \to 0$ when $|z| \to \infty$. It then follows from (4.21) that

$$c_n - b_n = 0 \quad \text{for } n \geq 0,$$

and

$$b_{-n} - c_{-n} = 0 \quad \text{for } n \geq 1.$$

This establishes the uniqueness of the constants in (4.18).   □

The series in (4.18) is called a *Laurent series*.

Example 4.30

Let us consider the function

$$f(z) = \frac{1}{z^2 - z} = \frac{1}{z(z-1)},$$

which is holomorphic in $\mathbb{C} \setminus \{0, 1\}$. In the ring $\{z \in \mathbb{C} : 0 < |z| < 1\}$, we have

$$f(z) = \frac{1}{z^2 - z} = -\frac{1}{z} \frac{1}{1-z}$$

$$= -\frac{1}{z} \sum_{n=0}^{\infty} z^n = \sum_{m=-1}^{\infty} -z^m, \tag{4.22}$$

while in the ring $\{z \in \mathbb{C} : 1 < |z| < \infty\} = \{z \in \mathbb{C} : |z| > 1\}$,

$$f(z) = \frac{1}{z} \cdot \frac{1/z}{1 - 1/z} = \frac{1}{z^2} \sum_{n=0}^{\infty} \left(\frac{1}{z}\right)^n = \sum_{m=-\infty}^{-2} z^{-m}.$$

By the uniqueness of the coefficients c_n in Theorem 4.29, these are necessarily the Laurent series of the function f in each of the rings.

Setting $r = 0$ in Theorem 4.29, we conclude that a holomorphic function in the ring

$$\{z \in \mathbb{C} : 0 < |z - a| < r\}$$

(such as any function with an isolated singularity at a, for some $r > 0$) has a (unique) representation as a Laurent series. We use this property in the following definition.

Definition 4.31

When a is an isolated singularity of f and the numbers c_n for $n \in \mathbb{Z}$ are the coefficients of the Laurent series in (4.18), we say that:

1. a is a *removable singularity* if $c_n = 0$ for every $n < 0$;
2. a is a *pole of order* $m \in \mathbb{N}$ if $c_{-m} \neq 0$ and $c_n = 0$ for every $n < -m$;
3. a is an *essential singularity* if $c_n \neq 0$ for infinitely many negative values of n.

Example 4.32

It follows from Example 4.26 that $z = 0$ is a removable singularity of the function $(\sin z)/z$.

Example 4.33

For the function

$$f(z) = \frac{1}{z^2 - z},$$

it follows from (4.22) that $z = 0$ is a pole of order 1. We also have

$$f(z) = \frac{1}{z(z-1)} = \frac{1}{z-1} \frac{1}{1 + (z-1)}$$

$$= \frac{1}{z-1} \sum_{n=0}^{\infty} [-(z-1)]^n = \sum_{n=0}^{\infty} (-1)^n (z-1)^{n-1},$$

and thus $z = 1$ is a pole of order 1.

More generally, one can show that if

$$f(z) = \frac{(z-a_1)^{n_1} \cdots (z-a_k)^{n_k}}{(z-b_1)^{m_1} \cdots (z-b_l)^{m_l}}, \tag{4.23}$$

with $k + l$ different numbers $a_j, b_j \in \mathbb{C}$, and with $k + l$ exponents $n_j, m_j \in \mathbb{N}$, then each point $z = b_j$ is a pole of order m_j (see Example 4.37).

Example 4.34

The function $f(z) = e^{1/z}$ has an essential singularity at $z = 0$. Indeed,

$$e^{1/z} = \sum_{n=0}^{\infty} \frac{1}{n!} \left(\frac{1}{z}\right)^n \tag{4.24}$$

for $z \neq 0$, and the Laurent series has infinitely many negative powers.

Now we give criteria for an isolated singularity to be a removable singularity or to be a pole.

Proposition 4.35

Let a be an isolated singularity of f.
  1. If

$$\lim_{z \to a} (z - a)f(z) = 0, \tag{4.25}$$

then a is a removable singularity.
  2. If there exists $m \in \mathbb{N}$ such that

$$\lim_{z \to a} (z - a)^m f(z) \neq 0, \tag{4.26}$$

then a is a pole of order m.

Proof

We first assume that condition (4.25) holds. By (4.19), given $n < 0$, we have

$$c_n = \frac{1}{2\pi i} \int_\gamma f(z)(z - a)^{-n-1} \, dz,$$

where the path $\gamma \colon [0, 2\pi] \to \mathbb{C}$ is given by $\gamma(t) = a + re^{it}$, for any sufficiently small r. It then follows from Proposition 2.42 that

$$|c_n| \leq \frac{L_\gamma}{2\pi} \sup \left\{ \left| f(z)(z - a)^{-n-1} \right| : z \in \gamma([0, 2\pi]) \right\}.$$

Now we observe that by (4.25), given $\varepsilon > 0$, there exists $r > 0$ such that

$$\left| (z - a)f(z) \right| < \varepsilon \quad \text{whenever } |z - a| \leq r.$$

Hence,

$$|c_n| \leq r\varepsilon \sup \left\{ \left| (z - a)^{-n-2} \right| : z \in \gamma([0, 2\pi]) \right\}$$
$$= \varepsilon r^{-n-1} \leq \varepsilon$$

for any sufficiently small $r \leq 1$, because $n < 0$. Since ε is arbitrary, we conclude that $c_n = 0$ for $n < 0$, and a is a removable singularity.

Now we assume that condition (4.26) holds. Then

$$\lim_{z \to a} (z - a)^{m+1} f(z) = 0,$$

and it follows from the former property that a is a removable singularity of the function $g(z) = (z - a)^m f(z)$. By Theorem 4.29, there exist unique constants $c_n \in \mathbb{C}$ for $n \in \mathbb{N} \cup \{0\}$ such that

$$g(z) = \sum_{n=0}^{\infty} c_n (z - a)^n, \tag{4.27}$$

for $z \in B_r(a) \setminus \{a\}$ and some $r > 0$. Thus,

$$f(z) = \frac{g(z)}{(z-a)^m} = \sum_{m=0}^{\infty} c_n (z - a)^{n-m}$$

for $z \in B_r(a) \setminus \{a\}$. On the other hand, by (4.26), we have

$$\lim_{z \to a} g(z) = \lim_{z \to a} (z - a)^m f(z) \neq 0.$$

Since power series define continuous functions, it follows from (4.27) that

$$c_0 = \lim_{z \to a} g(z) \neq 0,$$

and thus, a is a pole of order m. $\qquad\qquad\qquad\qquad\qquad\qquad\qquad\square$

We also give some examples.

Example 4.36

The function

$$f(z) = e^{1/z} + \frac{1}{(z+1)^2(z-2)}$$

has isolated singularities at the points -1, 0 and 2. Since

$$\lim_{z \to 2} (z - 2) f(z) = \lim_{z \to 2} \left((z-2)e^{1/z} + \frac{1}{(z+1)^2} \right) = \frac{1}{9} \neq 0,$$

it follows from Proposition 4.35 that 2 is a pole of order 1. Analogously, since

$$\lim_{z \to -1} (z+1)^2 f(z) = \lim_{z \to -1} \left((z+1)^2 e^{1/z} + \frac{1}{z-2} \right) = -\frac{1}{3} \neq 0,$$

it follows from Proposition 4.35 that -1 is a pole of order 2. Moreover, since the function

$$g(z) = \frac{1}{(z+1)^2(z-2)}$$

is holomorphic in some open ball centered at the origin, by Theorem 4.29 there exist unique constants $c_n \in \mathbb{C}$ for $n \in \mathbb{N} \cup \{0\}$ such that

$$g(z) = \sum_{n=0}^{\infty} c_n z^n,$$

for $z \in B_r(0)$ and some $r > 0$. It then follows from (4.24) that

$$f(z) = \sum_{n=0}^{\infty} \frac{1}{n!} \left(\frac{1}{z}\right)^n + \sum_{n=0}^{\infty} c_n z^n$$

for $z \in B_r(0) \setminus \{0\}$, and thus, 0 is an essential singularity.

Example 4.37

Let us consider the function f defined by (4.23), where the $k + l$ numbers a_j, $b_j \in \mathbb{C}$ are distinct. Since

$$\lim_{z \to b_j} (z - b_j)^{m_j} f(z) = \frac{\prod_{p=1}^{k}(z - a_p)^{n_p}}{\prod_{q \neq j}(z - b_q)^{m_q}} \neq 0,$$

it follows from Proposition 4.35 that $z = b_j$ is a pole of order m_j.

Example 4.38

Now we assume that a is an isolated singularity of f such that

$$M_r := \sup \left\{ |f(z)| : z \in B_r(a) \setminus \{a\} \right\} < +\infty$$

for some $r > 0$ (and thus for any sufficiently small r). We then have

$$\sup \left\{ |(z - a)f(z)| : z \in B_r(a) \setminus \{a\} \right\} \leq M_r r \to 0$$

when $r \to 0$, because the function $r \mapsto M_r$ is nondecreasing. Hence,

$$\lim_{z \to a} (z - a) f(z) = 0,$$

and it follows from Proposition 4.35 that a is a removable singularity.

Example 4.39

Let us assume that a is an isolated singularity of f and that the limit

$$w = \lim_{z \to a} f(z)$$

exists. Given $\varepsilon > 0$, there exists $r > 0$ such that

$$\left| f(z) - w \right| < \varepsilon \quad \text{whenever } |z - a| < r.$$

In particular,

$$\left| f(z) \right| \le \left| f(z) - w \right| + |w| < \varepsilon + |w|$$

whenever $z \in B_r(a)$, and it follows from Example 4.38 that a is a removable singularity.

Furthermore, the function

$$g(z) = \begin{cases} f(z) & \text{if } z \in B_r(a) \setminus \{a\}, \\ w & \text{if } z = a \end{cases}$$

is holomorphic. Indeed, by Theorem 4.29, there exist unique constants $c_n \in \mathbb{C}$ for $n \in \mathbb{N} \cup \{0\}$ such that

$$f(z) = \sum_{n=0}^{\infty} c_n (z - a)^n, \tag{4.28}$$

for $z \in B_r(a) \setminus \{a\}$ and some $r > 0$. Since the power series on the right-hand side of (4.28) defines an holomorphic function and

$$\sum_{n=0}^{\infty} c_n (z - a)^n \Big|_{z=a} = \lim_{z \to a} \sum_{n=0}^{\infty} c_n (z - a)^n$$

$$= \lim_{z \to a} f(z)$$

$$= w = g(a),$$

we conclude that

$$g(z) = \sum_{n=0}^{\infty} c_n (z - a)^n$$

for $z \in B_r(a)$, and thus the function g is holomorphic.

4.4 Residues

In order to compute in a somewhat expedited manner many integrals of non-holomorphic functions along closed paths, we introduce the notion of the residue at an isolated singularity.

Definition 4.40

Let a be an isolated singularity of f. The number

$$\mathrm{Res}(f,a) = \frac{1}{2\pi i} \int_\gamma f,$$

where the path $\gamma \colon [0, 2\pi] \to \mathbb{C}$ is given by $\gamma(t) = a + re^{it}$ for any sufficiently small r, is called the *residue* of f at a.

It follows from Theorem 2.63 that the residue is well defined. Moreover, it can be computed as follows.

Proposition 4.41

If a is an isolated singularity of f, then $\mathrm{Res}(f,a) = c_{-1}$, where c_{-1} is the coefficient of the term of degree -1 in the Laurent series in (4.18).

Proof

In a similar manner to that in the proof of Theorem 4.8, one can show that the Laurent series in (4.18) converges uniformly on the set

$$\{z \in \mathbb{C} : \rho_1 \leq |z - a| \leq \rho_2\},$$

for each $\rho_1, \rho_2 > 0$ such that $\rho_1 \leq \rho_2 < R$ (see (4.17)). It then follows from Proposition 3.38 that

$$\mathrm{Res}(f,a) = \frac{1}{2\pi i} \int_\gamma f = \frac{1}{2\pi i} \int_\gamma \sum_{n \in \mathbb{Z}} c_n (z - a)^n \, dz$$

$$= \frac{1}{2\pi i} \sum_{n \in \mathbb{Z}} c_n \int_\gamma (z - a)^n \, dz,$$

where the path $\gamma \colon [0, 2\pi] \to \mathbb{C}$ is given by $\gamma(t) = a + \rho e^{it}$, with $\rho \in (\rho_1, \rho_2)$ (we note that by Theorem 2.63, the integrals are independent of ρ). For $n \neq -1$ we have

$$\int_\gamma (z - a)^n \, dz = \left. \frac{(z - a)^{n+1}}{n+1} \right|_{z=\gamma(0)}^{z=\gamma(2\pi)} = 0. \tag{4.29}$$

Therefore,

$$\operatorname{Res}(f,a) = \frac{c_{-1}}{2\pi i} \int_\gamma \frac{dz}{z-a}$$

$$= \frac{c_{-1}}{2\pi i} \int_0^{2\pi} \frac{rie^{it}}{re^{it}}\, dt = c_{-1},$$

which yields the desired identity. □

Example 4.42

Let us compute the residue of the function

$$f(z) = \frac{z}{1+z^2} = \frac{z}{(z-i)(z+i)}$$

at the pole $z = i$. We have

$$f(z) = \frac{z}{z-i}\frac{1}{z+i} = \frac{z-i+i}{z-i}\frac{1}{2i+z-i}$$

$$= \left(1 + \frac{i}{z-i}\right)\frac{1}{2i} \cdot \frac{1}{1+(z-i)/(2i)}$$

$$= \left(1 + \frac{i}{z-i}\right)\frac{1}{2i}\left(1 - \frac{z-i}{2i} - \frac{(z-i)^2}{4} + \cdots\right)$$

$$= \frac{1}{2i} + \frac{1}{2}\cdot\frac{1}{z-i} + \frac{z-i}{4} - \frac{1}{4i} - \frac{(z-i)^2}{8i} - \frac{z-i}{8} + \cdots$$

$$= \frac{1}{2}(z-i)^{-1} + \frac{1}{4i} + \frac{1}{8}(z-i) + \cdots.$$

Therefore, $z = i$ is a pole of order 1 and $\operatorname{Res}(f,i) = 1/2$.

More generally, we have the following result.

Proposition 4.43

If a is a pole of order m for a function f, then

$$\operatorname{Res}(f,a) = \lim_{z\to a} \frac{1}{(m-1)!}\left[(z-a)^m f(z)\right]^{(m-1)}.$$

Proof

Writing

$$f(z) = \sum_{n=-m}^{\infty} c_n (z-a)^n,$$

we obtain

$$(z-a)^m f(z) = c_{-m} + c_{-m+1}(z-a) + \cdots,$$

and thus,

$$\left[(z-a)^m f(z) \right]^{(m-1)} = (m-1)! c_{-1} + \cdots.$$

This yields the desired result. $\qquad\square$

Example 4.44

For the function f in Example 4.42, we have

$$\mathrm{Res}(f, i) = \lim_{z \to i} \left[(z-i) f(z) \right]$$

$$= \lim_{z \to i} \frac{z}{z+i} = \frac{i}{2i} = \frac{1}{2}.$$

Example 4.45

For the function $f(z) = e^{1/z}$, we have already obtained the Laurent series in (4.24). It then follows from Proposition 4.41 that $\mathrm{Res}(f, 0) = 1$.

4.5 Meromorphic Functions

Now we introduce the notion of a meromorphic function.

Definition 4.46

A function f is said to be *meromorphic* in an open set $\Omega \subset \mathbb{C}$ if there exists $A \subset \Omega$ such that:

1. f is holomorphic in $\Omega \setminus A$;
2. f has a pole at each point of A;
3. A has no limit points in Ω.

Holomorphic functions are a particular case of meromorphic functions (they correspond to taking $A = \varnothing$ in Definition 4.46).

Example 4.47

It follows from Example 4.33 that the function $f(z) = 1/(z^2 - z)$ has only the isolated singularities $z = 0$ and $z = 1$, which are poles. Hence, f is a meromorphic function.

More generally, one can show that if $f(z) = P(z)/Q(z)$ for some polynomials P and Q, then f is meromorphic.

The following result allows one to compute in a somewhat expedited manner the integral of a meromorphic function along a closed path.

Theorem 4.48 (Residue theorem)

If f is a meromorphic function in a simply connected open set $\Omega \subset \mathbb{C}$ and γ is a closed piecewise regular path in $\Omega \setminus A$, where $A \subset \Omega$ is the set of the poles of f, then

$$\frac{1}{2\pi i} \int_\gamma f = \sum_{a \in A} \operatorname{Res}(f, a) \operatorname{Ind}_\gamma(a).$$

Proof

Let us consider the set

$$B = \big\{ a \in A : \operatorname{Ind}_\gamma(a) \neq 0 \big\}.$$

We recall that $\operatorname{Ind}_\gamma(a) = 0$ for any point a in the unbounded connected component U of the complement of the curve defined by γ. Hence, B is contained in the compact set $\mathbb{C} \setminus U \subset \Omega$. This implies that B is finite. Otherwise, there would exist a sequence $(a_n)_n$ of distinct points in B. Being bounded, this sequence would have a limit point in $\mathbb{C} \setminus U$, and thus also in Ω. But then this would be a limit point of A in Ω, which does not exist, since the function f is meromorphic.

Now let P_a be the sum of the negative powers of the Laurent series of f in the ring

$$R = \big\{ z \in \mathbb{C} : 0 < |z - a| < r \big\},$$

for some sufficiently small r such that $R \subset \Omega \setminus A$. Then the function

$$g = f - \sum_{a \in B} P_a$$

has a removable singularity at each point of B. Since Ω is simply connected, it follows from Theorem 2.64 that

$$0 = \int_\gamma g = \int_\gamma f - \sum_{a \in B} \int_\gamma P_a. \tag{4.30}$$

Now we proceed as in (4.29), for the path γ. Integrating P_a term by term, by Proposition 4.41 we obtain

$$\frac{1}{2\pi i} \int_\gamma P_a = \frac{1}{2\pi i} \int_\gamma \frac{\mathrm{Res}(f, a)}{z - a} \, dz = \mathrm{Res}(f, a) \, \mathrm{Ind}_\gamma(a).$$

It then follows from (4.30) that

$$\frac{1}{2\pi i} \int_\gamma f = \sum_{a \in B} \mathrm{Res}(f, a) \, \mathrm{Ind}_\gamma(a)$$

$$= \sum_{a \in A} \mathrm{Res}(f, a) \, \mathrm{Ind}_\gamma(a),$$

which yields the desired identity. $\qquad \square$

Example 4.49

Let us consider the function $f(z) = z/(1 + z^2)$, and let $\gamma \colon [0, 2\pi] \to \mathbb{C}$ be the path given by $\gamma(t) = i + e^{it}$. By Example 4.42, we have $\mathrm{Res}(f, i) = 1/2$. Since $\mathrm{Ind}_\gamma(i) = 1$ and $\mathrm{Ind}_\gamma(-i) = 0$, we then obtain

$$\int_\gamma f = 2\pi i \big[\mathrm{Res}(f, i) \, \mathrm{Ind}_\gamma(i) + \mathrm{Res}(f, -i) \, \mathrm{Ind}_\gamma(-i)\big] = \pi i.$$

Example 4.50

Let us consider the integral

$$\int_0^\infty \frac{dx}{1 + x^2}.$$

Since a primitive of $1/(1+x^2)$ is $\tan^{-1} x$, we have

$$\int_0^\infty \frac{dx}{1+x^2} = \lim_{a\to\infty} \int_0^a \frac{dx}{1+x^2}$$

$$= \lim_{a\to\infty} \left(\tan^{-1} a - \tan^{-1} 0\right) = \frac{\pi}{2}.$$

Now we show how the integral can be computed using the Residue theorem. Let us consider the path $\gamma = \gamma_1 + \gamma_2$ with $\gamma_1 \colon [-R, R] \to \mathbb{C}$ and $\gamma_2 \colon [0, \pi] \to \mathbb{C}$ given respectively by

$$\gamma_1(t) = t \quad \text{and} \quad \gamma_2(t) = Re^{it}$$

(see Figure 4.4). For $R > 1$ we have $\mathrm{Ind}_\gamma(i) = 1$ and $\mathrm{Ind}_\gamma(-i) = 0$. Therefore, if $f(z) = 1/(1 + z^2)$, then since

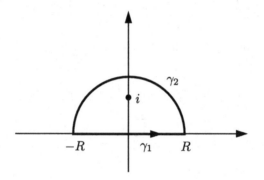

Figure 4.4   Path $\gamma = \gamma_1 + \gamma_2$

$$\mathrm{Res}(f, i) = \lim_{z\to i} \frac{z-i}{1+z^2} = \lim_{z\to i} \frac{1}{z+i} = \frac{1}{2i},$$

we obtain

$$\int_\gamma f = 2\pi i \, \mathrm{Res}(f, i) \, \mathrm{Ind}_\gamma(i) = 2\pi i \frac{1}{2i} = \pi. \tag{4.31}$$

On the other hand,

$$\int_\gamma f = \int_{-R}^R f(t)\, dt + \int_{\gamma_2} f = 2 \int_0^R f(t)\, dt + \int_{\gamma_2} f.$$

Since

$$\left| \int_{\gamma_2} f \right| \le L_{\gamma_2} \sup\left\{ \frac{1}{|z^2 + 1|} : z \in \gamma_2\big([0, \pi]\big) \right\}$$

(see Proposition 2.42), and

$$\left|z^2 + 1\right| \geq \left|z^2\right| - 1 = |z|^2 - 1$$

for $|z| > 1$, we obtain

$$\left|\int_{\gamma_2} f\right| \leq \pi R \frac{1}{R^2 - 1} \to 0$$

when $R \to \infty$. Therefore,

$$\int_{\gamma} f = 2 \int_0^R f(t)\, dt + \int_{\gamma_2} f \to 2 \int_0^\infty f(t)\, dt$$

when $R \to \infty$. It then follows from (4.31) that

$$\int_0^\infty \frac{dx}{1 + x^2} = \frac{\pi}{2}.$$

Example 4.51

Let us compute the integral

$$\int_0^\infty \frac{\sin x}{x}\, dx.$$

Given $r, R > 0$ with $r < R$, we consider the path

$$\gamma = \gamma_1 + \gamma_2 + \gamma_3 + \gamma_4,$$

where $\gamma_1 \colon [r, R] \to \mathbb{C}$, $\gamma_2 \colon [0, \pi] \to \mathbb{C}$, $\gamma_3 \colon [-R, -r] \to \mathbb{C}$ and $\gamma_4 \colon [0, \pi] \to \mathbb{C}$ are given by

$$\gamma_1(t) = t, \quad \gamma_2(t) = Re^{it}, \quad \gamma_3(t) = t \quad \text{and} \quad \gamma_4(t) = re^{i(\pi - t)}$$

(see Figure 4.5). By Theorem 4.48, for the function

$$f(z) = \frac{e^{iz}}{2iz}$$

we have $\int_{\gamma} f = 0$. Moreover,

$$\int_{\gamma} f - \int_{\gamma_2} f - \int_{\gamma_4} f = \int_r^R \frac{e^{ix}}{2ix}\, dx + \int_{-R}^{-r} \frac{e^{ix}}{2ix}\, dx$$

$$= \int_r^R \frac{e^{ix} - e^{-ix}}{2ix}\, dx = \int_r^R \frac{\sin x}{x}\, dx. \qquad (4.32)$$

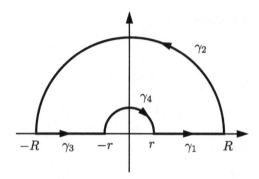

Figure 4.5   Path $\gamma = \gamma_1 + \gamma_2 + \gamma_3 + \gamma_4$

Now we observe that by Proposition 2.42,

$$\left| \int_{\gamma_2} f \right| \leq \int_0^\pi \left| f(Re^{it}) Rie^{it} \right| dt$$

$$= \frac{1}{2} \int_0^\pi e^{-R\sin t} \, dt = \int_0^{\pi/2} e^{-R\sin t} \, dt$$

$$\leq \int_0^{\pi/2} e^{-Rt/\pi} \, dt = \frac{1 - \pi e^{-R/2}}{R} \to 0$$

when $R \to +\infty$, since $\sin t \geq 2t/\pi$ for $t \in [0, \pi/2]$ (in order to obtain this inequality, it is sufficient to compare the graphs of $\sin t$ and $2t/\pi$). Moreover, since $z = 0$ is a removable singularity of $(e^{iz} - 1)/z$, we have

$$\left| \int_{\gamma_4} \frac{e^{iz} - 1}{z} \, dz \right| \leq \pi r \sup\left\{ \left| \frac{e^{iz} - 1}{z} \right| : |z| < 1 \right\}$$

for $r < 1$, and thus,

$$\lim_{r \to 0} \int_{\gamma_4} f = \lim_{r \to 0} \int_{\gamma_4} \left(\frac{e^{iz} - 1}{2iz} + \frac{1}{2iz} \right) dz$$

$$= 0 + \lim_{r \to 0} \int_{\gamma_4} \frac{dz}{2iz}$$

$$= \lim_{r \to 0} - \int_0^\pi \frac{rie^{i(\pi - t)}}{2ire^{i(\pi - t)}} \, dt = -\frac{\pi}{2}.$$

Hence, letting $R \to +\infty$ and $r \to 0$ in (4.32), we obtain

$$\int_0^\infty \frac{\sin x}{x} \, dx = - \lim_{R \to +\infty} \int_{\gamma_2} f - \lim_{r \to 0} \int_\gamma f = \frac{\pi}{2}.$$

Incidentally, we note that

$$\int_0^\infty \left| \frac{\sin x}{x} \right| dx = +\infty.$$

Indeed,

$$\int_0^\infty \left| \frac{\sin x}{x} \right| dx \geq \sum_{n=1}^\infty \int_{2\pi n+\pi/4}^{2\pi n+\pi/2} \left| \frac{\sin x}{x} \right| dx$$

$$\geq \sum_{n=1}^\infty \frac{\pi}{4\sqrt{2}} \frac{1}{2\pi n + \pi/2} = +\infty.$$

4.6 Solved Problems and Exercises

Problem 4.1

Compute the radius of convergence of the power series

$$\sum_{n=1}^\infty (z+i)^n/5^n \quad \text{and} \quad \sum_{n=1}^\infty (z+i/2)^n 2^n/n^2.$$

Solution

The radius of convergence of the first series is given by

$$1/\limsup_{n\to\infty} \sqrt[n]{1/5^n} = 5.$$

Alternatively, note that

$$\frac{1}{5^n} : \frac{1}{5^{n+1}} = 5 \to 5$$

when $n \to \infty$, and by Proposition 4.4 the radius of convergence is 5.
The radius of convergence of the second series is given by

$$1/\limsup_{n\to\infty} \sqrt[n]{2^n/n^2} = 1/2.$$

Alternatively, note that

$$\frac{2^n}{n^2} : \frac{2^{n+1}}{(n+1)^2} = \frac{n+1}{2n} \to \frac{1}{2}$$

when $n \to \infty$, and by Proposition 4.4 the radius of convergence is $1/2$.

Problem 4.2

Compute the radius of convergence of the power series $\sum_{n=1}^{\infty} [3 - (-1)^n]^n z^n$.

Solution

The radius of convergence is given by

$$1/\limsup_{n\to\infty} \sqrt[n]{[3 - (-1)^n]^n} = 1/\limsup_{n\to\infty} [3 - (-1)^n] = 1/4.$$

We note that the sequence

$$a_n = \frac{[3 - (-1)^n]^n}{[3 - (-1)^{n+1}]^{n+1}}$$

takes the values

$$a_n = \frac{2^n}{4^{n+1}} = \frac{2^n}{2^{2(n+1)}} = \frac{1}{2^{n+2}}$$

for n even, and

$$a_n = \frac{4^n}{2^{n+1}} = \frac{2^{2n}}{2^{n+1}} = 2^{n-1}$$

for n odd, and hence it does not converge. Therefore, in this case, it is impossible to find the radius of convergence by computing the limit of the sequence a_n.

Problem 4.3

Find a power series with radius of convergence equal to zero.

Solution

For example, the power series $\sum_{n=1}^{\infty} n^n z^n$ has radius of convergence

$$1/\limsup_{n\to\infty} \sqrt[n]{n^n} = 1/\limsup_{n\to\infty} n = 0.$$

Problem 4.4

Show that the function $f(z) = (z - \sin z)/z^3$ has a removable singularity at $z = 0$.

Solution

We have

$$\sin z = \sum_{n=0}^{\infty} \frac{(-1)^n}{(2n+1)!} z^{2n+1}, \quad z \in \mathbb{C}, \tag{4.33}$$

and hence, for $z \neq 0$,

$$\frac{z - \sin z}{z^3} = \frac{1}{z^3}\left(z - z + \frac{z^3}{6} - \frac{z^5}{120} + \cdots\right) = \frac{1}{6} - \frac{z^2}{120} + \cdots. \tag{4.34}$$

Since the power series in (4.33) has radius of convergence $+\infty$, the same happens with the last series in (4.34). Moreover, by Theorem 4.12, a power series defines a holomorphic function, and thus also a continuous function, in the interior of its domain of convergence. Hence, in order to compute the limit of the series $1/6 - z^2/120 + \cdots$ when $z \to 0$ it is sufficient to take $z = 0$. That is,

$$\lim_{z \to 0} \frac{z - \sin z}{z^3} = \frac{1}{6} - \frac{z^2}{120} + \cdots \bigg|_{z=0} = \frac{1}{6} \neq 0,$$

and it follows from Proposition 4.35 that $z = 0$ is a removable singularity of f.

Problem 4.5

Verify that if $f \colon \Omega \to \mathbb{C}$ is a holomorphic function in an open set $\Omega \subset \mathbb{C}$, and $f'(z_0) \neq 0$ for some point $z_0 \in \Omega$, then the function

$$g(z) = \frac{z - z_0}{f(z) - f(z_0)}$$

has a removable singularity at $z = z_0$.

Solution

Since

$$g(z) = \frac{1}{(f(z) - f(z_0))/(z - z_0)},$$

we have

$$\lim_{z \to z_0} g(z) = \frac{1}{f'(z_0)} \neq 0,$$

and it follows from Proposition 4.35 that $z = z_0$ is a removable singularity of g.

Problem 4.6

For the path $\gamma \colon [0, 4\pi] \to \mathbb{C}$ given by $\gamma(t) = e^{it}$, compute the integral

$$\int_\gamma \frac{az^8 + b\sin(z^4) + c\sin^4 z}{z^4} \, dz$$

for each $a, b, c \in \mathbb{C}$.

Solution

We have

$$\int_\gamma \frac{az^8 + b\sin(z^4) + c\sin^4 z}{z^4} \, dz = \int_\gamma \left[az^4 + b\frac{\sin(z^4)}{z^4} + c\left(\frac{\sin z}{z}\right)^4 \right] dz.$$

Clearly, the function az^4 is holomorphic. Moreover, both functions $\sin(z^4)/z^4$ and $((\sin z)/z)^4$ have a removable singularity at $z = 0$, since

$$\lim_{z \to 0} \frac{\sin(z^4)}{z^4} = 1 \quad \text{and} \quad \lim_{z \to 0} \left(\frac{\sin z}{z}\right)^4 = 1$$

(see Proposition 4.35). Therefore, by Cauchy's theorem (Theorem 2.55), we have

$$\int_\gamma \frac{az^8 + b\sin(z^4) + c\sin^4 z}{z^4} \, dz = 0.$$

Problem 4.7

Find the power series of the function $z^2 \sin z$ centered at $z = \pi$ up to the term of order 3.

Solution

Since

$$\sin(z - \pi) = \frac{e^{i(z-\pi)} - e^{-i(z-\pi)}}{2i} = \frac{-e^{iz} + e^{-iz}}{2i} = -\sin z,$$

it follows from (4.33) that

$$\sin z = -\sin(z - \pi) = \sum_{n=0}^{\infty} \frac{(-1)^{n+1}}{(2n+1)!} (z - \pi)^{2n+1}.$$

Therefore,

$$
\begin{aligned}
z^2 \sin z &= (z - \pi + \pi)^2 \sin z \\
&= (z - \pi)^2 \sin z + 2\pi(z - \pi) \sin z + \pi^2 \sin z \\
&= (z - \pi)^2 \left[-(z - \pi) + \cdots \right] \\
&\quad + 2\pi(z - \pi) \left[-(z - \pi) + \cdots \right] \\
&\quad + \pi^2 \left[-(z - \pi) + \frac{(z - \pi)^3}{6} - \cdots \right] \\
&= -\pi^2(z - \pi) - 2\pi(z - \pi)^2 + \frac{\pi^2 - 6}{6}(z - \pi)^3 + \cdots .
\end{aligned}
$$

Problem 4.8

Find and classify all isolated singularities of the function $\cos(1/z)$.

Solution

The function $\cos(1/z)$ is holomorphic in $\mathbb{C} \setminus \{0\}$, and thus it has only the isolated singularity $z = 0$. Since

$$
\cos z = \sum_{n=0}^{\infty} \frac{(-1)^n}{(2n)!} z^{2n}
$$

for $z \in \mathbb{C}$, we obtain

$$
\cos(1/z) = \sum_{n=0}^{\infty} \frac{(-1)^n}{(2n)!} \left(\frac{1}{z} \right)^{2n} \tag{4.35}
$$

for $z \neq 0$. Since the Laurent series in (4.35) has infinitely many negative powers, $z = 0$ is an essential singularity.

Problem 4.9

Find and classify all isolated singularities of the function $f(z) = e^z / \sin z$.

Solution

The function f has isolated singularities at the points where $\sin z = 0$, that is, at the points $z = k\pi$ with $k \in \mathbb{Z}$. By Problem 2.14, for each $k \in \mathbb{Z}$ we have

$$\lim_{z \to k\pi} (z - k\pi) \frac{e^z}{\sin z} = e^{k\pi} \lim_{z \to k\pi} \frac{z - k\pi}{\sin z}$$

$$= e^{k\pi} \lim_{z \to k\pi} \frac{1}{\cos z} = e^{k\pi} \cos(k\pi) \neq 0.$$

Therefore, by Proposition 4.35, $z = k\pi$ is a pole of order 1 for each $k \in \mathbb{Z}$.

Problem 4.10

Find and classify all isolated singularities of the function

$$f(z) = \frac{z(z + 1)}{(z - 3)^2(z - 1)} + e^{1/z^2}.$$

Solution

The function f has isolated singularities at $z = 1$ and $z = 3$ (since at these points the denominator of the first fraction vanishes), and also at $z = 0$ (since e^{1/z^2} is not defined at this point). The isolated singularity $z = 1$ is a pole of order 1, since

$$\lim_{z \to 1} (z - 1)f(z) = \lim_{z \to 1} \frac{z(z + 1)}{(z - 3)^2} + \lim_{z \to 1} (z - 1)e^{1/z} = \frac{1}{2} \neq 0,$$

while $z = 3$ is a pole of order 2, since

$$\lim_{z \to 3} (z - 3)^2 f(z) = \lim_{z \to 3} \frac{z(z + 1)}{z - 1} + \lim_{z \to 3} (z - 3)^2 e^{1/z} = 6 \neq 0$$

(see Proposition 4.35). Moreover, since the function

$$g(z) = \frac{z(z + 1)}{(z - 3)^2(z - 1)}$$

is holomorphic in some open ball centered at the origin, by Theorem 4.29 there exist unique constants $c_n \in \mathbb{C}$ for $n \in \mathbb{N} \cup \{0\}$ such that

$$g(z) = \sum_{n=0}^{\infty} c_n z^n,$$

for $z \in B_r(0)$ and some $r > 0$. It then follows from (4.24) that

$$f(z) = \sum_{n=0}^{\infty} \frac{1}{n!} \left(\frac{1}{z^2}\right)^n + \sum_{n=0}^{\infty} c_n z^n,$$

and thus, 0 is an essential singularity.

Problem 4.11

For the function $f(z) = z/(z^2 - 16)$, compute the residue $\text{Res}(f, 4)$.

Solution

Since

$$f(z) = \frac{z}{(z-4)(z+4)},$$

we have

$$\lim_{z \to 4}(z-4)f(z) = \frac{z}{z+4} = \frac{1}{2} \neq 0,$$

and it follows from Proposition 4.35 that $z = 4$ is a pole of order 1. Moreover, by Proposition 4.43, we have $\text{Res}(f, 4) = 1/2$.

Problem 4.12

Compute the residue $\text{Res}((\sin z)/z^{100}, 0)$.

Solution

It follows from (4.33) that

$$\frac{\sin z}{z^{100}} = \sum_{n=0}^{\infty} \frac{(-1)^n}{(2n+1)!} z^{2n-99}$$

for $z \neq 0$. We note that this is the Laurent series of f. Since the residue is the coefficient of the term of order -1 (see Proposition 4.41), we obtain

$$\text{Res}((\sin z)/z^{100}, 0) = (-1)^{49}/99! = -1/99!.$$

Problem 4.13

Let g and h be holomorphic functions. Show that the residue of the function

$$f(z) = \frac{g(z)}{h(z)}$$

at a pole z_0 of order 1, with $g(z_0) \neq 0$, $h(z_0) = 0$, and $h'(z_0) \neq 0$, is given by

$$\text{Res}(f, z_0) = \frac{g(z_0)}{h'(z_0)}.$$

Solution

Since z_0 is a pole of order 1, by Proposition 4.43 we have

$$\text{Res}(f, z_0) = \lim_{z \to z_0} (z - z_0) f(z)$$

$$= \lim_{z \to z_0} \frac{g(z)}{(h(z) - h(z_0))/(z - z_0)}$$

$$= \frac{g(z_0)}{h'(z_0)}.$$

Problem 4.14

Find whether the function $f(z) = (z + 1)/(z - 2)$ is analytic in some open set.

Solution

Given $a \in \mathbb{C} \setminus \{2\}$, we have

$$\frac{z + 1}{z - 2} = \frac{z + 1}{z - a - (2 - a)}$$

$$= \frac{z + 1}{a - 2} \cdot \frac{1}{1 - (z - a)/(2 - a)}$$

$$= \frac{z + 1}{a - 2} \sum_{n=0}^{\infty} \left(\frac{z - a}{2 - a} \right)^n$$

whenever $|z - a| < |2 - a|$. Hence,

$$f(z) = -(z - a + a + 1) \sum_{n=0}^{\infty} \frac{(z - a)^n}{(2 - a)^{n+1}}$$

$$= -\sum_{n=0}^{\infty} \frac{(z - a)^{n+1}}{(2 - a)^{n+1}} - (a + 1) \sum_{n=0}^{\infty} \frac{(z - a)^n}{(2 - a)^{n+1}}$$

$$= \frac{a + 1}{a - 2} - \sum_{n=1}^{\infty} \frac{3}{(2 - a)^{n+1}} (z - a)^n,$$

and the last power series has radius of convergence

$$R = 1/\limsup_{n \to \infty} \sqrt[n]{\left| 3/(2 - a)^{n+1} \right|} = |2 - a|.$$

We note that R coincides with the distance from a to the isolated singularity 2. This shows that in each ball $B(a,r) \subset \mathbb{C} \setminus \{2\}$ the function f can be represented by a power series centered at a, and thus f is analytic in $\mathbb{C} \setminus \{2\}$.

Problem 4.15

Find the Laurent series of the function $f(z) = z/(z^2 - 1)$ in the ring

$$\Omega = \{z \in \mathbb{C} : 0 < |z + 1| < 2\}.$$

Solution

The function f is holomorphic in $\mathbb{C} \setminus \{-1, 1\}$. For $z \in \Omega$, we have

$$
\begin{aligned}
\frac{z}{z^2 - 1} &= \frac{z}{(z-1)(z+1)} \\
&= \frac{1}{z+1} \cdot \frac{z}{z+1-2} \\
&= -\frac{z}{z+1} \cdot \frac{1}{1-(z+1)/2},
\end{aligned}
$$

and hence,

$$
\begin{aligned}
f(z) &= -\frac{z+1-1}{z+1} \sum_{n=0}^{\infty} \frac{1}{2^n} (z+1)^n \\
&= -\sum_{n=0}^{\infty} \frac{1}{2^n} (z+1)^n + \sum_{n=0}^{\infty} \frac{1}{2^n} (z+1)^{n-1} \\
&= -\sum_{n=0}^{\infty} \frac{1}{2^n} (z+1)^n + \sum_{n=-1}^{\infty} \frac{1}{2^{n+1}} (z+1)^n \\
&= \frac{1}{z+1} + \sum_{n=0}^{\infty} \left(\frac{1}{2^{n+1}} - \frac{1}{2^n} \right) (z+1)^n \\
&= \frac{1}{z+1} - \sum_{n=0}^{\infty} \frac{1}{2^{n+1}} (z+1)^n.
\end{aligned}
$$

This is the Laurent series of f in the ring Ω.

Problem 4.16

Consider the function

$$f(z) = \frac{z}{z^2 + \sin^3 z}.$$

Classify the isolated singularity of f at the origin and find the terms of order -1 and -2 of the Laurent series of f centered at $z = 0$.

Solution

Since

$$\lim_{z \to 0} \frac{\sin z}{z} = 1,$$

we have

$$\begin{aligned}
\lim_{z \to 0} z f(z) &= \lim_{z \to 0} \frac{z^2}{z^2 + \sin^3 z} \\
&= \lim_{z \to 0} \frac{1}{1 + z((\sin z)/z)^3} = 1 \neq 0,
\end{aligned} \tag{4.36}$$

and it follows from Proposition 4.35 that $z = 0$ is a pole of order 1. Hence, the term of order -2 of the Laurent series is zero. The term of order -1 is $1/z$, since the coefficient is the residue $\mathrm{Res}(f, 0)$, which in this case is given by the limit in (4.36).

Problem 4.17

For the path $\gamma \colon [0, 2\pi] \to \mathbb{C}$ given by $\gamma(t) = 5e^{it}$, compute the integral

$$\int_\gamma \frac{e^z}{(z+2)(z-3)} \, dz.$$

Solution

The function

$$f(z) = \frac{e^z}{(z+2)(z-3)}$$

has poles of order 1 at the points -2 and 3, with residues

$$\mathrm{Res}(f, -2) = \lim_{z \to -2} (z+2) f(z) = \left. \frac{e^z}{z-3} \right|_{z=-2} = -\frac{e^{-2}}{5}$$

and

$$\text{Res}(f,3) = \lim_{z\to 3}(z-3)f(z) = \frac{e^z}{z+2}\Big|_{z=3} = \frac{e^3}{5}.$$

Moreover,

$$\text{Ind}_\gamma(-2) = \text{Ind}_\gamma(3) = 1.$$

It then follows from the Residue theorem (Theorem 4.48) that

$$\int_\gamma f = 2\pi i\big[\text{Res}(f,-2) + \text{Res}(f,-3)\big]$$

$$= 2\pi i\left(-\frac{e^{-2}}{5} + \frac{e^3}{5}\right) = \frac{2\pi i(e^3 - e^{-2})}{5}.$$

Problem 4.18

Use the Residue theorem (Theorem 4.48) to compute the integrals $\int_\alpha f$ and $\int_\beta f$ of the function

$$f(z) = \frac{z+3}{z^2-1},$$

for the paths $\alpha, \beta \colon [0, 2\pi] \to \mathbb{C}$ given by $\alpha(t) = 1 + e^{it}$ and $\beta(t) = 3e^{-it}$ (see Figure 4.6).

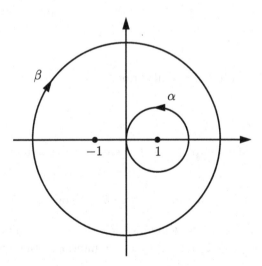

Figure 4.6 Paths α and β

Solution

The function f has poles of order 1 at the points 1 and -1, with residues

$$\text{Res}(f,1) = \lim_{z \to 1} \frac{(z-1)(z+3)}{(z-1)(z+1)} = \lim_{z \to 1} \frac{z+3}{z+1} = 2$$

and

$$\text{Res}(f,-1) = \lim_{z \to -1} \frac{(z+1)(z+3)}{(z-1)(z+1)} = \lim_{z \to -1} \frac{z+3}{z-1} = -1.$$

Since $\text{Ind}_\alpha(1) = 1$ and $\text{Ind}_\alpha(-1) = 0$, it follows from the Residue theorem that

$$\int_\alpha f = 2\pi i \big[\text{Res}(f,1)\,\text{Ind}_\alpha(1) + \text{Res}(f,-1)\,\text{Ind}_\alpha(-1)\big] = 4\pi i.$$

Similarly, since $\text{Ind}_\beta(1) = \text{Ind}_\beta(-1) = -1$, it follows again from the Residue theorem that

$$\int_\beta f = -2\pi i \big[\text{Res}(f,1) + \text{Res}(f,-1)\big] = -2\pi i.$$

Problem 4.19

For the function

$$f(z) = \frac{1}{(z^2-5)^2 - 16},$$

compute the integral $\int_\gamma f(z)\,dz$, where γ is a path looping once along the boundary of the square

$$Q = \big\{x + iy \in \mathbb{C} : |x| + |y| \le 2\big\} \tag{4.37}$$

(see Figure 4.7), in the positive direction.

Solution

The isolated singularities of f are the zeros of the polynomial $(z^2-5)^2 - 16$. We have

$$(z^2-5)^2 - 16 = 0 \quad \Leftrightarrow \quad z^2 - 5 = 4 \quad \text{or} \quad z^2 - 5 = -4,$$

and thus, $z = \pm 3$ or $z = \pm 1$. Being distinct, these four points are poles of f of order 1. Only the poles 1 and -1 are in the interior of the square Q, and thus,

$$\text{Ind}_\gamma(1) = \text{Ind}_\gamma(-1) = -1 \quad \text{and} \quad \text{Ind}_\gamma(3) = \text{Ind}_\gamma(-3) = 0.$$

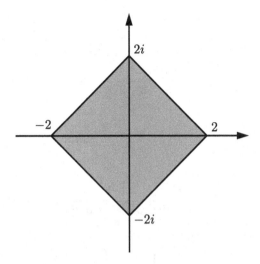

Figure 4.7 Square Q (4.37)

It then follows from the Residue theorem that

$$\int_\gamma f(z)\,dz = 2\pi i\big[\mathrm{Res}(f,-1)\,\mathrm{Ind}_\gamma(-1) + \mathrm{Res}(f,1)\,\mathrm{Ind}_\gamma(1)\big]$$

$$= -2\pi i\big[\mathrm{Res}(f,-1) + \mathrm{Res}(f,1)\big]. \qquad (4.38)$$

Since

$$f(z) = \frac{1}{(z-3)(z+3)(z-1)(z+1)},$$

we have

$$\mathrm{Res}(f,-1) = \lim_{z\to -1}(z+1)f(z) = \frac{1}{16}$$

and

$$\mathrm{Res}(f,1) = \lim_{z\to 1}(z-1)f(z) = -\frac{1}{16}.$$

Substituting in (4.38), we finally obtain

$$\int_\gamma f(z)\,dz = -2\pi i\left(\frac{1}{16} - \frac{1}{16}\right) = 0.$$

Problem 4.20

Compute the integral

$$\int_0^{2\pi} \frac{1}{3 + 2\cos t}\, dt.$$

Solution

We have

$$\int_0^{2\pi} \frac{1}{3 + 2\cos t}\, dt = \int_0^{2\pi} \frac{ie^{it}}{ie^{it}(3 + e^{it} + e^{-it})}\, dt = \int_\gamma f(z)\, dz, \qquad (4.39)$$

where

$$f(z) = \frac{1}{iz(3 + z + 1/z)} = \frac{1}{i(z^2 + 3z + 1)},$$

with the path $\gamma \colon [0, 2\pi] \to \mathbb{C}$ given by $\gamma(t) = e^{it}$. The function f has poles at the zeros of the polynomial $z^2 + 3z + 1$, that is, at

$$z_1 = (-3 + \sqrt{5})/2 \quad \text{and} \quad z_2 = (-3 - \sqrt{5})/2.$$

Since $|z_1| < 1$ and $|z_2| > 1$, we have $\operatorname{Ind}_\gamma(z_1) = 1$ and $\operatorname{Ind}_\gamma(z_2) = 0$. By the Residue theorem, we thus obtain

$$\int_\gamma f(z)\, dz = 2\pi i \sum_{j=1}^2 \operatorname{Res}(f, z_j)\operatorname{Ind}_\gamma(z_j) = 2\pi i \operatorname{Res}(f, z_1).$$

Since

$$\operatorname{Res}(f, z_1) = \lim_{z \to z_1} (z - z_1)f(z) = \frac{1}{i(z_1 - z_2)} = \frac{1}{i\sqrt{5}},$$

it follows from (4.39) that

$$\int_0^{2\pi} 1/(3 + 2\cos t)\, dt = 2\pi i \frac{1}{i\sqrt{5}} = \frac{2\pi}{\sqrt{5}}.$$

Problem 4.21

Compute the integral

$$\int_0^\infty \frac{1}{1 + x^4}\, dx. \qquad (4.40)$$

Solution

Take $R > 1$. We consider the path $\gamma = \gamma_1 + \gamma_2$, where $\gamma_1 \colon [-R, R] \to \mathbb{C}$ and $\gamma_2 \colon [0, \pi] \to \mathbb{C}$ are given respectively by

$$\gamma_1(t) = t \quad \text{and} \quad \gamma_2(t) = Re^{it}$$

(see Figure 4.8). We also consider the function

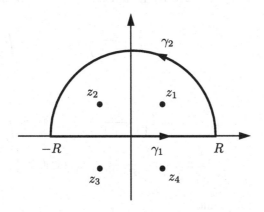

Figure 4.8   Path $\gamma = \gamma_1 + \gamma_2$

$$f(z) = \frac{1}{1 + z^4},$$

which has the poles

$$z_1 = (1 + i)/\sqrt{2}, \qquad z_2 = (-1 + i)/\sqrt{2},$$
$$z_3 = (-1 - i)/\sqrt{2}, \qquad z_4 = (1 - i)/\sqrt{2},$$

all of them of order 1. Since $R > 1$, we have

$$\mathrm{Ind}_\gamma(z_1) = \mathrm{Ind}_\gamma(z_2) = 1 \quad \text{and} \quad \mathrm{Ind}_\gamma(z_3) = \mathrm{Ind}_\gamma(z_4) = 0.$$

Moreover,

$$\mathrm{Res}(f, z_1) = \lim_{z \to z_1} \frac{1}{(z - z_2)(z - z_3)(z - z_4)} = \frac{-1 - i}{4\sqrt{2}}$$

and

$$\mathrm{Res}(f, z_2) = \lim_{z \to z_2} \frac{1}{(z - z_1)(z - z_3)(z - z_4)} = \frac{1 - i}{4\sqrt{2}}.$$

It then follows from the Residue theorem that

$$\int_\gamma f = 2\pi i \sum_{j=1}^4 \operatorname{Res}(f, z_j) \operatorname{Ind}_\gamma(z_j)$$

$$= 2\pi i \big[\operatorname{Res}(f, z_1) + \operatorname{Res}(f, z_2)\big] = \frac{\pi}{\sqrt{2}}. \qquad (4.41)$$

Now we use (4.41) to compute the integral in (4.40). We have

$$\int_\gamma f = \int_{-R}^R f(x)\,dx + \int_{\gamma_2} f = 2\int_0^R f(x)\,dx + \int_{\gamma_2} f.$$

Since

$$\left|\int_{\gamma_2} f\right| \le L_{\gamma_2} \sup\left\{\frac{1}{|z^4+1|} : z \in \gamma_2([0,\pi])\right\}$$

$$= \pi R \sup\left\{\frac{1}{|R^4 e^{4it}+1|} : t \in [0,\pi]\right\}$$

(see Proposition 2.42), and

$$\big|R^4 e^{4it} + 1\big| \ge \big|R^4 e^{4it}\big| - 1 = R^4 - 1,$$

we obtain

$$\left|\int_{\gamma_2} f\right| \le \pi R \frac{1}{R^4 - 1} \to 0$$

when $R \to +\infty$. Hence,

$$\int_\gamma f = 2\int_0^R f(x)\,dx + \int_{\gamma_2} f \to 2\int_0^\infty f(x)\,dx$$

when $R \to +\infty$, and thus,

$$\int_0^\infty f(x)\,dx = \frac{1}{2}\int_\gamma f.$$

It then follows from (4.41) that

$$\int_0^\infty \frac{1}{1+x^4}\,dx = \int_0^\infty f(x)\,dx = \frac{1}{2}\int_\gamma f = \frac{\pi}{2\sqrt{2}}.$$

Problem 4.22

Compute the integral

$$\int_0^\infty \frac{1}{(x^2+1)^2}\, dx.$$

Solution

We consider the path $\gamma = \gamma_1 + \gamma_2$ in Figure 4.8, and the function

$$f(z) = \frac{1}{(z^2+1)^2},$$

which has poles of order 2 at the points i and $-i$. Indeed,

$$f(z) = \frac{1}{[(z-i)(z+i)]^2} = \frac{1}{(z-i)^2(z+i)^2}.$$

We have

$$\mathrm{Ind}_\gamma(i) = 1 \quad \text{and} \quad \mathrm{Ind}_\gamma(-i) = 0,$$

and also

$$\mathrm{Res}(f,i) = \left(\frac{1}{(z+i)^2}\right)'\bigg|_{z=i} = -\frac{2}{(z+i)^3}\bigg|_{z=i} = \frac{1}{4i}.$$

It then follows from the Residue theorem that

$$\int_\gamma f = 2\pi i \,\mathrm{Res}(f,i) = \frac{\pi}{2}.$$

On the other hand,

$$\left|\int_{\gamma_2} f\right| \le \pi R \sup\left\{\frac{1}{|(z^2+1)^2|} : z \in \gamma_2([0,\pi])\right\}$$

$$\le \frac{\pi R}{(R^2-1)^2} \to 0$$

when $R \to +\infty$. Hence,

$$\frac{\pi}{2} = \int_\gamma f = \int_{-R}^R f(x)\, dx + \int_{\gamma_2} f$$

$$= 2\int_0^R f(x)\, dx + \int_{\gamma_2} f \to 2\int_0^\infty \frac{1}{(x^2+1)^2}\, dx$$

when $R \to +\infty$, and thus,

$$\int_0^\infty \frac{1}{(x^2+1)^2} \, dx = \frac{\pi}{4}.$$

Problem 4.23

Compute the integral

$$\int_0^\infty \frac{\cos x}{1+x^2} \, dx.$$

Solution

Take $R > 1$. Again we consider the path $\gamma = \gamma_1 + \gamma_2$ in Figure 4.8. We also consider the function

$$f(z) = \frac{e^{iz}}{1+z^2},$$

which has poles of order 1 at the points i and $-i$. Since $R > 1$, we have

$$\mathrm{Ind}_\gamma(i) = 1 \quad \text{and} \quad \mathrm{Ind}_\gamma(-i) = 0.$$

Moreover,

$$\mathrm{Res}(f, i) = \lim_{z \to i}(z - i)\frac{e^{iz}}{1+z^2} = \lim_{z \to i}\frac{e^{iz}}{z+i} = \frac{1}{2ei}.$$

It then follows from the Residue theorem that

$$\int_\gamma f = 2\pi i \, \mathrm{Res}(f, i) = \frac{\pi}{e}. \tag{4.42}$$

On the other hand,

$$\int_\gamma f = \int_{-R}^R f(x) \, dx + \int_{\gamma_2} f. \tag{4.43}$$

Since

$$\int_0^R f(x) \, dx = \int_0^R \frac{e^{ix}}{1+x^2} \, dx,$$

and

$$\int_{-R}^0 f(x) \, dx = -\int_R^0 f(-x) \, dx$$

$$= \int_0^R f(-x) \, dx = \int_0^R \frac{e^{-ix}}{1+x^2} \, dx,$$

it follows from (4.43) that

$$\int_\gamma f = \int_0^R \frac{e^{ix} + e^{-ix}}{1 + x^2} \, dx + \int_{\gamma_2} f$$

$$= 2 \int_0^R \frac{\cos x}{1 + x^2} \, dx + \int_{\gamma_2} f. \tag{4.44}$$

Moreover,

$$\left| \int_{\gamma_2} f \right| \le L_{\gamma_2} \sup \left\{ \left| \frac{e^{iz}}{1 + z^2} \right| : z \in \gamma_2([0, \pi]) \right\}$$

$$= \pi R \sup \left\{ \frac{e^{-\operatorname{Im} z}}{|z^2 + 1|} : z \in \gamma_2([0, \pi]) \right\}.$$

Since $e^{-\operatorname{Im} z} \le 1$ for $\operatorname{Im} z \ge 0$, and

$$|z^2 + 1| \ge |z^2| - 1 = |z|^2 - 1,$$

we obtain

$$\left| \int_{\gamma_2} f \right| \le \frac{\pi R}{R^2 - 1}. \tag{4.45}$$

Letting $R \to +\infty$, it follows from (4.44) and (4.45) that

$$\int_\gamma f = 2 \int_0^\infty \frac{\cos x}{1 + x^2} \, dx.$$

By (4.42), we finally obtain

$$\int_0^\infty \frac{\cos x}{1 + x^2} \, dx = \frac{1}{2} \int_\gamma f = \frac{\pi}{2e}.$$

Problem 4.24

For each $a \in (0, 1)$, compute the integral

$$\int_{-\infty}^{+\infty} \frac{e^{ax}}{1 + e^x} \, dx.$$

Solution

Take $R > 0$. We consider the path $\gamma = \gamma_1 + \gamma_2 + \gamma_3 + \gamma_4$, where

$$\gamma_1 \colon [-R, R] \to \mathbb{C} \quad \text{is given by } \gamma_1(t) = t,$$
$$\gamma_2 \colon [0, 2\pi] \to \mathbb{C} \quad \text{is given by } \gamma_2(t) = R + it,$$
$$\gamma_3 \colon [-R, R] \to \mathbb{C} \quad \text{is given by } \gamma_3(t) = -t + i2\pi,$$
$$\gamma_4 \colon [0, 2\pi] \to \mathbb{C} \quad \text{is given by } \gamma_4(t) = R + i(2\pi - t)$$

(see Figure 4.9). We also consider the function

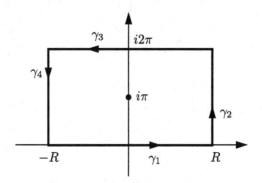

Figure 4.9   Path $\gamma = \gamma_1 + \gamma_2 + \gamma_3 + \gamma_4$

$$f(z) = \frac{e^{az}}{1 + e^z},$$

which has poles at the points $(2n + 1)\pi i$ for $n \in \mathbb{Z}$. We note that

$$\operatorname{Ind}_\gamma\big((2n+1)\pi i\big) = \begin{cases} 1 & \text{if } n = 0, \\ 0 & \text{if } n \neq 0. \end{cases}$$

Moreover,

$$\begin{aligned}
\operatorname{Res}(f, i\pi) &= \lim_{z \to i\pi} (z - i\pi)\frac{e^{az}}{1 + e^z} \\
&= e^{a\pi i} \lim_{z \to i\pi} \frac{z - i\pi}{1 + e^z} \\
&= e^{a\pi i} \lim_{z \to i\pi} \frac{1}{e^z} = -e^{a\pi i}.
\end{aligned}$$

It then follows from the Residue theorem that

$$\int_\gamma f = 2\pi i \operatorname{Res}(f, i\pi) = -2\pi i e^{a\pi i}. \tag{4.46}$$

On the other hand,

$$\int_{\gamma_1} f = \int_{-R}^{R} \frac{e^{at}}{1 + e^t} \, dt$$

and

$$\int_{\gamma_3} f = -\int_{-R}^{R} \frac{e^{a(-t+i2\pi)}}{1 + e^{-t+i2\pi}} \, dt$$

$$= -e^{2\pi i a} \int_{-R}^{R} \frac{e^{-at}}{1 + e^{-t}} \, dt = -e^{2\pi i a} \int_{-R}^{R} \frac{e^{at}}{1 + e^t} \, dt. \tag{4.47}$$

We also have

$$\left| \int_{\gamma_2} f \right| \le L_{\gamma_2} \sup \left\{ \left| \frac{e^{az}}{1 + e^z} \right| : z \in \gamma_2([0, 2\pi]) \right\}$$

$$= 2\pi \sup \left\{ \left| \frac{e^{a(R+it)}}{1 + e^{R+it}} \right| : t \in [0, 2\pi] \right\}$$

$$\le \frac{2\pi e^{aR}}{e^R - 1} \to 0$$

when $R \to +\infty$, and analogously,

$$\left| \int_{\gamma_4} f \right| \le L_{\gamma_4} \sup \left\{ \left| \frac{e^{az}}{1 + e^z} \right| : z \in \gamma_4([0, 2\pi]) \right\}$$

$$\le \frac{2\pi e^{aR}}{e^R - 1} \to 0$$

when $R \to +\infty$. It then follows from (4.46) and (4.47) that

$$-2\pi i e^{a\pi i} = \int_\gamma f = \sum_{j=1}^{4} \int_{\gamma_j} f$$

$$= (1 - e^{2\pi i a}) \int_{-R}^{R} \frac{e^{at}}{1 + e^t} \, dt + \int_{\gamma_2} f + \int_{\gamma_4} f$$

$$\to (1 - e^{2\pi i a}) \int_{-\infty}^{+\infty} \frac{e^{at}}{1 + e^t} \, dt$$

when $R \to +\infty$. We finally obtain

$$\int_{-\infty}^{+\infty} \frac{e^{at}}{1+e^t}\, dt = -\frac{2\pi i e^{a\pi i}}{1 - e^{2\pi i a}}$$

$$= \frac{2\pi i}{e^{\pi i a} - e^{-\pi i a}} = \frac{\pi}{\sin(\pi a)}.$$

Problem 4.25

For each $a > 0$, show that

$$\int_0^{\infty} \frac{\log x}{x^2 + a^2}\, dx = \frac{\pi}{2a} \log a. \tag{4.48}$$

Solution

Take $r, R > 0$ such that $r < a < R$. We consider the path $\gamma = \gamma_1 + \gamma_2 + \gamma_3 + \gamma_4$, where

$$\gamma_1 \colon [r, R] \to \mathbb{C} \qquad \text{is given by } \gamma_1(t) = t,$$
$$\gamma_2 \colon [0, \pi] \to \mathbb{C} \qquad \text{is given by } \gamma_2(t) = Re^{it},$$
$$\gamma_3 \colon [-R, -r] \to \mathbb{C} \quad \text{is given by } \gamma_3(t) = t,$$
$$\gamma_4 \colon [0, \pi] \to \mathbb{C} \qquad \text{is given by } \gamma_4(t) = re^{i(\pi - t)}$$

(see Figure 4.5). We also consider the function

$$f(z) = \frac{\log z}{z^2 + a^2},$$

which has poles of order 1 at ai and $-ai$. Since

$$\operatorname{Ind}_\gamma(ai) = 1 \quad \text{and} \quad \operatorname{Ind}_\gamma(-ai) = 0,$$

as well as

$$\operatorname{Res}(f, ai) = \lim_{z \to ai} (z - ai) f(z) = \lim_{z \to ai} \frac{\log z}{z + ai}$$

$$= \frac{\log(ai)}{2ai} = \frac{1}{2ai}\left(\log a + i\frac{\pi}{2}\right),$$

it follows from the Residue theorem that

$$\int_\gamma f(z)\, dz = 2\pi i \operatorname{Res}(f, ai) = \frac{\pi}{a}\left(\log a + i\frac{\pi}{2}\right). \tag{4.49}$$

On the other hand,

$$\left|\int_{\gamma_2} f\right| \leq L_{\gamma_2} \sup\left\{\left|\frac{\log R + it}{(Re^{it})^2 + a^2}\right| : t \in [0, \pi]\right\}$$

$$\leq \pi R \frac{\log R + \pi}{R^2 - a^2} \to 0$$

when $R \to +\infty$. Moreover,

$$\left|\int_{\gamma_4} f\right| \leq \pi r \frac{\log r + \pi}{a^2 - r^2} \to 0$$

when $r \to 0$. Taking limits when $R \to +\infty$ and $r \to 0$, it follows from (4.49) that

$$\frac{\pi}{a}\left(\log a + i\frac{\pi}{2}\right) = \int_{\gamma_1} f + \int_{\gamma_2} f + \int_{\gamma_3} f + \int_{\gamma_4} f$$

$$= \int_r^R \frac{\log x}{x^2 + a^2} \, dx + \int_{\gamma_2} f + \int_{-R}^{-r} \frac{\log x}{x^2 + a^2} \, dx + \int_{\gamma_4} f$$

$$\to \int_{-\infty}^{\infty} \frac{\log x}{x^2 + a^2} \, dx.$$

For $x < 0$ we have $\log x = \log|x| + i\pi$, and thus,

$$\frac{\pi}{a}\left(\log a + i\frac{\pi}{2}\right) = \int_{-\infty}^{\infty} \frac{\log|x|}{x^2 + a^2} \, dx + i\pi \int_{-\infty}^0 \frac{1}{x^2 + a^2} \, dx$$

$$= 2\int_0^{\infty} \frac{\log x}{x^2 + a^2} \, dx + i\pi \int_{-\infty}^0 \frac{1}{x^2 + a^2} \, dx. \qquad (4.50)$$

Taking the real part, we finally obtain identity (4.48).

Problem 4.26

Verify that if $f: \Omega \to \mathbb{C}$ is a nonvanishing holomorphic function, then the function $g(z) = 1/f(z)$ is holomorphic in Ω.

Solution

We need to show that g has derivatives at all points of Ω. We first note that

$$\frac{g(z) - g(z_0)}{z - z_0} = \frac{1/f(z) - 1/f(z_0)}{z - z_0}$$

$$= \frac{f(z_0) - f(z)}{(z - z_0)f(z)f(z_0)}.$$

Moreover, since the function f is continuous (because it is holomorphic), we have

$$\lim_{z \to z_0} \frac{1}{f(z)} = \frac{1}{f(z_0)}.$$

Finally, since

$$f'(z_0) = \lim_{z \to z_0} \frac{f(z) - f(z_0)}{z - z_0},$$

we obtain

$$\lim_{z \to z_0} \frac{g(z) - g(z_0)}{z - z_0} = \lim_{z \to z_0} \frac{f(z_0) - f(z)}{(z - z_0) f(z) f(z_0)}$$

$$= \lim_{z \to z_0} \frac{f(z_0) - f(z)}{z - z_0} \cdot \lim_{z \to z_0} \frac{1}{f(z) f(z_0)}$$

$$= -\frac{f'(z_0)}{f(z_0)^2}.$$

Therefore, the function g is holomorphic in Ω and $g' = -f'/f^2$.

Problem 4.27

Show that if $f \colon \mathbb{C} \to \mathbb{C}$ is an analytic function and $|f^{(n)}(0)| \le 2^n$ for every $n \in \mathbb{N} \cup \{0\}$, then $|f(z)| \le e^{2|z|}$ for $z \in \mathbb{C}$.

Solution

Since f is analytic in $\mathbb{C}$, by Theorem 4.12 (see (4.4) and (4.5)), we have

$$f(z) = \sum_{n=0}^{\infty} c_n z^n, \quad \text{with } c_n = \frac{f^{(n)}(0)}{n!}, \tag{4.51}$$

for each $z \in \mathbb{C}$. Moreover, since

$$|c_n| \le \frac{|f^{(n)}(0)|}{n!} \le \frac{2^n}{n!},$$

it follows from (4.51) that

$$|f(z)| \leq \sum_{n=0}^{\infty} |c_n||z|^n$$

$$\leq \sum_{n=0}^{\infty} \frac{2^n}{n!}|z|^n$$

$$= \sum_{n=0}^{\infty} \frac{(2|z|)^n}{n!} = e^{2|z|}.$$

Problem 4.28

Show that if the function f is holomorphic in an open set $\Omega \subset \mathbb{C}$, then the function

$$g(z) = \overline{f(\bar{z})}$$

is holomorphic in the open set $\Omega' = \{\bar{z} : z \in \Omega\}$.

Solution

Writing $z = x + iy$ and $f = u + iv$, we obtain

$$g(x + iy) = \overline{f(x - iy)}$$

$$= u(x, -y) - iv(x, -y)$$

$$= \bar{u}(x, y) + i\bar{v}(x, y),$$

where

$$\bar{u}(x, y) = u(x, -y) \quad \text{and} \quad \bar{v}(x, y) = -v(x, -y).$$

Since f is holomorphic in Ω, the functions u and v, and thus also the functions $\bar{u}$ and $\bar{v}$ are of class C^1. Their partial derivatives are given by

$$\frac{\partial \bar{u}}{\partial x} = \frac{\partial u}{\partial x}(x, -y), \qquad \frac{\partial \bar{u}}{\partial y} = -\frac{\partial u}{\partial y}(x, -y),$$

and

$$\frac{\partial \bar{v}}{\partial x} = -\frac{\partial v}{\partial x}(x, -y), \qquad \frac{\partial \bar{v}}{\partial y} = \frac{\partial v}{\partial y}(x, -y).$$

On the other hand, it follows from the Cauchy–Riemann equations (for f) that

$$\frac{\partial \bar{u}}{\partial x} = \frac{\partial \bar{v}}{\partial y} \quad \text{and} \quad \frac{\partial \bar{u}}{\partial y} = -\frac{\partial \bar{v}}{\partial x} \tag{4.52}$$

in Ω'. Since Ω' is an open set, and the functions $\bar{u}$ and $\bar{v}$ are of class C^1, it follows from Theorem 2.23 and (4.52) that the function g is holomorphic in Ω'.

Problem 4.29

Let f be a holomorphic function in an open set $\Omega \subset \mathbb{C}$, with $f'(z_0) \neq 0$ at some point $z_0 \in \Omega$. Show that for the path $\gamma \colon [0, 2\pi] \to \mathbb{C}$ given by $\gamma(t) = z_0 + re^{it}$, we have

$$\frac{2\pi i}{f'(z_0)} = \int_\gamma \frac{1}{f(z) - f(z_0)} \, dz$$

for any sufficiently small $r > 0$.

Solution

By Problem 4.5, the function

$$g(z) = \frac{z - z_0}{f(z) - f(z_0)}$$

has a removable singularity at $z = z_0$, and

$$\lim_{z \to z_0} g(z) = \frac{1}{f'(z_0)} \neq 0.$$

Hence, the function $z \mapsto 1/(f(z) - f(z_0))$ has a pole of order 1 at $z = z_0$, with residue $1/f'(z_0)$. It then follows from the Residue theorem that

$$\int_\gamma \frac{1}{f(z) - f(z_0)} \, dz = 2\pi i \frac{1}{f'(z_0)} \, \mathrm{Ind}_\gamma(z_0) = \frac{2\pi i}{f'(z_0)}$$

for any sufficiently small $r > 0$.

Problem 4.30

Compute the radius of convergence of the power series

$$f(z) = \sum_{n=0}^{\infty} \frac{(-1)^n}{(n!)^2} \cdot \frac{(z^2)^n}{2^{2n}},$$

and verify that

$$z f''(z) + f'(z) + z f(z) = 0.$$

Solution

By Proposition 4.4, the radius of convergence is given by

$$\lim_{n\to\infty}\left|\frac{(-1)^n}{n!2^{2n}}:\frac{(-1)^{n+1}}{(n+1)!2^{2(n+1)}}\right|=\lim_{n\to\infty}\frac{n+1}{4}=+\infty.$$

In particular, in view of Theorem 4.16, the function f is holomorphic in $\mathbb{C}$. Since power series can be differentiated term by term in the interior of their domain of convergence (see Theorem 4.12), we obtain

$$f'(z)=\sum_{n=1}^{\infty}\frac{(-1)^n}{(n!)^2}\cdot\frac{nz^{2n-1}}{2^{2n-1}}$$

and

$$f''(z)=\sum_{n=1}^{\infty}\frac{(-1)^n}{(n!)^2}\cdot\frac{n(2n-1)z^{2(n-1)}}{2^{2n-1}},$$

for $z\in\mathbb{C}$. Therefore,

$$zf''(z)+f'(z)+zf(z)$$

$$=\sum_{n=1}^{\infty}\frac{(-1)^n}{(n!)^2}\cdot\frac{n(2n-1)z^{2n-1}}{2^{2n-1}}$$

$$+\sum_{n=1}^{\infty}\frac{(-1)^n}{(n!)^2}\cdot\frac{nz^{2n-1}}{2^{2n-1}}+\sum_{n=0}^{\infty}\frac{(-1)^n}{(n!)^2}\frac{z^{2n+1}}{2^{2n}}$$

$$=\sum_{n=1}^{\infty}\left(\frac{(-1)^n}{(n!)^2}\cdot\frac{n(2n-1)+n}{2^{2n-1}}+\frac{(-1)^{n-1}}{((n-1)!)^2}\cdot\frac{1}{2^{2n-2}}\right)z^{2n-1}$$

$$=\sum_{n=1}^{\infty}\frac{(-1)^{n-1}}{((n-1)!)^2}\left(-\frac{2n^2}{n^22^{2n-1}}+\frac{1}{2^{2n-2}}\right)z^{2n-1}=0.$$

Problem 4.31

Find a holomorphic function in $\mathbb{C}$ whose set of zeros is $\{2n:n\in\mathbb{Z}\}\setminus\{0\}$.

Solution

Let us consider the function

$$f(z)=\begin{cases}\sin(\pi z/2)/z & \text{if }z\neq 0,\\ \pi/2 & \text{if }z=0.\end{cases}$$

It follows from (4.33) that

$$\frac{\sin(\pi z/2)}{z} = \frac{1}{z} \sum_{n=0}^{\infty} \frac{(-1)^n}{(2n+1)!} \left(\frac{\pi z}{2}\right)^{2n+1}$$

$$= \frac{\pi}{2} \sum_{n=0}^{\infty} \frac{(-1)^n}{(2n+1)!} \left(\frac{\pi z}{2}\right)^{2n}$$

for $z \neq 0$, with radius of convergence $+\infty$. Hence, the last series defines a holomorphic function g in $\mathbb{C}$. Since $g(0) = \pi/2$, we conclude that $f = g$, and thus f is holomorphic in $\mathbb{C}$.

Now we solve the equation

$$\sin(\pi z/2) = \frac{e^{i\pi z/2} - e^{-i\pi z/2}}{2i} = 0,$$

that is, $e^{i\pi z/2} = e^{-i\pi z/2}$, which is equivalent to $e^{i\pi z} = 1$. Writing $z = x + iy$, with $x, y \in \mathbb{R}$, we obtain

$$e^{i\pi z} = e^{-\pi y} e^{i\pi x} = 1,$$

and hence $y = 0$ and $x = 2k$ with $k \in \mathbb{Z}$. The solutions of the equation $\sin(\pi z/2) = 0$ are thus $z = 2k$ with $k \in \mathbb{Z}$, and the zeros of f are $z = 2k$ with $k \in \mathbb{Z} \setminus \{0\}$, since $f(0) = \pi/2 \neq 0$ and $f(z) = \sin(\pi z/2)/z$ for $z \neq 0$.

Problem 4.32

Show that if a holomorphic function $f \colon \mathbb{C} \to \mathbb{C}$ vanishes at all points of the form $r + ir$ with $r \in \mathbb{Q}$, then $f = 0$.

Solution

By Theorem 4.21, the zeros of a nonzero holomorphic function are isolated. Since the set $\{r + ir : r \in \mathbb{Q}\}$ has points that are not isolated (in fact, it has no isolated points), we conclude that $f = 0$ in $\mathbb{C}$.

Problem 4.33

Show that if $f \colon \mathbb{C} \to \mathbb{C}$ is a holomorphic function satisfying

$$|f(z)| \leq \log(1 + |z|) \tag{4.53}$$

for every $z \in \mathbb{C}$, then f is constant.

Solution

Given $z \in \mathbb{C}$, let $\gamma \colon [0, 2\pi] \to \mathbb{C}$ be the path defined by $\gamma(t) = z + re^{it}$. Since $\mathrm{Ind}_\gamma(z) = 1$, it follows from Cauchy's integral formula for the first derivative (see Theorem 4.17) that

$$f'(z) = \frac{1}{2\pi i} \int_\gamma \frac{f(w)}{(w - z)^2} \, dw.$$

Therefore,

$$\left| f'(z) \right| = \frac{1}{2\pi} \left| \int_\gamma \frac{f(w)}{(w - z)^2} \, dw \right|$$

$$\leq \frac{1}{2\pi} L_\gamma \sup \left\{ \frac{|f(\gamma(t))|}{|\gamma(t) - z|^2} : t \in [0, 2\pi] \right\}. \tag{4.54}$$

On the other hand, by (4.53), we have

$$\left| f(\gamma(t)) \right| \leq \log\left(1 + |z + re^{it}|\right) \leq \log\left(1 + |z| + r\right).$$

Since $L_\gamma = 2\pi r$ and

$$\left| \gamma(t) - z \right|^2 = \left| re^{it} \right|^2 = r^2,$$

it then follows from (4.54) that

$$\left| f'(z) \right| \leq \frac{\log(1 + |z| + r)}{r} \to 0$$

when $r \to +\infty$. Hence, $f'(z) = 0$ for every $z \in \mathbb{C}$, and thus f is constant.

Problem 4.34

Show that if a holomorphic function f in $\mathbb{C}$ has the periods 1 and i, then it is constant.

Solution

Since f has the periods 1 and i, it is sufficient to know its values in the compact set

$$K = \left\{ a + ib : a, b \in [0, 1] \right\}.$$

Indeed,

$$f\big((a + n) + i(b + m)\big) = f(a + ib)$$

for every $a, b \in [0,1]$ and $n, m \in \mathbb{Z}$. In particular,

$$\sup\left\{|f(z)| : z \in \mathbb{C}\right\} = \sup\left\{|f(z)| : z \in K\right\}. \tag{4.55}$$

On the other hand, since f is holomorphic, the function $z \mapsto |f(z)|$ is continuous, and thus its supremum in K is finite. It then follows from (4.55) that f is bounded, and we conclude from Liouville's theorem (Theorem 4.18) that the function f is constant.

Problem 4.35

Let $f \colon \Omega \to \mathbb{C}$ be a meromorphic function in an open set $\Omega \subset \mathbb{C}$, and let $\gamma \colon [a,b] \to \Omega$ be a closed piecewise regular path without intersections, looping once in the positive direction, such that f has neither zeros nor poles in $\gamma([a,b])$. Show that

$$\frac{1}{2\pi i} \int_\gamma \frac{f'}{f} = Z - P,$$

where Z and P are respectively the number of zeros and the number of poles of f in the interior of γ, counted with their multiplicities.

Solution

If $z = z_0$ is a zero of f of multiplicity n, then

$$f(z) = (z - z_0)^n g(z),$$

where g is a holomorphic function in some open ball centered at z_0 such that $g(z_0) \neq 0$. On the other hand, if $z = z_0$ is a pole of f of order n, then

$$f(z) = g(z)/(z - z_0)^n,$$

where g is a holomorphic function in some open ball centered at z_0. In the first case, we have

$$f'(z) = n(z - z_0)^{n-1}g(z) + (z - z_0)^n g'(z),$$

and thus,

$$\frac{f'(z)}{f(z)} = \frac{n}{z - z_0} + \frac{g'(z)}{g(z)}. \tag{4.56}$$

In the second case, we have

$$f'(z) = \frac{g'(z)(z - z_0)^n - n(z - z_0)^{n-1}g(z)}{(z - z_0)^{2n}}$$

$$= \frac{g'(z)}{(z - z_0)^n} - \frac{ng(z)}{(z - z_0)^{n+1}},$$

and thus,

$$\frac{f'(z)}{f(z)} = -\frac{n}{z - z_0} + \frac{g'(z)}{g(z)}. \tag{4.57}$$

We conclude that f'/f is a meromorphic function whose poles are exactly the zeros and the poles of f. It then follows from the Residue theorem that

$$\int_\gamma \frac{f'}{f} = 2\pi i \left[\sum_{j=1}^{Z'} \operatorname{Res}\left(\frac{f'}{f}, p_j\right) + \sum_{j=1}^{P'} \operatorname{Res}\left(\frac{f'}{f}, q_j\right) \right], \tag{4.58}$$

where p_j and q_j are respectively the zeros and the poles of f in the interior of γ, counted with their multiplicities, and where Z' and P' are respectively the number of zeros and the number of poles of f, also in the interior of γ, but now counted without their multiplicities. If p_j has multiplicity n_j, then it follows from (4.56) that

$$\operatorname{Res}\left(\frac{f'}{f}, p_j\right) = \operatorname{Res}\left(\frac{n_j}{z - p_j}, p_j\right) = n_j.$$

Moreover, if q_j has multiplicity m_j, then it follows from (4.57) that

$$\operatorname{Res}\left(\frac{f'}{f}, q_j\right) = \operatorname{Res}\left(-\frac{m_j}{z - q_j}, q_j\right) = -m_j.$$

We also have

$$\sum_{j=1}^{Z'} n_j = Z \quad \text{and} \quad \sum_{j=1}^{P'} m_j = P.$$

It then follows from (4.58) that

$$\int_\gamma \frac{f'}{f} = 2\pi i \left(\sum_{j=1}^{Z'} n_j - \sum_{j=1}^{P'} m_j \right) = 2\pi i(Z - P).$$

Problem 4.36

Let f and g be holomorphic functions in a simply connected open set $\Omega \subset \mathbb{C}$. Show that if

$$|f(z)| > |g(z) - f(z)| \tag{4.59}$$

for every $z \in \gamma([a, b])$, where $\gamma \colon [a, b] \to \Omega$ is a closed piecewise regular path without intersections, then f and g have the same number of zeros in the interior of γ, counted with their multiplicities.

Solution

It follows from (4.59) that f and g do not vanish on the set $\gamma([a, b])$. Hence, this set contains neither poles nor zeros of the function

$$F(z) = \frac{g(z)}{f(z)}.$$

It also follows from (4.59) that

$$\left| \frac{g(z)}{f(z)} - 1 \right| < 1$$

for $z \in \gamma([a, b])$, and thus,

$$\left| F(\gamma(t)) - 1 \right| < 1 \quad \text{for } t \in [a, b].$$

Hence, the closed piecewise regular path $F \circ \gamma$ is contained in the disk of radius 1 centered at 1. In particular, 0 is not in its interior. Therefore,

$$\int_{F \circ \gamma} \frac{1}{w} \, dw = 0.$$

On the other hand, it follows from Problem 4.35 that

$$0 = \int_{F \circ \gamma} \frac{1}{w} \, dw = \int_a^b \frac{F'(\gamma(t))}{F(\gamma(t))} \gamma'(t) \, dt = \int_\gamma \frac{F'(z)}{F(z)} \, dz = Z - P,$$

where Z and P are respectively the number of zeros and the number of poles of F in the interior of γ, counted with their multiplicities. These are respectively the zeros of g and f, also counted with their multiplicities.

Problem 4.37

Find the number of roots of the equation $z^3 + 3z + 1 = 0$ in the interior of the circle $|z| = 2$, counted with their multiplicities.

Solution

Let $f(z) = 3z$ and $g(z) = z^3 + 3z + 1$. We have

$$\left| g(z) - f(z) \right| \leq |z|^3 + 1 < |3z| = \left| f(z) \right|$$

for $|z| = 1$. Since f has a single zero in the interior of $|z| = 1$, it follows from Problem 4.36 that g also has a single zero in the interior of $|z| = 1$. On the other hand, if $f(z) = z^3$, then

$$\left| g(z) - f(z) \right| = |3z + 1| \leq 3|z| + 1 = 7 < 8 = |z|^3$$

for $|z| = 2$. Since f has three zeros in the interior of $|z| = 2$, counted with their multiplicities, it follows from Problem 4.36 that g also has three zeros in the interior of $|z| = 2$, again counted with their multiplicities. Two of them are in the ring $1 < |z| < 2$, since there are no zeros with $|z| = 1$.

EXERCISES

4.1. Write the function f as a power series centered at zero, indicating the radius of convergence:
  (a) $f(z) = z/(1 + z^2)$;
  (b) $f(z) = (z + 1)/(z - 1)$;
  (c) $f(z) = \sin z \cos z$.

4.2. Write the function f as a power series, indicating the radius of convergence:
  (a) $f(z) = 1/z$ at $a = 3$;
  (b) $f(z) = z/[(z - 1)(z - 3)]$ at $a = 2$;
  (c) $f(z) = z^3$ at $a = 1$.

4.3. Compute the radius of convergence of the power series:
  (a) $\sum_{n=1}^{\infty} z^n/(n^2)^n$;
  (b) $\sum_{n=1}^{\infty} n z^{n!}$;
  (c) $\sum_{n=0}^{\infty} z^{2n}/[2^{2n}(n!)^2]$.

4.4. Compute the radius of convergence of the power series of the function:
  (a) $z^2/(z^2 + 2z + 1)$ centered at $z = 3 + i$;
  (b) $1/(\cos z + 1)$ centered at $z = 1$.

4.5. Compute explicitly the function

$$\sum_{n=1}^{\infty} \left[3 - (-1)^n \right]^n z^n \quad \text{for } |z| < 1/4.$$

4.6. Find a power series with radius of convergence $\sqrt{2}$.

4.7. Verify that the power series

$$\sum_{n=0}^{\infty} c_n z^n \quad \text{and} \quad \sum_{n=0}^{\infty} (n^2+1)c_n z^{n+1}$$

have the same radius of convergence.

4.8. Write the function $(\sin z)/(z-\pi)$ as a power series centered at $z = \pi$.

4.9. For the function $z/\sin z$, find the Laurent series centered at $z = 0$ up to the term of order 4.

4.10. Find the term of order 4 of the power series of the function

$$\frac{\cos z \log(1+z)}{1-z}$$

centered at $z = 0$.

4.11. Let $u: \mathbb{R}^2 \to \mathbb{R}$ be the function $u(x,y) = e^{-y}\cos x + y(x-1)$.
   (a) Find v such that $f(x+iy) = u(x,y) + iv(x,y)$ is holomorphic in $\mathbb{C}$ and satisfies $f(0) = 1$.
   (b) Compute $\int_\gamma f(z)/(z-i)^2\, dz$ along the path $\gamma: [0, 4\pi] \to \mathbb{C}$ given by $\gamma(t) = 2e^{it}$.

4.12. Find and classify all isolated singularities of the function f, and compute the radius of convergence of its power series centered at the point a:
   (a) $f(z) = 1/(z^2+1)$, $a = 1$;
   (b) $f(z) = z^2/(z^2 - z - 2)$, $a = 0$;
   (c) $f(z) = e^{-z^2}/(z-2)$, $a = 0$.

4.13. Classify the isolated singularity at the origin for the function:
   (a) $(e^z+1)/(e^z-1)$;
   (b) $z\sin(1/z)$;
   (c) $\cos(1/z) - 1/\cos z$;
   (d) $(\sin z)/z^2$.

4.14. For the function

$$f(z) = \frac{\sin z - z + z^3/6}{z^5},$$

compute $\lim_{z\to 0} f(z)$ and verify that the origin is a removable singularity.

4.15. Find the limit of the function when $z \to 0$:
   (a) $\log(1-z)/z$;
   (b) $(e^{2z}-1)/z$;
   (c) $(e^{2z}-e^{-z})/z$.

4.16. Find and classify all isolated singularities of the function:

(a) $\dfrac{z^2+i}{(z-2)^2(z+1)} + \cos\left(\dfrac{1}{z}\right)$;

(b) $\dfrac{1}{z^2-1} + ze^{1/(z+3)}$;

(c) $\dfrac{z}{(e^z-1)^2} + e^{1/(z-4)}$.

4.17. Consider the function $f(z) = z/(e^z - 1)^2$.

(a) Find and classify all isolated singularities of f.

(b) Compute the integral of f along the circle $|z| = 2$ looping once in the positive direction.

4.18. Consider the function $f(z) = z/(z^2 + \sin^3 z)$.

(a) Classify the isolated singularity $z = 0$ of f.

(b) Find the terms of order -1 and -2 of the Laurent series of f at $z = 0$.

4.19. Compute the residue $\mathrm{Res}(e^z/z^{20}, 0)$.

4.20. Find the Laurent series of the function:

(a) $(\cos z)/z$ for $|z| > 0$;

(b) $z/(z^2 - 1)$ for $0 < |z+1| < 2$;

(c) $(\sin z)/z^2 + 1/(3 - z^2)$ for $0 < |z| < 3$ and for $|z| > 3$.

4.21. Compute the integral:

(a) $\displaystyle\int_\gamma \dfrac{1}{z^{20}}\, dz$, with $\gamma\colon [0, 2\pi] \to \mathbb{C}$ given by $\gamma(t) = 4e^{2it}$;

(b) $\displaystyle\int_\gamma \dfrac{1}{(z+2)(z-7)}\, dz$, with $\gamma\colon [0, 2\pi] \to \mathbb{C}$ given by $\gamma(t) = 6e^{it}$;

(c) $\displaystyle\int_\gamma \dfrac{\sin z}{z^3}\, dz$, with $\gamma\colon [0, 2\pi] \to \mathbb{C}$ given by $\gamma(t) = 3e^{it}$;

(d) $\displaystyle\int_\gamma \dfrac{\sinh z}{z^2-1}\, dz$, with $\gamma\colon [0, 2\pi] \to \mathbb{C}$ given by $\gamma(t) = 2e^{it}$.

4.22. For the path $\gamma\colon [0, 4\pi] \to \mathbb{C}$ given by $\gamma(t) = e^{it}$, compute the integral

$$\int_\gamma \frac{az^3 + bz^2 + cz + d}{z^3}\, dz$$

for each $a, b, c, d \in \mathbb{C}$.

4.23. Let $u\colon \mathbb{R}^2 \to \mathbb{R}$ be the function $u(x,y) = e^x \sin y$.

(a) Find $v\colon \mathbb{R}^2 \to \mathbb{R}$ such that $f = u + iv$ is holomorphic in $\mathbb{C}$ and satisfies $f(0) = -i$.

(b) Find explicitly the function f.

(c) Compute the integral

$$\int_\gamma \frac{f(z)}{z+i}\, dz$$

along the path $\gamma\colon [0, 6\pi] \to \mathbb{C}$ given by $\gamma(t) = 2e^{-it}$.

4.24. Let $u\colon \mathbb{R}^2 \to \mathbb{R}$ be the function $u(x, y) = e^{-y}\cos x + y(x - 1)$.

(a) Find $v\colon \mathbb{R}^2 \to \mathbb{R}$ such that $f = u + iv$ is holomorphic in $\mathbb{C}$ and satisfies $f(0) = 1$.

(b) Find explicitly the function f.

(c) Compute the integral

$$\int_\gamma \frac{f(z)}{(z-i)^2}\, dz$$

along the path $\gamma\colon [0, 4\pi] \to \mathbb{C}$ given by $\gamma(t) = 3e^{it}$.

4.25. Identify each statement as true or false.

(a) If f is a holomorphic function in $\mathbb{C}$ and $|f(z)| \le 1$ for every $z \in \mathbb{C}$, then f is a polynomial.

(b) There exists an analytic function in some open set $\Omega \subset \mathbb{C}$ that is not holomorphic in Ω.

(c) The derivative of $\sum_{n=0}^\infty c_n(z - a)^n$ has the same radius of convergence as this power series.

(d) All zeros of an analytic function are isolated.

(e) Two holomorphic functions $f, g\colon \mathbb{C} \to \mathbb{C}$ are equal if $f(z) = g(z)$ for every $z \in \mathbb{R}$.

(f) The function $z/(e^z - 1) + e^{1/(z-5)}$ has no essential singularities.

(g) The function $e^{e^{1/z}}$ has a pole.

(h) The residue of the function $z/(e^z - 1) + e^{1/(z-5)}$ at $z = 0$ is 0.

(i) The function $z/\sin z$ can be written as a power series in the ring $0 < |z| < \pi$.

(j) The boundary of the set $\{z \in \mathbb{C} : \sum_{n=0}^\infty (1 + 1/z)^n$ is convergent$\}$ is a straight line.

4.26. If γ is a closed path in $\mathbb{C} \setminus \{1, 2\}$ without intersections, find all possible values for the integral

$$\int_\gamma \frac{1}{(z-1)(z-2)}\, dz.$$

4.27. Let f be a function with a pole of order m at z_0, and let g be a function with a pole of order n at z_0. Show that fg has a pole of order $m + n$ at $z = z_0$.

4.28. Use the Residue theorem to compute the integral:

(a) $\displaystyle\int_0^{2\pi} \frac{1}{3 + \cos t}\, dt$;

(b) $\displaystyle\int_0^{2\pi} \frac{\cos(2t)}{5 - 4\cos t}\, dt$.

4.29. Given $a > 0$, use identity (4.50) to show that

$$\int_0^\infty \frac{1}{x^2 + a^2}\, dx = \frac{\pi}{2a}.$$

4.30. For $a > 1$, show that:

(a) $\displaystyle\frac{1}{2\pi}\int_0^{2\pi} \frac{\cos t}{a + \cos t}\, dt = 1 - \frac{a}{\sqrt{a^2 - 1}}$;

(b) $\displaystyle\frac{1}{2\pi}\int_0^{2\pi} \frac{\sin^2 t}{a + \cos t}\, dt = a - \sqrt{a^2 - 1}$.

4.31. Compute the integral:

(a) $\displaystyle\int_0^\infty \frac{1}{(x^2 + 2)^2}\, dx$;

(b) $\displaystyle\int_0^\infty \frac{1}{1 + x^6}\, dx$;

(c) $\displaystyle\int_0^\infty \frac{1}{(1 + x^2)(1 + x^4)}\, dx$;

(d) $\displaystyle\int_0^\infty \frac{x^2}{1 + x^4}\, dx$;

(e) $\displaystyle\int_0^\infty \frac{1}{(x^2 + a)^2}\, dx$ for $a > 0$.

4.32. For $a, b > 0$, show that

$$\int_0^\infty \frac{1}{(x^2 + a^2)(x^2 + b^2)}\, dx = \frac{\pi}{2ab(a + b)}.$$

4.33. Compute the integral:

(a) $\displaystyle\int_{-\infty}^\infty \frac{\sin x}{1 + x^2}\, dx$;

(b) $\displaystyle\int_0^\infty \frac{\cos x}{(1 + x^2)^2}\, dx$.

4.34. For $a > 0$, show that

$$\int_0^\infty \frac{\log x}{(x^2 + a^2)^2}\, dx = \frac{\pi}{4a^3}(\log a - 1).$$

4.35. Show that if $f, g: \Omega \to \mathbb{C}$ are holomorphic functions in a connected open set $\Omega \subset \mathbb{C}$, and $fg = 0$, then at least one of the functions f and g is zero in Ω.

4.36. Show that if $f = u + iv$ is a holomorphic function in an open set $\Omega \subset \mathbb{C}$, then $\Delta u = \Delta v = 0$ in Ω.

Hint: show that u and v are of class C^2.

4.37. Show that if f is a holomorphic function in $\mathbb{C}$, and there exists a polynomial p such that

$$\left| f(z) \right| \leq \log \left(1 + \left| p(z) \right| \right)$$

for every $z \in \mathbb{C}$, then f is constant.

4.38. Show that if f is a holomorphic function in $\mathbb{C}$, and there exist $c > 0$ and $n \in \mathbb{N}$ such that

$$\left| f(z) \right| < c \left(1 + \left| z \right|^n \right)$$

for every $z \in \mathbb{C}$, then f is a polynomial of degree at most n.

4.39. Show that if f is a holomorphic function in some open set containing the closed ball $\overline{B_r(z_0)}$, and

$$M = \sup \left\{ \left| f(z) \right| : z \in \overline{B_r(z_0)} \right\} < +\infty,$$

then

$$\left| f^{(n)}(z_0) \right| \leq \frac{n! M}{r^n} \quad \text{for } n \in \mathbb{N}.$$

Part II
Differential Equations

Ordinary Differential Equations

In this chapter we introduce the basic notions of the theory of ordinary differential equations. Besides establishing the existence and uniqueness of solutions, we study the class of linear differential equations with constant coefficients, as well as their perturbations. In particular, we show how to solve linear differential equations by computing exponentials of matrices, and we establish the Variation of parameters formula for the perturbations of these equations.

5.1 Basic Notions

In this chapter we consider ordinary differential equations of the form

$$x' = f(t, x), \tag{5.1}$$

where $f\colon D \to \mathbb{R}^n$ is a continuous function in some open set $D \subset \mathbb{R} \times \mathbb{R}^n$. We first introduce the notion of a solution.

Definition 5.1

We say that a function $x\colon (a, b) \to \mathbb{R}^n$ of class C^1 is a *solution* of the differential equation (5.1) if (see Figure 5.1):

L. Barreira, C. Valls, *Complex Analysis and Differential Equations*, Springer Undergraduate Mathematics Series, DOI 10.1007/978-1-4471-4008-5_5, © Springer-Verlag London 2012

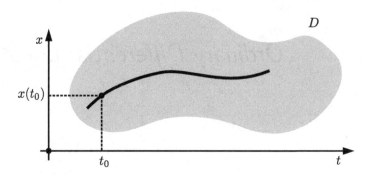

Figure 5.1   A solution of the equation $x' = f(t, x)$

(a) $(t, x(t)) \in D$ for each $t \in (a, b)$;

(b) $x'(t) = f(t, x(t))$ for each $t \in (a, b)$.

Example 5.2

Let us consider the equation $x' = x$ in $\mathbb{R}$. We note that $x(t)$ is a solution if and only if

$$\left(e^{-t}x(t)\right)' = -e^{-t}x(t) + e^{-t}x'(t)$$
$$= e^{-t}\left(-x(t) + x'(t)\right) = 0.$$

In order to simplify the notation, it is common to avoid writing explicitly the dependence of the solutions on t, thus writing instead

$$\left(e^{-t}x\right)' = -e^{-t}x + e^{-t}x'$$
$$= e^{-t}\left(-x + x'\right) = 0.$$

This shows that there exists $k \in \mathbb{R}$ such that $e^{-t}x(t) = k$, that is,

$$x(t) = ke^t, \quad t \in \mathbb{R}. \tag{5.2}$$

The solutions of the equation $x' = x$ are thus the functions in (5.2).

Example 5.3

We consider again the equation $x' = x$ in $\mathbb{R}$, and we describe an alternative method to find its solutions. Namely, if $x(t)$ is a nonvanishing solution, then

$$\frac{x'}{x} = 1 \quad \Leftrightarrow \quad \left(\log|x|\right)' = 1,$$

and hence,

$$\log|x(t)| = t + c$$

for some constant $c \in \mathbb{R}$. We thus obtain

$$|x(t)| = e^c e^t, \quad t \in \mathbb{R}.$$

Since $x(t)$ does not vanish and is continuous (since it is of class C^1), it is always positive or always negative. Hence,

$$x(t) = k e^t, \quad t \in \mathbb{R}$$

with $k \neq 0$, since by varying $c \in \mathbb{R}$ the function e^c takes all values of $\mathbb{R}^+$. By direct substitution in the equation, one can verify that the zero function is also a solution.

Example 5.4

Let us consider the equation $(x, y)' = (y, -x)$ in $\mathbb{R}^2$, which can be written in the form

$$\begin{cases} x' = y, \\ y' = -x. \end{cases}$$

If $(x(t), y(t))$ is a solution, then

$$\left(x^2 + y^2\right)' = 2xx' + 2yy'$$

$$= 2xy + 2y(-x) = 0.$$

Therefore, there exists $r \geq 0$ such that

$$x(t)^2 + y(t)^2 = r^2.$$

Writing

$$x(t) = r \cos \theta(t) \quad \text{and} \quad y(t) = r \sin \theta(t),$$

since $x' = y$, we obtain

$$x'(t) = -r \sin \theta(t) \cdot \theta'(t) = r \sin \theta(t)$$

(we note that since x is differentiable, the function θ must also be differentiable). Therefore, $\theta'(t) = -1$ and there exists $c \in \mathbb{R}$ such that $\theta(t) = -t + c$. We thus obtain

$$\left(x(t), y(t)\right) = \left(r \cos(-t + c), r \sin(-t + c)\right), \quad t \in \mathbb{R}.$$

Example 5.5

Now we consider the equation

$$x' = 2tx + t \tag{5.3}$$

in $\mathbb{R}$. One can write

$$\frac{x'}{2x + 1} = t \tag{5.4}$$

for the solutions x not taking the value $-1/2$. It follows from (5.4) that there exists $c \in \mathbb{R}$ such that

$$\frac{1}{2}\log|2x(t) + 1| = \frac{1}{2}t^2 + c,$$

that is,

$$|2x(t) + 1| = e^{t^2 + 2c}.$$

Proceeding as in Example 5.3, we then obtain

$$x(t) = -\frac{1}{2} + ke^{t^2}, \quad t \in \mathbb{R}$$

with $k \in \mathbb{R}$. These are the solutions of equation (5.3).

5.2 Existence and Uniqueness of Solutions

In the same way as it is important to know that a polynomial has roots even if one is not able to compute them, it is also important to know when a differential equation has solutions even if one is not able to compute them.

We have the following result concerning the existence and uniqueness of solutions of a differential equation.

Theorem 5.6

If $f\colon D \to \mathbb{R}^n$ is a function of class C^1 in some open set $D \subset \mathbb{R} \times \mathbb{R}^n$, then for each $(t_0, x_0) \in D$ there exists a unique solution of the equation $x' = f(t, x)$ with $x(t_0) = x_0$ in some open interval containing t_0.

More generally, one can consider functions f that are not necessarily of class C^1.

Definition 5.7

A function $f\colon D \to \mathbb{R}^n$ is said to be *locally Lipschitz* in (the variable) x if, for each compact set $K \subset D$, there exists $L > 0$ such that

$$\big\| f(t,x) - f(t,y) \big\| \leq L \|x - y\|$$

for every $(t,x), (t,y) \in K$.

Using the Mean value theorem, one can show that all functions of class C^1 are locally Lipschitz in x. But there are many other functions that are locally Lipschitz.

Example 5.8

For each $x, y \in \mathbb{R}$, we have

$$\big| |x| - |y| \big| \leq |x - y|.$$

This shows that the function $f(t,x) = |x|$ is locally Lipschitz in x, with $L = 1$.

We also give an example of a function that is not locally Lipschitz.

Example 5.9

For the function $f\colon \mathbb{R}^2 \to \mathbb{R}$ given by $f(t,x) = \sqrt{|x|}$, we have

$$\big| f(x,t) - f(t,0) \big| = \sqrt{|x|} = \frac{1}{\sqrt{|x|}} |x - 0|.$$

Since $1/\sqrt{|x|} \to +\infty$ when $x \to 0$, the function f is not locally Lipschitz in x in any open set $D \subset \mathbb{R} \times \mathbb{R}$ intersecting the line $\mathbb{R} \times \{0\}$.

The following result includes Theorem 5.6 as a particular case.

Theorem 5.10 (Picard–Lindelöf theorem)

If the function $f\colon D \to \mathbb{R}^n$ is continuous and locally Lipschitz in x in some open set $D \subset \mathbb{R} \times \mathbb{R}^n$, then for each $(t_0, x_0) \in D$ there exists a unique solution of the equation $x' = f(t,x)$ with $x(t_0) = x_0$ in some open interval containing t_0.

Proof

We first observe that there exists a solution of the problem

$$\begin{cases} x' = f(t, x), \\ x(t_0) = x_0 \end{cases} \tag{5.5}$$

if and only if there exists a continuous function $x \colon (a, b) \to \mathbb{R}^n$ in some open interval containing t_0 such that

$$x(t) = x_0 + \int_{t_0}^t f\big(s, x(s)\big) \, ds \tag{5.6}$$

for every $t \in (a, b)$. Indeed, it follows from (5.5) that

$$x(t) = x(t_0) + \int_{t_0}^t f\big(s, x(s)\big) \, ds = x_0 + \int_{t_0}^t f\big(s, x(s)\big) \, ds,$$

and the function x is continuous (since it is of class C^1). On the other hand, if x is a continuous function satisfying (5.6), then the function $s \mapsto f(s, x(s))$ is also continuous, since it is a composition of continuous functions. Therefore, $t \mapsto \int_{t_0}^t f(s, x(s)) \, ds$ is of class C^1. Taking derivatives with respect to t in (5.6), we then obtain

$$x'(t) = f\big(t, x(t)\big)$$

for every $t \in (a, b)$. It also follows from (5.6) that $x(t_0) = x_0$.

Take $a < t_0 < b$ and $\beta > 0$ such that $[a, b] \times \overline{B_\beta(x_0)} \subset D$, where

$$\overline{B_\beta(x_0)} = \big\{ y \in \mathbb{R}^n : \|y - x_0\| \le \beta \big\}.$$

Now let X be the family of continuous functions $x \colon (a, b) \to \mathbb{R}^n$ such that $\|x(t) - x_0\| \le \beta$ for every $t \in (a, b)$. We also consider the transformation T defined by

$$(Tx)(t) = x_0 + \int_{t_0}^t f\big(s, x(s)\big) \, ds$$

for each $x \in X$. We note that Tx is a continuous function, and that

$$\big\| (Tx)(t) - x_0 \big\| \le \left\| \int_{t_0}^t f\big(s, x(s)\big) \, ds \right\|$$

$$\le |t - t_0| M \le (b - a) M,$$

where

$$M = \sup\big\{ \big\| f(t, x) \big\| : t \in [a, b], \; x \in \overline{B_\beta(x_0)} \big\}.$$

Since the function $(t, x) \mapsto \|f(t, x)\|$ is continuous and the set $[a, b] \times \overline{B_\beta(x_0)}$ is compact, it follows from Weierstrass' theorem that M is finite. Moreover, for $b - a$ sufficiently small, we have $(b - a)M \leq \beta$, and thus $T(X) \subset X$.

We also note that

$$\|(Tx)(t) - (Ty)(t)\| \leq \left\| \int_{t_0}^t [f(s, x(s)) - f(s, y(s))] \, ds \right\|$$

$$\leq \left| \int_{t_0}^t L \|x(s) - y(s)\| \, ds \right|$$

$$\leq (b - a) L \|x - y\|_\infty$$

for every $x, y \in X$, where

$$\|x - y\|_\infty = \sup\{\|x(t) - y(t)\| : t \in (a, b)\}.$$

Therefore,

$$\|Tx - Ty\|_\infty \leq (b - a) L \|x - y\|_\infty. \tag{5.7}$$

Now we consider the sequence

$$x_m = T x_{m-1} = T^m x_0, \quad m \in \mathbb{N},$$

where $x_0 \in X$ denotes the constant function equal to x_0. The sequence is well defined, since $T(X) \subset X$. We note that if necessary one can rechoose a and b so that $c = (b - a) L < 1$. Then

$$\|x_p - x_q\|_\infty \leq \sum_{j=q}^{p-1} \|x_{j+1} - x_j\|_\infty \leq \sum_{j=q}^{\infty} \|T^j(Tx_0) - T^j x_0\|_\infty$$

$$\leq \sum_{j=q}^{\infty} c^j \|Tx_0 - x_0\|_\infty = \frac{c^q}{1 - c} \|Tx_0 - x_0\|_\infty$$

for every $p \geq q$. For each $t \in (a, b)$, we have

$$\|x_p(t) - x_q(t)\| \leq \|x_p - x_q\|_\infty \leq \frac{c^q}{1 - c} \|Tx_0 - x_0\|_\infty, \tag{5.8}$$

and thus, $(x_m(t))_m$ is a Cauchy sequence in $\mathbb{R}^n$. Therefore, the sequence is convergent and one can define

$$x(t) = \lim_{m \to \infty} x_m(t).$$

Letting $p \to \infty$ in (5.8), we obtain

$$\|x(t) - x_q(t)\| \le \frac{c^q}{1-c}\|Tx_0 - x_0\|_\infty. \tag{5.9}$$

Now we show that $x \in X$ and that it satisfies (5.6). We have

$$\|x(t) - x_0\| = \lim_{m\to\infty}\|x_m(t) - x_0\| \le \beta.$$

Moreover, for each $t, s \in (a, b)$,

$$\|x(t) - x(s)\| \le \|x(t) - x_m(t)\| + \|x_m(t) - x_m(s)\| + \|x_m(s) - x(s)\|. \tag{5.10}$$

Given $\varepsilon > 0$, it follows from (5.9) that there exists $p \in \mathbb{N}$ such that

$$\|x_m(t) - x(t)\| < \varepsilon$$

for every $t \in (a, b)$ and $m \ge p$. Taking $m = p$, it follows from (5.10) that

$$\|x(t) - x(s)\| < 2\varepsilon + \|x_p(t) - x_p(s)\|. \tag{5.11}$$

On the other hand, since x_p is continuous, there exists $\delta > 0$ such that

$$\|x_p(t) - x_p(s)\| < \varepsilon \quad \text{whenever } |t - s| < \delta,$$

and thus, it follows from (5.11) that

$$\|x(t) - x(s)\| < 3\varepsilon \quad \text{whenever } |t - s| < \delta.$$

This shows that x is continuous, and hence $x \in X$. In order to show that x also satisfies (5.6), we note that

$$x(t) = \lim_{m\to\infty} x_m(t) = \lim_{m\to\infty}\left(x_0 + \int_{t_0}^t f(s, x_m(s))\, ds\right), \tag{5.12}$$

and

$$\left|\int_{t_0}^t [f(s, x_m(s)) - f(s, x(s))]\, ds\right| \le \left|\int_{t_0}^t L\|x_m(s) - x(s)\|\, ds\right|$$

$$\le (b-a)L\|x_m - x\|_\infty \to 0$$

when $m \to \infty$. It then follows from (5.12) that x satisfies (5.6).

It remains to show that the solution x is unique. Let us assume that $y \in X$ is also a solution. In view of the identities $Tx = x$ and $Ty = y$, it follows from (5.7) that

$$\|x - y\|_\infty = \|Tx - Ty\|_\infty \le c\|x - y\|_\infty.$$

But since $c < 1$, we must have $\|x - y\|_\infty = 0$, and thus $x = y$. $\qquad\square$

Example 5.11

Let us consider the equation $x' = |x|$. By Example 5.8 and the Picard–Lindelöf theorem (Theorem 5.10), for each $(t_0, x_0) \in \mathbb{R}^2$ there exists a unique solution of the equation with $x(t_0) = x_0$ in some open interval containing t_0.

Clearly, $x(t) = 0$ is a solution. On the other hand, when $x > 0$ we obtain the equation $x' = x$, which by Example 5.3 has the solutions

$$x(t) = ke^t, \quad t \in \mathbb{R},$$

now with $k > 0$ (so that $x(t)$ is positive). When $x < 0$ we obtain the equation $x' = -x$ whose solutions satisfy

$$\left(e^t x\right)' = e^t x + e^t x' = e^t \left(x + x'\right) = 0,$$

and hence,

$$x(t) = ke^{-t}, \quad t \in \mathbb{R},$$

with $k < 0$. We thus obtain the solutions

$$x(t) = \begin{cases} ke^t & \text{with } k > 0, \\ 0, & \\ ke^{-t} & \text{with } k < 0. \end{cases} \tag{5.13}$$

One can easily verify that for each $(t_0, x_0) \in \mathbb{R}^2$ there exists exactly one solution with $x(t_0) = x_0$. On the other hand, by Theorem 5.10, all these solutions are unique. Therefore, each solution of the equation $x' = |x|$ takes one of the forms in (5.13).

The following example shows that for functions $f(t, x)$ that are not locally Lipschitz the solutions may not be unique.

Example 5.12

Let us consider the continuous function $f(t, x) = \sqrt{|x|}$ in Example 5.9. One can easily verify that both $x(t) = 0$ and

$$x(t) = \begin{cases} t^2/4 & \text{if } t \geq 0, \\ 0 & \text{if } t \leq 0 \end{cases}$$

are solutions of the equation $x' = \sqrt{|x|}$ with $x(0) = 0$.

We also show that each solution given by Theorem 5.10 can be extended to a maximal interval in a unique manner.

Theorem 5.13

If the function $f: D \to \mathbb{R}^n$ is continuous and locally Lipschitz in x in some open set $D \subset \mathbb{R} \times \mathbb{R}^n$, then for each $(t_0, x_0) \in D$ there exists a unique solution $\phi: (a, b) \to \mathbb{R}^n$ of problem (5.5) such that, for any solution $x: I_x \to \mathbb{R}^n$ of the same problem, we have $I_x \subset (a, b)$ and $x(t) = \phi(t)$ for every $t \in I_x$.

Proof

We note that $J = \bigcup_x I_x$ is an open interval, since the union of any family of open intervals containing t_0 is still an open interval (containing t_0). Now we define a function $\phi: J \to \mathbb{R}^n$ as follows. For each $t \in I_x$, let us take $\phi(t) = x(t)$. We show that the function ϕ is well defined, that is, $\phi(t)$ does not depend on the function x. Let $x: I_x \to \mathbb{R}^n$ and $y: I_y \to \mathbb{R}^n$ be solutions of problem (5.5). Let also I be the largest open interval containing t_0 where $x = y$. We want to show that $I = I_x \cap I_y$. Otherwise, the interval I would have an endpoint s that is not an endpoint of $I_x \cap I_y$. Since x and y are continuous in the interval $I_x \cap I_y$, we have

$$p := \lim_{t \to s} x(t) = \lim_{t \to s} y(t).$$

Moreover, by Theorem 5.10 with the pair (t_0, x_0) replaced by (s, p), there would exist an open interval

$$(s - \alpha, s + \alpha) \subset I_x \cap I_y \quad \text{where } x = y.$$

But since $(s - \alpha, s + \alpha) \setminus I \neq \emptyset$, this contradicts the fact that I is the largest interval containing t_0 where $x = y$. Therefore, $I = I_x \cap I_y$ and $x = y$ in $I_x \cap I_y$. Clearly, the function $\phi: J \to \mathbb{R}^n$ is a solution of problem (5.5). This yields the desired result. $\square$

By Theorem 5.13, one can introduce the notion of maximal interval (of existence) of a solution as follows.

Definition 5.14

Under the assumptions of Theorem 5.13, the *maximal interval (of existence)* of a solution $x: I \to \mathbb{R}^n$ of the equation $x' = f(t, x)$ is the largest open interval where there exists a solution coinciding with x in I.

Example 5.15

By Example 5.2, the solutions of the equation $x' = x$ are given by (5.2). All solutions have maximal interval $\mathbb{R}$, since they are defined for all $t \in \mathbb{R}$.

Example 5.16

By Example 5.11, the solutions of the equation $x' = |x|$ are given by (5.13). All solutions have maximal interval $\mathbb{R}$, since they are defined for all $t \in \mathbb{R}$.

Example 5.17

Let us consider the equation $x' = x^2$. Besides the solution $x(t) = 0$, which has maximal interval $\mathbb{R}$, the nonvanishing solutions are obtained writing

$$\frac{x'}{x^2} = 1 \quad \Leftrightarrow \quad \left(-\frac{1}{x}\right)' = 1.$$

Therefore,

$$-\frac{1}{x(t)} = t + c \quad \Leftrightarrow \quad x(t) = -\frac{1}{t+c}, \tag{5.14}$$

for some constant $c \in \mathbb{R}$. The maximal interval is thus $(-\infty, -c)$ or $(-c, +\infty)$, depending on whether the initial time t_0 is contained in the first or second intervals. For example, the solution of the equation $x' = x^2$ with $x(2) = 3$ is obtained substituting $t = 2$ in (5.14): we obtain

$$x(2) = -\frac{1}{2+c} = 3 \quad \Leftrightarrow \quad c = -\frac{7}{3},$$

and thus,

$$x(t) = -\frac{1}{t - 7/3} = \frac{3}{7 - 3t}$$

for $t \in (-\infty, 7/3)$, since $2 \in (-\infty, 7/3)$.

5.3 Linear Equations: Scalar Case

In this section we consider the particular case of equations in $\mathbb{R}$ of the form

$$x' = a(t)x + b(t), \tag{5.15}$$

where $a, b: \mathbb{R} \to \mathbb{R}$ are continuous functions. The solutions are obtained as follows.

Theorem 5.18

For each $(t_0, x_0) \in \mathbb{R}^2$, the (unique) solution of equation (5.15) with $x(t_0) = x_0$ is given by

$$x(t) = e^{\int_{t_0}^{t} a(s)\, ds} x_0 + \int_{t_0}^{t} e^{\int_{u}^{t} a(s)\, ds} b(u)\, du$$

for $t \in \mathbb{R}$ (and thus has maximal interval $\mathbb{R}$).

Proof

We first note that the function $f(t,x) = a(t)x + b(t)$ is continuous and locally Lipschitz in x. To verify that it is locally Lipschitz, we write

$$\bigl| f(t,x) - f(t,y) \bigr| = |a(t)| \cdot |x - y|.$$

Since the function $(t,x) \mapsto |a(t)|$ is continuous, it has a maximum in each compact set, and hence f is locally Lipschitz in x. By the Picard–Lindelöf theorem (Theorem 5.10), for each $(t_0, x_0) \in \mathbb{R}^2$ there exists a unique solution of equation (5.15) with $x(t_0) = x_0$ in some open interval containing t_0.

Now we note that if $x(t)$ is a solution with $x(t_0) = x_0$, then

$$\left(x(t) e^{-\int_{t_0}^{t} a(s)\, ds} \right)' = e^{-\int_{t_0}^{t} a(s)\, ds} \bigl[x'(t) - a(t)x(t) \bigr]$$

$$= e^{-\int_{t_0}^{t} a(s)\, ds} b(t).$$

Integrating over t, we obtain

$$x(t) e^{-\int_{t_0}^{t} a(s)\, ds} - x_0 = \int_{t_0}^{t} e^{-\int_{t_0}^{u} a(s)\, ds} b(u)\, du$$

and

$$x(t) = e^{\int_{t_0}^{t} a(s)\, ds} x_0 + \int_{t_0}^{t} e^{\int_{u}^{t} a(s)\, ds} b(u)\, du.$$

Since the integrands are continuous, the solution $x(t)$ is defined for every $t \in \mathbb{R}$, and thus has maximal interval $\mathbb{R}$. □

Example 5.19

It follows from Theorem 5.18 with $b(t) = 0$ that the (unique) solution of the equation $x' = a(t)x$ with $x(t_0) = x_0$ is given by

$$x(t) = e^{\int_{t_0}^{t} a(s)\, ds} x_0 \quad \text{for } t \in \mathbb{R}. \tag{5.16}$$

Example 5.20

Let us consider the equation $x' = x \cos t$. The solution with $x(0) = 2$ is given by (5.16), that is,

$$x(t) = e^{\int_0^t \cos s \, ds} 2 = 2e^{\sin t} \quad \text{for } t \in \mathbb{R}.$$

Example 5.21

Now we consider the equation

$$x' = 3x + t$$

with the condition $x(1) = 0$. By Theorem 5.18, the solution is given by

$$x(t) = e^{3(t-1)}0 + \int_1^t e^{3(t-s)} s \, ds$$

$$= -e^{3(t-s)} \left(\frac{s}{3} + \frac{1}{9} \right) \Big|_{s=1}^{s=t}$$

$$= -\frac{t}{3} - \frac{1}{9} + \frac{4}{9} e^{3(t-1)}$$

for $t \in \mathbb{R}$.

Example 5.22

Finally, we consider the equation

$$x' = \frac{t}{t^2 + 1} x + t$$

with the condition $x(0) = 1$. We note that the functions

$$a(t) = \frac{t}{t^2 + 1} \quad \text{and} \quad b(t) = t$$

are continuous. By Theorem 5.18, since

$$\exp\left(\int_{t_0}^t a(s) \, ds \right) = \exp\left(\frac{1}{2} \log(s^2 + 1) \Big|_{s=t_0}^{s=t} \right) = \sqrt{\frac{t^2 + 1}{t_0^2 + 1}},$$

the solution is given by

$$
\begin{aligned}
x(t) &= \sqrt{\frac{t^2 + 1}{0^2 + 1}}\, x(0) + \int_0^t \sqrt{\frac{t^2 + 1}{s^2 + 1}}\, s \, ds \\
&= \sqrt{t^2 + 1} + \sqrt{t^2 + 1} \int_0^t \frac{s}{\sqrt{s^2 + 1}} \, ds \\
&= \sqrt{t^2 + 1} + \sqrt{t^2 + 1}\sqrt{s^2 + 1}\,\Big|_{s=0}^{s=t} \\
&= \sqrt{t^2 + 1} + t^2 + 1 - \sqrt{t^2 + 1} = t^2 + 1
\end{aligned}
$$

for $t \in \mathbb{R}$.

5.4 Linear Equations: General Case

Now we consider equations in $\mathbb{R}^n$ of the form

$$ x' = Ax + b(t), \tag{5.17} $$

where A is an $n \times n$ matrix with real entries and $b \colon \mathbb{R} \to \mathbb{R}^n$ is a continuous function. On purpose, we do not consider the more general case of equations in $\mathbb{R}^n$ of the form

$$ x' = A(t)x + b(t), $$

where $A(t)$ is an $n \times n$ matrix varying continuously with t. In spite of their importance, these equations fall outside the scope of the book.

We start our study with the particular case when $b(t) = 0$, that is, with the equation

$$ x' = Ax, \tag{5.18} $$

where A is an $n \times n$ matrix with real entries. Since the function $f(t, x) = Ax$ is of class C^1, it is also continuous and locally Lipschitz in x.

Example 5.23

Let us consider the equation

$$ \begin{pmatrix} x \\ y \end{pmatrix}' = \begin{pmatrix} 0 & 1 \\ -1 & 0 \end{pmatrix} \begin{pmatrix} x \\ y \end{pmatrix}, \tag{5.19} $$

which can be written in the form

$$\begin{cases} x' = y, \\ y' = -x. \end{cases}$$

By Example 5.4, its solutions are

$$\begin{pmatrix} x(t) \\ y(t) \end{pmatrix} = \begin{pmatrix} r\cos(-t+c) \\ r\sin(-t+c) \end{pmatrix},$$

with $r \geq 0$ and $c \in [0, 2\pi)$. We note that

$$\begin{pmatrix} x(t) \\ y(t) \end{pmatrix} = \begin{pmatrix} r\cos c\cos t + r\sin c\sin t \\ -r\cos c\sin t + r\sin c\cos t \end{pmatrix}$$

$$= r\cos c \begin{pmatrix} \cos t \\ -\sin t \end{pmatrix} + r\sin c \begin{pmatrix} \sin t \\ \cos t \end{pmatrix}.$$

Therefore, the set of solutions of equation (5.19) is a linear space of dimension 2, generated by the vectors

$$\begin{pmatrix} \cos t \\ -\sin t \end{pmatrix} \quad \text{and} \quad \begin{pmatrix} \sin t \\ \cos t \end{pmatrix}.$$

In order to solve equation (5.18) for an arbitrary matrix A, we introduce the notion of the exponential of a matrix.

Definition 5.24

We define the *exponential* of a square matrix A by

$$e^A = \sum_{k=0}^{\infty} \frac{1}{k!} A^k, \tag{5.20}$$

with the convention that A^0 is the identity matrix Id.

We show that the series converges.

Proposition 5.25

The series in (5.20) is convergent, that is, there exists an $n \times n$ matrix B such that

$$\sum_{k=0}^{m} \frac{1}{k!} A^k \to B$$

entry by entry when $m \to \infty$.

Proof

Let

$$\|A\| = \sup_{x \neq 0} \frac{\|Ax\|}{\|x\|},$$

with the norm in $\mathbb{R}^n$ given by

$$\|(x_1, \dots, x_n)\| = \left(\sum_{i=1}^{n} x_i^2 \right)^{1/2}.$$

We note that

$$\|A^k\| \leq \|A\|^k. \tag{5.21}$$

Indeed,

$$\|A^k\| = \sup_{x \neq 0} \frac{\|A^k x\|}{\|x\|}$$

$$= \sup_{x \neq 0,\ A^{k-1}x \neq 0} \frac{\|A(A^{k-1})x\|}{\|x\|}$$

$$= \sup_{x \neq 0,\ A^{k-1}x \neq 0} \left(\frac{\|A(A^{k-1}x)\|}{\|A^{k-1}x\|} \cdot \frac{\|A^{k-1}x\|}{\|x\|} \right)$$

$$\leq \sup_{y \neq 0} \frac{\|Ay\|}{\|y\|} \cdot \sup_{x \neq 0} \frac{\|A^{k-1}x\|}{\|x\|}$$

$$= \|A\| \cdot \|A^{k-1}\|,$$

and inequality (5.21) follows by induction. Therefore,

$$\sum_{k=0}^{\infty} \left\| \frac{1}{k!} A^k \right\| = \sum_{k=0}^{\infty} \frac{1}{k!} \|A^k\|$$

$$\leq \sum_{k=0}^{\infty} \frac{1}{k!} \|A\|^k = e^{\|A\|} < \infty. \tag{5.22}$$

Now we observe that if the entries of an $n \times n$ matrix B are b_{ij}, for $i, j = 1, \dots, n$, and $e_1, \dots, e_n$ is the standard basis of $\mathbb{R}^n$, then

$$\|B\| \geq \frac{\|Be_j\|}{\|e_j\|} = \|Be_j\|$$

$$= \left(\sum_{i=1}^{n} b_{ij}^2 \right)^{1/2} \geq |b_{ij}|.$$

It thus follows from (5.22) that the series $\sum_{k=0}^{\infty}|a_{ij}^{(k)}|/k!$, where $a_{ij}^{(k)}$ are the entries of A^k, is convergent for $i, j = 1, \ldots, n$. This shows that the series $\sum_{k=0}^{\infty} A^k/k!$ is convergent entry by entry. $\square$

Example 5.26

Let 0 be the $n \times n$ matrix with all entries equal to zero. Since $0^k = 0$ for each $k \in \mathbb{N}$, we obtain

$$e^0 = \sum_{k=0}^{\infty} \frac{1}{k!} 0^k = \frac{1}{0!} 0^0 = \mathrm{Id}.$$

Example 5.27

We have

$$e^{\mathrm{Id}} = \sum_{k=0}^{\infty} \frac{1}{k!} \mathrm{Id}^k = \sum_{k=0}^{\infty} \frac{1}{k!} \mathrm{Id} = e\mathrm{Id}.$$

In Section 5.5 we describe a method to compute the exponential of a matrix A. Here, we show how the exponential can be used to solve equations (5.18) and (5.17). We start with an auxiliary result.

Proposition 5.28

$(e^{At})' = Ae^{At}$ for every $t \in \mathbb{R}$, with the derivative computed entry by entry.

Proof

By Proposition 5.25, the exponential

$$e^{At} = \sum_{k=0}^{\infty} \frac{1}{k!} t^k A^k$$

is well defined for every $t \in \mathbb{R}$. Hence, each entry of e^{At} is a power series in t with radius of convergence $+\infty$. Since power series can be differentiated term by term in the interior of their domain of convergence, we obtain

$$\left(e^{At}\right)' = \sum_{k=1}^{\infty} \frac{1}{k!} k t^{k-1} A^k$$

$$= A \sum_{k=1}^{\infty} \frac{1}{(k-1)!} t^{k-1} A^{k-1}$$

$$= A e^{At}.$$

This yields the desired result. □

Example 5.29

We show that there exists no 2×2 matrix A with

$$e^{At} = \begin{pmatrix} \cos t & e^t \\ 0 & 1 \end{pmatrix}.$$

Otherwise, we would have

$$e^{At}\big|_{t=0} = \begin{pmatrix} 1 & 1 \\ 0 & 1 \end{pmatrix},$$

but by Example 5.26, we always have $e^{A0} = e^0 = \mathrm{Id}$.

Example 5.30

Let us find a 2×2 matrix A such that

$$e^{At} = \begin{pmatrix} e^{2t} & te^{2t} \\ 0 & e^{2t} \end{pmatrix}.$$

Taking derivatives with respect to t, we obtain

$$\left(e^{At}\right)'\big|_{t=0} = \begin{pmatrix} 2e^{2t} & e^{2t} + 2te^{2t} \\ 0 & 2e^{2t} \end{pmatrix} = \begin{pmatrix} 2 & 1 \\ 0 & 2 \end{pmatrix}.$$

On the other hand, it follows from Proposition 5.28 that

$$\left(e^{At}\right)'\big|_{t=0} = A e^{At}\big|_{t=0} = A e^0 = A\,\mathrm{Id} = A.$$

Hence,

$$A = \begin{pmatrix} 2 & 1 \\ 0 & 2 \end{pmatrix}.$$

Now we obtain all solutions of equation (5.17).

Theorem 5.31 (Variation of parameters formula)

For each $(t_0, x_0) \in \mathbb{R} \times \mathbb{R}^n$, the (unique) solution of the equation $x' = Ax + b(t)$ with $x(t_0) = x_0$ is given by

$$x(t) = e^{A(t-t_0)}x_0 + \int_{t_0}^t e^{A(t-s)}b(s)\,ds \tag{5.23}$$

for $t \in \mathbb{R}$ (and thus has maximal interval $\mathbb{R}$).

Proof

It is sufficient to verify that the function $x(t)$ defined by (5.23) satisfies

$$x(t_0) = e^{A0}x_0 = e^0 x_0 = \mathrm{Id}x_0 = x_0$$

and

$$x'(t) = Ae^{A(t-t_0)}x_0 + \int_{t_0}^t Ae^{A(t-s)}b(s)\,ds + e^{A(t-t)}b(t)$$

$$= A\left(e^{A(t-t_0)}x_0 + \int_{t_0}^t e^{A(t-s)}b(s)\,ds\right) + e^0 b(t)$$

$$= Ax(t) + b(t).$$

Since the integrands are continuous, the function $x(t)$ is defined for $t \in \mathbb{R}$.   □

The solutions of the equation $x' = Ax$ can be obtained as a particular case of Theorem 5.31.

Proposition 5.32

For each $(t_0, x_0) \in \mathbb{R} \times \mathbb{R}^n$, the (unique) solution of the equation $x' = Ax$ with $x(t_0) = x_0$ is given by

$$x(t) = e^{A(t-t_0)}x_0$$

for $t \in \mathbb{R}$ (and thus has maximal interval $\mathbb{R}$). Moreover, the set of solutions of the equation $x' = Ax$ is a linear space of dimension n.

Proof

The first statement follows from Theorem 5.31 by setting $b(t) = 0$. For the second statement, we note that any linear combination of solutions of the equation

$x' = Ax$ is still a solution of this equation. Therefore, the set of solutions is a linear space, generated by the columns of the matrix $e^{A(t-t_0)}$. Since

$$e^{A(t-t_0)}\big|_{t=t_0} = e^0 = \mathrm{Id},$$

these columns are linearly independent, because they are linearly independent for a particular value of t. Hence, the space of the solutions has dimension n. $\square$

Example 5.33

Let us consider the equation

$$\begin{cases} x' = 2x + y, \\ y' = 2y + t. \end{cases} \tag{5.24}$$

By Example 5.30, for the matrix

$$A = \begin{pmatrix} 2 & 1 \\ 0 & 2 \end{pmatrix}$$

we have

$$e^{At} = \begin{pmatrix} e^{2t} & te^{2t} \\ 0 & e^{2t} \end{pmatrix}.$$

Hence, the solutions of equation (5.24) are given by

$$\begin{pmatrix} x(t) \\ y(t) \end{pmatrix} = \begin{pmatrix} e^{2(t-t_0)} & (t-t_0)e^{2(t-t_0)} \\ 0 & e^{2(t-t_0)} \end{pmatrix} \begin{pmatrix} x(t_0) \\ y(t_0) \end{pmatrix}$$

$$+ \int_{t_0}^t \begin{pmatrix} e^{2(t-s)} & (t-s)e^{2(t-s)} \\ 0 & e^{2(t-s)} \end{pmatrix} \begin{pmatrix} 0 \\ s \end{pmatrix} ds.$$

For example, when $t_0 = 0$, after some computations we obtain

$$\begin{pmatrix} x(t) \\ y(t) \end{pmatrix} = \begin{pmatrix} e^{2t}x(0) + te^{2t}y(0) \\ e^{2t}y(0) \end{pmatrix} + \begin{pmatrix} \int_0^t (t-s)e^{2(t-s)}s\,ds \\ \int_0^t e^{2(t-s)}s\,ds \end{pmatrix}$$

$$= \begin{pmatrix} e^{2t}x(0) + te^{2t}y(0) + \frac{1}{4}(1 + t - e^{2t} + te^{2t}) \\ e^{2t}y(0) - \frac{1}{4}(1 + 2t - e^{2t}) \end{pmatrix}.$$

Example 5.34

Now we show that

$$e^{A(t-s)} = e^{At}e^{-As} \tag{5.25}$$

for every $t, s \in \mathbb{R}$. Given $v \in \mathbb{R}^n$, we consider the functions

$$x(t) = e^{A(t-s)}v \quad \text{and} \quad y(t) = e^{At}e^{-As}v.$$

We note that

$$x(s) = e^{A0}v = \text{Id}\, v = v$$

and

$$y(s) = e^{As}e^{-As}v.$$

Moreover,

$$x'(t) = Ae^{A(t-s)}v = Ax(t)$$

and

$$y'(t) = Ae^{At}e^{-As}v = Ay(t).$$

Hence, if we show that

$$e^{As}e^{-As} = \text{Id}, \tag{5.26}$$

then it follows from the uniqueness of the solutions of the equation $x' = Ax$ that $x(t) = y(t)$ for every $t \in \mathbb{R}$, that is,

$$e^{A(t-s)}v = e^{At}e^{-As}v$$

for every $t, s \in \mathbb{R}$ and $v \in \mathbb{R}^n$, which establishes (5.25). In order to show that identity (5.26) holds, we first note that

$$\frac{d}{ds}\left(e^{As}e^{-As}\right) = Ae^{As}e^{-As} - e^{As}Ae^{-As}.$$

We also have

$$Ae^{As} = A\left(\sum_{k=0}^{\infty}\frac{1}{k!}s^k A^k\right)$$

$$= \left(\sum_{k=0}^{\infty}\frac{1}{k!}s^k A^k\right)A = e^{As}A,$$

since $AA^k = A^k A$ for each k. Therefore,

$$\frac{d}{ds}\left(e^{As}e^{-As}\right) = 0.$$

Finally, since

$$e^{As}e^{-As}\big|_{s=0} = e^0 e^{-0} = \text{Id}^2 = \text{Id},$$

we conclude that identity (5.26) holds for every $s \in \mathbb{R}$.

In particular, it follows from (5.25) that

$$e^{At}e^{As} = e^{As}e^{At}$$

for every $t, s \in \mathbb{R}$.

5.5 Computing Exponentials of Matrices

In this section we describe a method to compute the exponential of a matrix. We first recall an important result from linear algebra.

Theorem 5.35 (Complex Jordan form)

For each $n \times n$ matrix A there exists an invertible $n \times n$ matrix S with entries in $\mathbb{C}$ such that

$$S^{-1}AS = \begin{pmatrix} R_1 & & 0 \\ & \ddots & \\ 0 & & R_k \end{pmatrix}, \tag{5.27}$$

where each block R_j is an $n_j \times n_j$ matrix, for some $n_j \leq n$, of the form

$$R_j = \begin{pmatrix} \lambda_j & 1 & & 0 \\ & \ddots & \ddots & \\ & & \ddots & 1 \\ 0 & & & \lambda_j \end{pmatrix},$$

where each complex number λ_j is an eigenvalue of A.

We give several examples.

Example 5.36

When $n_j = 1$ we obtain the 1×1 matrix $R_j = [\lambda_j]$. In this case there are no 1s above the main diagonal.

Example 5.37

We recall that if an $n \times n$ matrix A has distinct eigenvalues $\mu_1, \ldots, \mu_n$, then it can be diagonalized, that is, there exists an invertible $n \times n$ matrix S (with entries in $\mathbb{C}$) such that

$$S^{-1}AS = \begin{pmatrix} \mu_1 & & 0 \\ & \ddots & \\ 0 & & \mu_n \end{pmatrix}.$$

Example 5.38

Let us consider the matrix

$$A = \begin{pmatrix} 0 & -2 \\ 1 & 0 \end{pmatrix}. \tag{5.28}$$

The eigenvalues of A are $i\sqrt{2}$ and $-i\sqrt{2}$. For example, taking the eigenvectors $(\sqrt{2}, -i)$ and $(\sqrt{2}, i)$ associated respectively to $i\sqrt{2}$ and $-i\sqrt{2}$, we consider the matrix

$$S = \begin{pmatrix} \sqrt{2} & \sqrt{2} \\ -i & i \end{pmatrix}. \tag{5.29}$$

Since

$$S^{-1} = \frac{1}{i2\sqrt{2}} \begin{pmatrix} i & -\sqrt{2} \\ i & \sqrt{2} \end{pmatrix}, \tag{5.30}$$

we obtain

$$S^{-1}AS = \begin{pmatrix} i\sqrt{2} & 0 \\ 0 & -i\sqrt{2} \end{pmatrix}. \tag{5.31}$$

This is the complex Jordan form of the matrix A.

Example 5.39

For a 3×3 matrix A with all eigenvalues equal to λ, the complex Jordan form is one of the matrices

$$\begin{pmatrix} \lambda & 0 & 0 \\ 0 & \lambda & 0 \\ 0 & 0 & \lambda \end{pmatrix}, \quad \begin{pmatrix} \lambda & 1 & 0 \\ 0 & \lambda & 0 \\ 0 & 0 & \lambda \end{pmatrix} \quad \text{and} \quad \begin{pmatrix} \lambda & 1 & 0 \\ 0 & \lambda & 1 \\ 0 & 0 & \lambda \end{pmatrix}. \tag{5.32}$$

We note that if

$$S = \begin{pmatrix} 0 & 0 & 1 \\ 0 & 1 & 0 \\ 1 & 0 & 0 \end{pmatrix},$$

then

$$S^{-1} \begin{pmatrix} \lambda & 1 & 0 \\ 0 & \lambda & 0 \\ 0 & 0 & \lambda \end{pmatrix} S = \begin{pmatrix} \lambda & 0 & 0 \\ 0 & \lambda & 1 \\ 0 & 0 & \lambda \end{pmatrix}.$$

This explains why we did not include the last matrix in (5.32).

When A has a single block R_j in its complex Jordan from, one can easily compute the exponential e^{At}. More precisely, let A be the $n \times n$ matrix given by

$$A = \begin{pmatrix} \lambda & 1 & & 0 \\ & \ddots & \ddots & \\ & & \ddots & 1 \\ 0 & & & \lambda \end{pmatrix}. \tag{5.33}$$

We write

$$A = \lambda \mathrm{Id} + N,$$

where N is the $n \times n$ matrix given by

$$N = \begin{pmatrix} 0 & 1 & & 0 \\ & \ddots & \ddots & \\ & & \ddots & 1 \\ 0 & & & 0 \end{pmatrix}.$$

One can easily verify that $N^n = 0$.

Proposition 5.40

For the $n \times n$ matrix A in (5.33), we have

$$e^{At} = e^{\lambda t} \left(\mathrm{Id} + tN + \frac{t^2}{2!} N^2 + \cdots + \frac{t^{n-1}}{(n-1)!} N^{n-1} \right) \tag{5.34}$$

for each $t \in \mathbb{R}$.

Proof

Let $B(t)$ be the matrix on the right-hand side of (5.34). Taking derivatives with respect to t, we obtain

$$B'(t) = \lambda e^{\lambda t}\left(\mathrm{Id} + tN + \cdots + \frac{t^{n-1}}{(n-1)!}N^{n-1}\right)$$
$$+ e^{\lambda t}\left(N + tN^2 + \cdots + \frac{t^{n-2}}{(n-2)!}N^{n-1}\right)$$
$$= \lambda e^{\lambda t}\left(\mathrm{Id} + tN + \cdots + \frac{t^{n-1}}{(n-1)!}N^{n-1}\right)$$
$$+ e^{\lambda t}\left(N + tN^2 + \cdots + \frac{t^{n-1}}{(n-1)!}N^n\right)$$
$$= (\lambda + N)e^{\lambda t}\left(\mathrm{Id} + tN + \cdots + \frac{t^{n-1}}{(n-1)!}N^{n-1}\right),$$

since $N^n = 0$. Therefore,

$$B'(t) = AB(t).$$

Since $B(0) = \mathrm{Id}$, we conclude that for each $v \in \mathbb{R}^n$ the function $x(t) = B(t)v$ is the solution of the equation $x' = Ax$ with $x(0) = v$. But this solution is also given by $e^{At}v$. Hence, $B(t)v = e^{At}v$ for every $v \in \mathbb{R}$, that is, $B(t) = e^{At}$. □

Example 5.41

For the matrix

$$A = \begin{pmatrix} 2 & 1 & 0 \\ 0 & 2 & 1 \\ 0 & 0 & 2 \end{pmatrix},$$

we have $A = 2\mathrm{Id} + N$, with

$$N = \begin{pmatrix} 0 & 1 & 0 \\ 0 & 0 & 1 \\ 0 & 0 & 0 \end{pmatrix}, \quad N^2 = \begin{pmatrix} 0 & 0 & 1 \\ 0 & 0 & 0 \\ 0 & 0 & 0 \end{pmatrix} \quad \text{and} \quad N^3 = 0.$$

Hence, by Proposition 5.40,

$$e^{At} = e^{2t}\left(\mathrm{Id} + tN + \frac{t^2}{2}N\right) = \begin{pmatrix} e^{2t} & te^{2t} & \frac{1}{2}t^2e^{2t} \\ 0 & e^{2t} & te^{2t} \\ 0 & 0 & e^{2t} \end{pmatrix}.$$

For example, the solution of the equation $x' = Ax$ with $x(0) = (3, 0, 1)$ is given by

$$x(t) = e^{At} \begin{pmatrix} 3 \\ 0 \\ 1 \end{pmatrix} = \begin{pmatrix} 3e^{2t} + \frac{1}{2}t^2 e^{2t} \\ te^{2t} \\ e^{2t} \end{pmatrix}.$$

Now we consider arbitrary matrices.

Proposition 5.42

If A is a square matrix with the complex Jordan form in (5.27), then

$$e^{At} = Se^{(S^{-1}AS)t}S^{-1} = S \begin{pmatrix} e^{R_1 t} & & 0 \\ & \ddots & \\ 0 & & e^{R_k t} \end{pmatrix} S^{-1} \tag{5.35}$$

for each $t \in \mathbb{R}$.

Proof

Let us consider the change of variables $y = S^{-1}x$. If $x = x(t)$ is a solution of the equation $x' = Ax$, then the function $y = y(t) = S^{-1}x(t)$ satisfies

$$y' = S^{-1}x' = S^{-1}Ax = S^{-1}ASy.$$

Hence, $y' = By$, where $B = S^{-1}AS$. Therefore,

$$y(t) = e^{Bt}y(0) = e^{(S^{-1}AS)t}y(0). \tag{5.36}$$

On the other hand, since $x(0) = Sy(0)$, we also have

$$y(t) = S^{-1}x(t) = S^{-1}e^{At}x(0) = S^{-1}e^{At}Sy(0). \tag{5.37}$$

Comparing (5.36) and (5.37), we conclude that

$$e^{(S^{-1}AS)t} = S^{-1}e^{At}S.$$

This establishes the first identity in (5.35).

Moreover, since $S^{-1}AS$ is the complex Jordan form, we obtain

$$
e^{(S^{-1}AS)t} = \exp\left\{ \begin{pmatrix} R_1 & & 0 \\ & \ddots & \\ 0 & & R_k \end{pmatrix} t \right\}
$$

$$
= \sum_{m=0}^{\infty} \frac{1}{m!} t^m \begin{pmatrix} R_1 & & 0 \\ & \ddots & \\ 0 & & R_k \end{pmatrix}^m
$$

$$
= \sum_{m=0}^{\infty} \frac{1}{m!} t^m \begin{pmatrix} R_1^m & & 0 \\ & \ddots & \\ 0 & & R_k^m \end{pmatrix},
$$

and thus,

$$
e^{(S^{-1}AS)t} = \begin{pmatrix} \sum_{m=0}^{\infty} \frac{1}{m!} t^m R_1^m & & 0 \\ & \ddots & \\ 0 & & \sum_{m=0}^{\infty} \frac{1}{m!} t^m R_k^m \end{pmatrix}
$$

$$
= \begin{pmatrix} e^{R_1 t} & & 0 \\ & \ddots & \\ 0 & & e^{R_k t} \end{pmatrix}.
$$

This completes the proof of the proposition.   □

Example 5.43

Let us consider the matrix

$$
A = \begin{pmatrix} 5 & 1 & 0 \\ 0 & 5 & 0 \\ 0 & 0 & 3 \end{pmatrix}.
$$

Since

$$
e^{\left(\begin{smallmatrix} 5 & 1 \\ 0 & 5 \end{smallmatrix} \right)t} = e^{5t} \left(\mathrm{Id} + t \begin{pmatrix} 0 & 1 \\ 0 & 0 \end{pmatrix} \right) = \begin{pmatrix} e^{5t} & te^{5t} \\ 0 & e^{5t} \end{pmatrix},
$$

it follows from Proposition 5.42 that

$$
e^{At} = \begin{pmatrix} e^{\left(\begin{smallmatrix} 5 & 1 \\ 0 & 5 \end{smallmatrix} \right)t} & 0 \\ 0 & e^{3t} \end{pmatrix} = \begin{pmatrix} e^{5t} & te^{5t} & 0 \\ 0 & e^{5t} & 0 \\ 0 & 0 & e^{3t} \end{pmatrix}.
$$

Example 5.44

Now we consider the matrix A in (5.28). It follows from (5.29), (5.30) and (5.31) that

$$e^{At} = Se^{(S^{-1}AS)t}S^{-1} = S\begin{pmatrix} e^{i\sqrt{2}t} & 0 \\ 0 & e^{-i\sqrt{2}t} \end{pmatrix}S^{-1}.$$

Using the formulas

$$\cos(\sqrt{2}t) = \frac{e^{i\sqrt{2}t} + e^{-i\sqrt{2}t}}{2} \quad \text{and} \quad \sin(\sqrt{2}t) = \frac{e^{i\sqrt{2}t} - e^{-i\sqrt{2}t}}{2i},$$

we then obtain

$$e^{At} = \frac{1}{i2\sqrt{2}}\begin{pmatrix} \sqrt{2} & \sqrt{2} \\ -i & i \end{pmatrix}\begin{pmatrix} e^{i\sqrt{2}t} & 0 \\ 0 & e^{-i\sqrt{2}t} \end{pmatrix}\begin{pmatrix} i & -\sqrt{2} \\ i & \sqrt{2} \end{pmatrix}$$

$$= \begin{pmatrix} (e^{i\sqrt{2}t} + e^{-i\sqrt{2}t})/2 & -(e^{i\sqrt{2}t} - e^{-i\sqrt{2}t})/(\sqrt{2}i) \\ (e^{i\sqrt{2}t} - e^{-i\sqrt{2}t})/(2\sqrt{2}i) & (e^{i\sqrt{2}t} + e^{-i\sqrt{2}t})/2 \end{pmatrix}$$

$$= \begin{pmatrix} \cos(\sqrt{2}t) & -\sqrt{2}\sin(\sqrt{2}t) \\ \sin(\sqrt{2}t)/\sqrt{2} & \cos(\sqrt{2}t) \end{pmatrix}.$$

5.6 Solved Problems and Exercises

Problem 5.1

Verify that te^t is a solution of the equation $x'' - 2x' + x = 0$.

Solution

Let $x(t) = te^t$. We have

$$x' = e^t + te^t \quad \text{and} \quad x'' = 2e^t + te^t.$$

Hence,

$$x'' - 2x' + x = 2e^t + te^t - 2e^t - 2te^t + te^t = 0,$$

and thus, x is a solution of the equation.

Problem 5.2

Find all solutions of the equation $x''' = 0$.

Solution

Writing $x'' = y$, we obtain $y' = x''' = 0$. Thus, $y(t) = a$ for some $a \in \mathbb{R}$, that is, $x'' = a$. Writing $z = x'$, we obtain $z' = x'' = a$. Thus, $z(t) = at + b$ for some $b \in \mathbb{R}$, that is, $x'(t) = at + b$. Finally, integrating on both sides we obtain

$$x(t) = \frac{a}{2}t^2 + bt + c, \quad \text{with } a, b, c \in \mathbb{R},$$

or equivalently

$$x(t) = kt^2 + bt + c, \quad \text{with } k, b, c \in \mathbb{R}.$$

In particular, all solutions have maximal interval $\mathbb{R}$.

Problem 5.3

Find all solutions of the equation $x' = x^3$.

Solution

For a nonvanishing solution x, one can write

$$\frac{x'}{x^3} = 1 \quad \Leftrightarrow \quad \left(-\frac{1}{2x^2} \right)' = 1,$$

and thus,

$$\frac{1}{x(t)^2} = -2t + c \quad \Leftrightarrow \quad x(t) = \pm \frac{1}{\sqrt{-2t + c}},$$

for some constant $c \in \mathbb{R}$. In order that a solution x is well defined, it is necessary that $-2t + c > 0$, which is the same as $t \in (-\infty, c/2)$. The nonvanishing solutions are thus

$$x(t) = \frac{1}{\sqrt{-2t + c}} \quad \text{and} \quad x(t) = -\frac{1}{\sqrt{-2t + c}}$$

with $c \in \mathbb{R}$. Both have maximal interval $(-\infty, c/2)$. By direct substitution in the equation one can verify that the zero function is also a solution, thus with maximal interval $\mathbb{R}$.

Problem 5.4

Find all solutions of the equation $x' + te^t x = 0$.

Solution

For a nonvanishing solution x, one can write

$$\frac{x'}{x} = -te^t \quad \Leftrightarrow \quad \left(\log|x|\right)' = -te^t,$$

and thus,

$$\log|x(t)| = e^t(1 - t) + c \quad \Leftrightarrow \quad |x(t)| = e^{e^t(1-t)+c},$$

for some constant $c \in \mathbb{R}$. We note that x must be continuous (since by definition the solutions of a differential equation are of class C^1). Hence, it is always positive or always negative. Therefore,

$$x(t) = ke^{e^t(1-t)} \quad \text{for } t \in \mathbb{R}, \tag{5.38}$$

with $k \neq 0$, since by varying $c \in \mathbb{R}$ the function e^c takes all values of $\mathbb{R}^+$. By direct substitution in the equation, one can verify that the zero function is also a solution. In conclusion, the solutions are given by (5.38) with $k \in \mathbb{R}$, and all have maximal interval $\mathbb{R}$.

Problem 5.5

Find the solution of the equation

$$x' + (t\sin t)x = 0 \quad \text{with } x(0) = 1. \tag{5.39}$$

Solution

For a nonvanishing solution x, one can write

$$\frac{x'}{x} = -t\sin t.$$

Integrating over t, we obtain

$$\log|x(t)| = t\cos t - \sin t + c \quad \Leftrightarrow \quad |x(t)| = e^{t\cos t - \sin t + c},$$

for some constant $c \in \mathbb{R}$. Proceeding in a similar manner to that in Problem 5.4, we conclude that

$$x(t) = ke^{t\cos t - \sin t} \quad \text{for } t \in \mathbb{R}, \tag{5.40}$$

with $k \neq 0$. By direct substitution in the equation, one can verify that the zero function is also a solution. Therefore, the solutions are given by (5.40) with $k \in \mathbb{R}$, and all have maximal interval $\mathbb{R}$. For $x(0) = 1$, we obtain $1 = ke^0 = k$, and thus, the solution of problem (5.39) is $x(t) = e^{t\cos t - \sin t}$ for $t \in \mathbb{R}$.

Problem 5.6

Find all solutions of the equation $x' = -|x|$.

Solution

For each $x, y \in \mathbb{R}$, we have

$$\big||x| - |y|\big| \le |x - y| \tag{5.41}$$

(see Figure 5.2), and thus, the continuous function $f(t, x) = -|x|$ is locally Lipschitz in x. It follows from the Picard–Lindelöf theorem (Theorem 5.10) that for each $(t_0, x_0) \in \mathbb{R}^2$ there exists a unique solution of the equation $x' = -|x|$ with $x(t_0) = x_0$ in some open interval containing t_0.

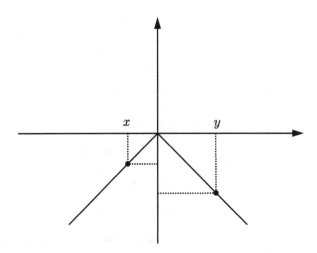

Figure 5.2 Graph of the function $-|x|$

Clearly, $x(t) = 0$ is a solution. When $x > 0$ we obtain the equation $x' = -x$, which has the solutions

$$x(t) = ke^{-t}, \quad t \in \mathbb{R},$$

with $k > 0$ (so that $x(t)$ is positive). Finally, when $x < 0$ we obtain the equation $x' = x$, which has the solutions

$$x(t) = ke^{t}, \quad t \in \mathbb{R},$$

with $k < 0$ (so that $x(t)$ is negative). In conclusion, the solutions of the equation

$x' = -|x|$ are

$$x(t) = \begin{cases} ke^{-t} & \text{with } k > 0, \\ 0, \\ ke^{t} & \text{with } k < 0, \end{cases}$$

and all have maximal interval $\mathbb{R}$.

Problem 5.7

Verify that the function $f\colon \mathbb{R}^2 \to \mathbb{R}$ defined by

$$f(t, x) = tx + |t + x|$$

is locally Lipschitz in x.

Solution

For each $t, x, y \in \mathbb{R}$, it follows from (5.41) that

$$\begin{aligned} \big| f(t, x) - f(t, y) \big| &= \big| tx + |t + x| - ty - |t + y| \big| \\ &\le |tx - ty| + \big| |t + x| - |t + y| \big| \\ &\le |t||x - y| + |t + x - (t + y)| \\ &= \big(|t| + 1 \big)|x - y|. \end{aligned} \tag{5.42}$$

Now let $K \subset \mathbb{R}^2$ be a compact set (that is, a closed bounded set). Since K is bounded, there exists $M = M(K) > 0$ such that $|t| < M$ for every $(t, x) \in K$. It then follows from (5.42) that

$$\big| f(t, x) - f(t, y) \big| \le (M + 1)|x - y|$$

for each $(t, x), (t, y) \in K$. This shows that f is locally Lipschitz in x.

Problem 5.8

Find all solutions of the equation

$$xx'' = \big(x' \big)^2. \tag{5.43}$$

Solution

We note that $x(t) = 0$ is a solution. For the nonvanishing solutions, equation (5.43) can be written in the form

$$\frac{d}{dt}\left(\frac{x'}{x}\right) = \frac{x''x - (x')^2}{x^2} = 0.$$

We then obtain

$$\frac{x'}{x} = a \quad \text{for some } a \in \mathbb{R}.$$

Since

$$\left(\log|x(t)|\right)' = \frac{x'}{x} = a,$$

we conclude that

$$\log|x(t)| = at + b$$

for some $a, b \in \mathbb{R}$. This is equivalent to $x(t) = ke^{at}$, with $a \in \mathbb{R}$ and $k \in \mathbb{R} \setminus \{0\}$. Therefore, the solutions of equation (5.43) are $x(t) = ke^{at}$, with $a, k \in \mathbb{R}$, and all have maximal interval $\mathbb{R}$.

Problem 5.9

Letting $x = ty$, solve the equation

$$x' = \frac{t + x}{t - x} \tag{5.44}$$

as explicitly as possible.

Solution

Since

$$x' = \frac{t + x}{t - x} = \frac{1 + x/t}{1 - x/t},$$

letting $x = ty$ (that is, $x(t) = ty(t)$) we obtain

$$y + ty' = \frac{1 + y}{1 - y},$$

or equivalently

$$ty' = \frac{1 + y}{1 - y} - y = \frac{1 + y^2}{1 - y}.$$

Writing this equation in the form

$$\frac{1-y}{1+y^2}y' = \frac{1}{t},$$

and integrating on both sides with respect to t, we obtain

$$\tan^{-1}y - \frac{1}{2}\log(1+y^2) = \log|t| + c$$

for some constant $c \in \mathbb{R}$. This shows that each solution of equation (5.44) satisfies

$$\tan^{-1}\left(\frac{x(t)}{t}\right) - \frac{1}{2}\log\left(1+\frac{x(t)^2}{t^2}\right) = \log|t| + c$$

for some $c \in \mathbb{R}$.

Problem 5.10

Letting $y = x^4$, solve the equation

$$x' = -\frac{x}{2t} + \frac{t}{x^3}.$$

Solution

Letting $y = x^4$, we obtain

$$y' = 4x^3x' = 4x^3\left(-\frac{x}{2t} + \frac{t}{x^3}\right)$$

$$= -\frac{2x^4}{t} + 4t = -\frac{2}{t}y + 4t.$$

Therefore,

$$(t^2y)' = 2ty + t^2y' = 2ty - 2ty + 4t^3 = 4t^3,$$

and hence,

$$t^2y(t) = t^4 + c$$

for some constant $c \in \mathbb{R}$, which yields

$$x(t) = \pm\sqrt[4]{t^2 + c/t^2}.$$

Each solution $x(t)$ is defined when $t^2 + c/t^2 > 0$ (and not $t^2 + c/t^2 \geq 0$, so that $x(t)$ is of class C^1). Hence, $t^4 > -c$, and the maximal interval of each solution is $\mathbb{R}^+$ or $\mathbb{R}^-$ for $c \geq 0$, and is $(-\infty, -\sqrt[4]{|c|})$ or $(\sqrt[4]{|c|}, +\infty)$ for $c < 0$.

Problem 5.11

Find the solution of the equation $x' = 2x + t$ with $x(1) = 3$.

Solution

By Theorem 5.18, the solution is given by

$$x(t) = e^{\int_1^t 2\,ds}3 + \int_1^t e^{\int_u^t 2\,ds}u\,du$$

$$= 3e^{2(t-1)} + \int_1^t e^{2(t-u)}u\,du$$

$$= 3e^{2(t-1)} - \frac{1}{4}e^{2(t-u)}(1+2u)\Big|_{u=1}^{u=t}$$

$$= -\frac{1}{4} - \frac{t}{2} + \frac{15}{4}e^{2(t-1)}$$

for $t \in \mathbb{R}$.

Problem 5.12

Given $a \in \mathbb{R}$, compute the exponential e^{At} for the matrix

$$A = \begin{pmatrix} a & 1 \\ 0 & a \end{pmatrix}.$$

Solution

We have

$$A = \begin{pmatrix} a & 0 \\ 0 & a \end{pmatrix} + \begin{pmatrix} 0 & 1 \\ 0 & 0 \end{pmatrix} = a\mathrm{Id} + N,$$

where

$$N = \begin{pmatrix} 0 & 1 \\ 0 & 0 \end{pmatrix}.$$

Since $N^2 = 0$, for each $k \geq 0$ we have

$$A^k = (a\mathrm{Id} + N)^k$$

$$= a^k\mathrm{Id}^k + ka^{k-1}\mathrm{Id}^{k-1}N$$

$$= a^k\mathrm{Id} + ka^{k-1}N.$$

Therefore, by (5.20), we obtain

$$e^{At} = \sum_{k=0}^{\infty} \frac{1}{k!} A^k t^k$$

$$= \sum_{k=0}^{\infty} \frac{1}{k!} (at)^k \mathrm{Id} + \sum_{k=1}^{\infty} \frac{1}{(k-1)!} (at)^{k-1} tN$$

$$= e^{at} (\mathrm{Id} + tN)$$

$$= e^{at} \left(\mathrm{Id} + t \begin{pmatrix} 0 & 1 \\ 0 & 0 \end{pmatrix} \right)$$

$$= \begin{pmatrix} e^{at} & te^{at} \\ 0 & e^{at} \end{pmatrix}.$$

Problem 5.13

Find the solution of the equation

$$\begin{cases} x' = 2x + y, \\ y' = 2y + 1 \end{cases} \tag{5.45}$$

with $x(0) = y(0) = 0$.

Solution

Equation (5.45) can be written in the form

$$\begin{pmatrix} x' \\ y' \end{pmatrix} = A \begin{pmatrix} x \\ y \end{pmatrix} + \begin{pmatrix} 0 \\ 1 \end{pmatrix}, \quad \text{where } A = \begin{pmatrix} 2 & 1 \\ 0 & 2 \end{pmatrix}.$$

By Problem 5.12, we have

$$e^{At} = \begin{pmatrix} e^{2t} & te^{2t} \\ 0 & e^{2t} \end{pmatrix}.$$

Hence, it follows from the Variation of parameters formula (Theorem 5.31) that the solution of equation (5.45) with $x(0) = y(0) = 0$ is given by

$$\begin{pmatrix} x(t) \\ y(t) \end{pmatrix} = e^{At} \begin{pmatrix} 0 \\ 0 \end{pmatrix} + \int_0^t e^{A(t-s)} \begin{pmatrix} 0 \\ 1 \end{pmatrix} ds$$

$$= \int_0^t \begin{pmatrix} (t-s)e^{2(t-s)} \\ e^{2(t-s)} \end{pmatrix} ds$$

$$= \left(\frac{1}{4} + \frac{1}{2}te^{2t} - \frac{1}{4}e^{2t}, \frac{1}{2}e^{2t} - \frac{1}{2} \right)$$

for $t \in \mathbb{R}$.

Problem 5.14

Find all solutions of the equation

$$\begin{cases} x' = 3x + y, \\ y' = 3y - t \end{cases}$$

in terms of the initial condition $(x(0), y(0)) = (x_0, y_0)$.

Solution

We write the equation in the matrix form

$$\begin{pmatrix} x \\ y \end{pmatrix}' = A \begin{pmatrix} x \\ y \end{pmatrix} + \begin{pmatrix} 0 \\ -t \end{pmatrix}, \quad \text{where } A = \begin{pmatrix} 3 & 1 \\ 0 & 3 \end{pmatrix}.$$

By Problem 5.12, we have

$$e^{At} = \begin{pmatrix} e^{3t} & te^{3t} \\ 0 & e^{3t} \end{pmatrix}.$$

Therefore, by the Variation of parameters formula (Theorem 5.31), the solutions are given by

$$\begin{pmatrix} x(t) \\ y(t) \end{pmatrix} = e^{A(t-t_0)} \begin{pmatrix} x(t_0) \\ y(t_0) \end{pmatrix} + \int_{t_0}^t e^{A(t-s)} \begin{pmatrix} 0 \\ -s \end{pmatrix} ds$$

for $t \in \mathbb{R}$. Taking $t_0 = 0$, we obtain

$$\begin{pmatrix} x(t) \\ y(t) \end{pmatrix} = \begin{pmatrix} e^{3t} & te^{3t} \\ 0 & e^{3t} \end{pmatrix} \begin{pmatrix} x_0 \\ y_0 \end{pmatrix} + \int_0^t \begin{pmatrix} e^{3(t-s)} & (t-s)e^{3(t-s)} \\ 0 & e^{3(t-s)} \end{pmatrix} \begin{pmatrix} 0 \\ -s \end{pmatrix} ds$$

$$= \begin{pmatrix} e^{3t}x_0 + te^{3t}y_0 - \int_0^t s(t-s)e^{3(t-s)}\, ds \\ e^{3t}y_0 - \int_0^t se^{3(t-s)}\, ds \end{pmatrix}$$

$$= \begin{pmatrix} e^{3t}x_0 + te^{3t}y_0 - (e^{3t}(3t-2) + 3t + 2)/27 \\ e^{3t}y_0 - (e^{3t} - 3t - 1)/9 \end{pmatrix}$$

for $t \in \mathbb{R}$.

Problem 5.15

Diagonalize the matrix

$$A = \begin{pmatrix} 0 & -3 \\ 1 & 0 \end{pmatrix}.$$

Solution

It follows from

$$\det(A - \lambda \mathrm{Id}) = \lambda^2 + 3 = 0$$

that the eigenvalues of A are $i\sqrt{3}$ and $-i\sqrt{3}$. For example, taking the eigenvectors $(i\sqrt{3}, 1)$ and $(-i\sqrt{3}, 1)$ associated respectively to $i\sqrt{3}$ and $-i\sqrt{3}$, we consider the matrix

$$S = \begin{pmatrix} i\sqrt{3} & -i\sqrt{3} \\ 1 & 1 \end{pmatrix},$$

whose columns are these eigenvectors. Then

$$S^{-1} = \frac{1}{2\sqrt{3}} \begin{pmatrix} -i & \sqrt{3} \\ i & \sqrt{3} \end{pmatrix},$$

and the diagonal matrix

$$S^{-1}AS = \begin{pmatrix} i\sqrt{3} & 0 \\ 0 & -i\sqrt{3} \end{pmatrix}$$

is the complex Jordan form of A.

Problem 5.16

Find the complex Jordan form of the matrix

$$A = \begin{pmatrix} 2 & 3 \\ 1 & 0 \end{pmatrix} \tag{5.46}$$

and compute e^{At}.

Solution

It follows from

$$\det(A - \lambda \mathrm{Id}) = \lambda^2 - 2\lambda - 3 = 0$$

that the eigenvalues of A are 3 and -1. For example, taking the eigenvectors $(3,1)$ and $(-1,1)$ associated respectively to 3 and -1, we consider the matrix

$$S = \begin{pmatrix} 3 & -1 \\ 1 & 1 \end{pmatrix}.$$

Then

$$S^{-1} = \frac{1}{4} \begin{pmatrix} 1 & 1 \\ -1 & 3 \end{pmatrix},$$

and the matrix

$$J = S^{-1}AS = \begin{pmatrix} 3 & 0 \\ 0 & -1 \end{pmatrix}$$

is the (real and) complex Jordan form of A. Hence,

$$S^{-1}e^{At}S = e^{Jt} = \begin{pmatrix} e^{3t} & 0 \\ 0 & e^{-t} \end{pmatrix},$$

and it follows from Proposition 5.42 that

$$e^{At} = e^{SJS^{-1}t} = Se^{Jt}S^{-1}$$

$$= \begin{pmatrix} 3 & -1 \\ 1 & 1 \end{pmatrix} \begin{pmatrix} e^{3t} & 0 \\ 0 & e^{-t} \end{pmatrix} \frac{1}{4} \begin{pmatrix} 1 & 1 \\ -1 & 3 \end{pmatrix}$$

$$= \frac{1}{4} \begin{pmatrix} 3e^{3t} + e^{-t} & 3e^{3t} - 3e^{-t} \\ e^{3t} - e^{-t} & e^{3t} + 3e^{-t} \end{pmatrix}. \tag{5.47}$$

Problem 5.17

Find the solution of the equation

$$x' = \begin{pmatrix} 2 & 3 \\ 1 & 0 \end{pmatrix} x \quad \text{with } x(0) = \begin{pmatrix} 1 \\ 2 \end{pmatrix}.$$

Solution

By Proposition 5.32, the solution is given by

$$\begin{pmatrix} x(t) \\ y(t) \end{pmatrix} = e^{At} \begin{pmatrix} x(0) \\ y(0) \end{pmatrix} = e^{At} \begin{pmatrix} 1 \\ 2 \end{pmatrix},$$

where A is the matrix in (5.46). It then follows from (5.47) that

$$\begin{pmatrix} x(t) \\ y(t) \end{pmatrix} = \frac{1}{4} \begin{pmatrix} 3e^{3t} + e^{-t} & 3e^{3t} - 3e^{-t} \\ e^{3t} - e^{-t} & e^{3t} + 3e^{-t} \end{pmatrix} \begin{pmatrix} 1 \\ 2 \end{pmatrix}$$

$$= \frac{1}{4} \begin{pmatrix} 9e^{3t} - 5e^{-t} \\ 3e^{3t} + 5e^{-t} \end{pmatrix}$$

for $t \in \mathbb{R}$.

Problem 5.18

Compute e^{At} for the matrix

$$A = \begin{pmatrix} 4 & -1 & 0 & 0 \\ 0 & 4 & 0 & 0 \\ 0 & 0 & 0 & 5 \\ 0 & 0 & -5 & 0 \end{pmatrix}.$$

Solution

We note that A is in block form, that is,

$$A = \begin{pmatrix} A_1 & 0 \\ 0 & A_2 \end{pmatrix},$$

where

$$A_1 = \begin{pmatrix} 4 & -1 \\ 0 & 4 \end{pmatrix} \quad \text{and} \quad A_2 = \begin{pmatrix} 0 & 5 \\ -5 & 0 \end{pmatrix}.$$

Hence,

$$e^{At} = \begin{pmatrix} e^{A_1 t} & 0 \\ 0 & e^{A_2 t} \end{pmatrix}.$$ (5.48)

By Proposition 5.40, for the block A_1 we have

$$e^{A_1 t} = e^{(-A_1)(-t)}$$

$$= e^{(-4)(-t)} \left(\mathrm{Id} + (-t) \begin{pmatrix} 0 & 1 \\ 0 & 0 \end{pmatrix} \right)$$

$$= e^{4t} \begin{pmatrix} 1 & -t \\ 0 & 1 \end{pmatrix} = \begin{pmatrix} e^{4t} & -te^{4t} \\ 0 & e^{4t} \end{pmatrix}.$$

On the other hand, the eigenvalues of A_2 are $5i$ and $-5i$. For example, taking the eigenvectors $(-i, 1)$ and $(i, 1)$ associated respectively to $5i$ and $-5i$, we consider the matrix

$$S = \begin{pmatrix} -i & i \\ 1 & 1 \end{pmatrix}.$$

Then

$$J = S^{-1} A_2 S = \begin{pmatrix} 5i & 0 \\ 0 & -5i \end{pmatrix}$$

is the complex Jordan form of A_2, and by Proposition 5.42 we obtain

$$e^{A_2 t} = S e^{Jt} S^{-1}$$

$$= \begin{pmatrix} -i & i \\ 1 & 1 \end{pmatrix} \begin{pmatrix} e^{5it} & 0 \\ 0 & e^{-5it} \end{pmatrix} \frac{1}{2} \begin{pmatrix} i & 1 \\ -i & 1 \end{pmatrix}$$

$$= \frac{1}{2} \begin{pmatrix} e^{5it} + e^{-5it} & -ie^{5it} + ie^{-5it} \\ ie^{5it} - ie^{-5it} & e^{5it} + e^{-5it} \end{pmatrix}$$

$$= \begin{pmatrix} \cos(5t) & \sin(5t) \\ -\sin(5t) & \cos(5t) \end{pmatrix}.$$

It follows from (5.48) that

$$e^{At} = \begin{pmatrix} e^{4t} & -te^{4t} & 0 & 0 \\ 0 & e^{4t} & 0 & 0 \\ 0 & 0 & \cos(5t) & \sin(5t) \\ 0 & 0 & -\sin(5t) & \cos(5t) \end{pmatrix}.$$

Problem 5.19

For the matrix

$$A = \begin{pmatrix} 2 & 1 & 0 & 0 \\ 0 & 4 & 0 & 0 \\ 0 & 0 & 5 & -1 \\ 0 & 0 & 0 & 5 \end{pmatrix}$$

and the vector $x = (0, 0, 1, 2)$, compute

$$\limsup_{t \to +\infty} \frac{1}{t} \log \|e^{At} x\|.$$

Solution

We have

$$e^{At} = \begin{pmatrix} e^{A_1 t} & 0 \\ 0 & e^{A_2 t} \end{pmatrix},$$

where

$$A_1 = \begin{pmatrix} 2 & 1 \\ 0 & 4 \end{pmatrix} \quad \text{and} \quad A_2 = \begin{pmatrix} 5 & -1 \\ 0 & 5 \end{pmatrix}.$$

For the first block, we consider the eigenvectors $(1, 0)$ and $(1, 2)$, and the matrix

$$S = \begin{pmatrix} 1 & 1 \\ 0 & 2 \end{pmatrix}.$$

We then obtain

$$S^{-1} \begin{pmatrix} 2 & 1 \\ 0 & 4 \end{pmatrix} S = \begin{pmatrix} 2 & 0 \\ 0 & 4 \end{pmatrix},$$

and it follows from Proposition 5.42 that

$$\begin{aligned}
e^{A_1 t} &= S \begin{pmatrix} e^{2t} & 0 \\ 0 & e^{4t} \end{pmatrix} S^{-1} \\
&= \begin{pmatrix} 1 & 1 \\ 0 & 2 \end{pmatrix} \begin{pmatrix} e^{2t} & 0 \\ 0 & e^{4t} \end{pmatrix} \begin{pmatrix} 1 & -1/2 \\ 0 & 1/2 \end{pmatrix} \\
&= \begin{pmatrix} e^{2t} & e^{4t}/2 - e^{-2t}/2 \\ 0 & e^{4t} \end{pmatrix}.
\end{aligned}$$

For the second block, we have

$$e^{A_2 t} = e^{\left(\begin{smallmatrix} -5 & 1 \\ 0 & -5 \end{smallmatrix}\right)(-t)}$$

$$= \begin{pmatrix} e^{-5(-t)} & -te^{-5(-t)} \\ 0 & e^{-5(-t)} \end{pmatrix}$$

$$= \begin{pmatrix} e^{5t} & -te^{5t} \\ 0 & e^{5t} \end{pmatrix}.$$

Therefore,

$$e^{At} = \begin{pmatrix} e^{2t} & e^{4t}/2 - e^{-2t}/2 & 0 & 0 \\ 0 & e^{4t} & 0 & 0 \\ 0 & 0 & e^{5t} & -te^{5t} \\ 0 & 0 & 0 & e^{5t} \end{pmatrix}$$

and

$$\limsup_{t \to +\infty} \frac{1}{t} \log \|e^{At} x\| = \limsup_{t \to +\infty} \frac{1}{t} \log \|(0, 0, e^{5t}, 2(1-t)e^{5t})\|$$

$$= \limsup_{t \to +\infty} \frac{1}{t} \log\left(e^{5t} \sqrt{1 + 4(1-t)^2}\right) = 5.$$

Problem 5.20

Find all solutions of the equation

$$x'' + 2x' + x = 0. \tag{5.49}$$

Solution

Taking $x' = y$, one can write the equation in the matrix form

$$\begin{pmatrix} x \\ y \end{pmatrix}' = A \begin{pmatrix} x \\ y \end{pmatrix}, \tag{5.50}$$

where

$$A = \begin{pmatrix} 0 & 1 \\ -1 & -2 \end{pmatrix}.$$

By Proposition 5.32, the solutions of equation (5.50) are given by

$$\begin{pmatrix} x(t) \\ y(t) \end{pmatrix} = e^{A(t-t_0)} \begin{pmatrix} x(t_0) \\ y(t_0) \end{pmatrix} \tag{5.51}$$

for $t \in \mathbb{R}$. Now we compute the exponential $e^{A(t-t_0)}$. Since

$$\det(A - \lambda \mathrm{Id}) = \lambda^2 + 2\lambda + 1,$$

the matrix A has only the eigenvalue -1. Moreover, one can easily verify that there exists no basis formed by eigenvectors, and thus one must consider the root space. An eigenvector is $(1, -1)$, and, for example, the vector $(0, 1)$ satisfies $(A - \lambda \mathrm{Id})(0, 1) = (1, -1)$. Taking

$$S = \begin{pmatrix} 1 & 0 \\ -1 & 1 \end{pmatrix},$$

we then obtain

$$J = S^{-1}AS$$

$$= \begin{pmatrix} 1 & 0 \\ 1 & 1 \end{pmatrix} \begin{pmatrix} 0 & 1 \\ -1 & -2 \end{pmatrix} \begin{pmatrix} 1 & 0 \\ -1 & 1 \end{pmatrix} = \begin{pmatrix} -1 & 1 \\ 0 & -1 \end{pmatrix}.$$

Therefore, by Proposition 5.42,

$$e^{A(t-t_0)} = S e^{J(t-t_0)} S^{-1}$$

$$= \begin{pmatrix} 1 & 0 \\ -1 & 1 \end{pmatrix} \begin{pmatrix} e^{-(t-t_0)} & (t-t_0)e^{-(t-t_0)} \\ 0 & e^{-(t-t_0)} \end{pmatrix} \begin{pmatrix} 1 & 0 \\ 1 & 1 \end{pmatrix}$$

$$= e^{-(t-t_0)} \begin{pmatrix} 1+t-t_0 & t-t_0 \\ -t+t_0 & 1-t+t_0 \end{pmatrix}.$$

For example, taking $t_0 = 0$, it follows from (5.51) that the solutions of equation (5.50) are given by

$$\begin{pmatrix} x(t) \\ y(t) \end{pmatrix} = e^{-t} \begin{pmatrix} 1+t & t \\ -t & 1-t \end{pmatrix} \begin{pmatrix} x(0) \\ y(0) \end{pmatrix}$$

for $t \in \mathbb{R}$. The solutions of equation (5.49) are given by the first component, that is,

$$x(t) = e^{-t}(1+t)x(0) + e^{-t}ty(0)$$

$$= e^{-t}x(0) + e^{-t}t[x(0) + x'(0)].$$

Problem 5.21

Find whether there exists a matrix A such that

$$e^{At} = \begin{pmatrix} e^t & e^{2t} - 1 \\ -1 + \cos t & 1 \end{pmatrix}. \tag{5.52}$$

Solution

We show that there exists no matrix A satisfying (5.52). Otherwise, taking derivatives with respect to t, we would have

$$\left(e^{At}\right)'\big|_{t=0} = \begin{pmatrix} e^t & 2e^{2t} \\ -\sin t & 0 \end{pmatrix}\bigg|_{t=0} = \begin{pmatrix} 1 & 2 \\ 0 & 0 \end{pmatrix}.$$

On the other hand, by Proposition 5.28, we have $\left(e^{At}\right)'|_{t=0} = A$, and hence, we should have

$$A = \begin{pmatrix} 1 & 2 \\ 0 & 0 \end{pmatrix}.$$

Since this matrix has eigenvalues 1 and 0, there would exist an invertible matrix S such that

$$S^{-1}AS = \begin{pmatrix} 1 & 0 \\ 0 & 0 \end{pmatrix},$$

and hence,

$$e^{At} = Se^{S^{-1}ASt}S^{-1} = S\begin{pmatrix} e^t & 0 \\ 0 & 1 \end{pmatrix}S^{-1}.$$

In particular, the entries of e^{At} would be linear combinations of the functions e^t and 1, and thus, it is impossible to obtain the entries $e^{2t} - 1$ and $-1 + \cos t$ in (5.52). Therefore, there exists no matrix A satisfying (5.52).

Problem 5.22

Verify that the identity

$$e^{(A+B)t} = e^{At}e^{Bt}$$

is not always satisfied for every $t \in \mathbb{R}$.

Solution

Let

$$A = \begin{pmatrix} 0 & 1 \\ 0 & 0 \end{pmatrix} \quad \text{and} \quad B = \begin{pmatrix} 0 & 0 \\ 0 & 1 \end{pmatrix}.$$

We have

$$e^{At} = \begin{pmatrix} 1 & t \\ 0 & 1 \end{pmatrix} \quad \text{and} \quad e^{Bt} = \begin{pmatrix} 1 & 0 \\ 0 & e^t \end{pmatrix}.$$

Hence,

$$e^{At}e^{Bt} = \begin{pmatrix} 1 & te^t \\ 0 & e^t \end{pmatrix}.$$

On the other hand, the matrix

$$A + B = \begin{pmatrix} 0 & 1 \\ 0 & 1 \end{pmatrix}$$

has eigenvalues 0 and 1. Taking

$$S = \begin{pmatrix} 1 & 1 \\ 0 & 1 \end{pmatrix},$$

we then obtain

$$S^{-1}(A + B)S = \begin{pmatrix} 0 & 0 \\ 0 & 1 \end{pmatrix}$$

and thus,

$$e^{(A+B)t} = Se^{S^{-1}(A+B)St}S^{-1}$$

$$= \begin{pmatrix} 1 & 1 \\ 0 & 1 \end{pmatrix} \begin{pmatrix} 1 & 0 \\ 0 & e^t \end{pmatrix} \begin{pmatrix} 1 & -1 \\ 0 & 1 \end{pmatrix}$$

$$= \begin{pmatrix} 1 & e^t - 1 \\ 0 & e^t \end{pmatrix} \neq \begin{pmatrix} 1 & te^t \\ 0 & e^t \end{pmatrix}.$$

Problem 5.23

Verify that the equation $x'' + x = \sqrt[5]{t}$ has solutions.

Solution

Let $y = x'$ and $z = (x, y)$. Since $z' = (x', y')$, the equation can be written in the matrix form

$$z' = Az + f(t), \quad \text{where } A = \begin{pmatrix} 0 & 1 \\ -1 & 0 \end{pmatrix} \text{ and } f(t) = \begin{pmatrix} 0 \\ \sqrt[5]{t} \end{pmatrix}.$$

Now we consider the continuous function $F(t, z) = Az + f(t)$. For $z_1, z_2 \in \mathbb{R}^2$, we have

$$\|F(t, z_1) - F(t, z_2)\| = \|Az_1 - Az_2 + f(t) - f(t)\|$$

$$= \|A(z_1 - z_2)\| \leq \|A\| \cdot \|z_1 - z_2\|,$$

where

$$\|A\| = \sup_{z \neq 0} \frac{\|Az\|}{\|z\|},$$

and thus, the function F is locally Lipschitz in z. It then follows from the Picard–Lindelöf theorem (Theorem 5.10) that for each $(t_0, z_0) \in \mathbb{R} \times \mathbb{R}^2$ there exists a unique solution of the equation with $z(t_0) = z_0$ in some open interval containing t_0. Moreover, since

$$z(t_0) = \big(x(t_0), y(t_0)\big) = \big(x(t_0), x'(t_0)\big),$$

for each $t_0, x_0, y_0 \in \mathbb{R}$ there exists a unique solution of the equation $x'' + x = \sqrt[5]{t}$ with $x(t_0) = x_0$ and $x'(t_0) = y_0$ in some open interval containing t_0.

Problem 5.24

Verify that the problem

$$x' = 2\sqrt{|x|} \quad \text{with } x(0) = 0 \tag{5.53}$$

has more than one solution.

Solution

A solution of problem (5.53) is $x(t) = 0$. To obtain another solution we consider separately the cases $x > 0$ and $x < 0$. For $x > 0$ we obtain the equation $x' = 2\sqrt{x}$, that is,

$$(\sqrt{x})' = \frac{x'}{2\sqrt{x}} = 1,$$

and thus,

$$\sqrt{x(t)} = t + c \quad \text{for some } c \in \mathbb{R}.$$

Taking $t = 0$, since $x(0) = 0$, we obtain $c = 0$, and hence $x(t) = t^2$. On the other hand, for $x < 0$ we obtain the equation $x' = 2\sqrt{-x}$, that is,

$$(\sqrt{-x})' = \frac{-x'}{2\sqrt{-x}} = -1,$$

and thus,

$$\sqrt{-x(t)} = -t + d \quad \text{for some } d \in \mathbb{R}.$$

Taking $t = 0$, since $x(0) = 0$, we obtain $d = 0$, and hence, $x(t) = -t^2$. Therefore, one can consider, for example, the function

$$
x(t) = \begin{cases} t^2 & \text{if } t > 0, \\ 0 & \text{if } t = 0, \\ -t^2 & \text{if } t < 0, \end{cases}
$$

that is, $x(t) = t|t|$. We note that x is of class C^1. Indeed, outside the origin we have

$$
x'(t) = \begin{cases} 2t & \text{if } t > 0, \\ -2t & \text{if } t < 0, \end{cases}
$$

and at the origin,

$$
x'(0) = \lim_{t \to 0} \frac{x(t) - x(0)}{t} = \lim_{t \to 0} \frac{t|t|}{t} = 0.
$$

Hence, $x'(t) = 2|t|$, which is a continuous function, and $x(t)$ is of class C^1. We thus have the solutions $x(t) = 0$ and $x(t) = t|t|$.

Problem 5.25

Verify that if a matrix A has at least one eigenvalue in $\mathbb{R}^+$, then the equation $x' = Ax$ has at least one solution not converging to zero when $t \to +\infty$.

Solution

We recall that for each $v \in \mathbb{R}^n$ the function $t \mapsto e^{At}v$ is a solution of the equation $x' = Ax$ (see Proposition 5.32). Now let $v \in \mathbb{R}^n \setminus \{0\}$ be an eigenvector associated to an eigenvalue $\lambda \in \mathbb{R}^+$. We have $Av = \lambda v$, and thus, $A^k v = \lambda^k v$ for each $k \in \mathbb{N}$. Therefore,

$$
x(t) = e^{At}v = \sum_{k=0}^{\infty} \frac{1}{k!} t^k A^k v = \sum_{k=0}^{\infty} \frac{1}{k!} t^k \lambda^k v = e^{\lambda t} v.
$$

Since $\lambda \in \mathbb{R}^+$, the solution $x(t)$ does not converge to zero when $t \to +\infty$.

Problem 5.26

Find all power series $x(t) = \sum_{n=0}^{\infty} c_n t^n$, with $t \in \mathbb{R}$, that are a solution of the equation $x'' + x = 0$.

Solution

By Theorem 4.12, the power series can be differentiated term by term in the interior of its domain of convergence to obtain

$$x'(t) = \sum_{n=1}^{\infty} n c_n t^{n-1}$$

and

$$x''(t) = \sum_{n=2}^{\infty} n(n-1) c_n t^{n-2}.$$

We thus have

$$x''(t) + x(t) = \sum_{n=0}^{\infty} \left[(n+2)(n+1)c_{n+2} + c_n \right] t^n.$$

This series is the zero function if and only if

$$(n+2)(n+1)c_{n+2} + c_n = 0$$

for every $n \in \mathbb{N} \cup \{0\}$. Hence,

$$c_{n+2} = -\frac{c_n}{(n+2)(n+1)},$$

that is, given $c_0, c_1 \in \mathbb{R}$, we have

$$c_{2n} = \frac{(-1)^n c_0}{(2n)!} \quad \text{and} \quad c_{2n+1} = \frac{(-1)^n c_1}{(2n+1)!}$$

for each $n \in \mathbb{N}$. Therefore,

$$x(t) = \sum_{n=0}^{\infty} c_n t^n = \sum_{n=0}^{\infty} c_{2n} t^{2n} + \sum_{n=0}^{\infty} c_{2n+1} t^{2n+1}$$

$$= c_0 \sum_{n=0}^{\infty} \frac{(-1)^n c_0}{(2n)!} t^{2n} + c_1 \sum_{n=0}^{\infty} \frac{(-1)^n c_1}{(2n+1)!} t^{2n+1}$$

$$= c_0 \cos t + c_1 \sin t.$$

Problem 5.27

Verify that if $x(t)$ and $y(t)$ are solutions respectively of the equations $x' = Ax$ and $y' = -A^* y$, then

$$\langle x(t), y(t) \rangle = \langle x(0), y(0) \rangle, \quad t \in \mathbb{R}, \tag{5.54}$$

where $\langle \cdot, \cdot \rangle$ is the standard inner product in $\mathbb{R}^n$.

Solution

We have

$$\frac{d}{dt}\langle x(t), y(t)\rangle = \lim_{h\to 0} \frac{1}{h}\left[\langle x(t+h), y(t+h)\rangle - \langle x(t), y(t)\rangle\right]$$

$$= \lim_{h\to 0} \frac{1}{h}\left[\langle x(t+h) - x(t), y(t+h)\rangle + \langle x(t), y(t+h) - y(t)\rangle\right]$$

$$= \lim_{h\to 0} \left\langle \frac{x(t+h) - x(t)}{h}, y(t+h)\right\rangle + \lim_{h\to 0} \left\langle x(t), \frac{y(t+h) - y(t)}{h}\right\rangle$$

$$= \langle x'(t), y(t)\rangle + \langle x(t), y'(t)\rangle.$$

Therefore,

$$\frac{d}{dt}\langle x(t), y(t)\rangle = \langle Ax(t), y(t)\rangle + \langle x(t), -A^*y(t)\rangle$$

$$= \langle Ax(t), y(t)\rangle - \langle Ax(t), y(t)\rangle = 0,$$

which yields identity (5.54).

Problem 5.28

Given an $n \times n$ matrix A, we assume that the function $q\colon \mathbb{R}^n \to \mathbb{R}_0^+$ given by

$$q(x) = \int_0^\infty \left\|e^{At}x\right\|^2 dt$$

is well defined. Show that q is a polynomial of degree 2 without terms of degree 0 or 1, and that the function $F(s) = q(e^{As}x)$ has derivative $F'(s) = -\|e^{As}x\|^2$.

Solution

We note that

$$q(x) = \int_0^\infty \left(e^{At}x\right)^* e^{At}x\, dt$$

$$= \int_0^\infty x^*\left(e^{At}\right)^* e^{At}x\, dt = x^*Cx, \tag{5.55}$$

where C is the $n \times n$ matrix given by

$$C = \int_0^\infty \left(e^{At}\right)^* e^{At}\, dt$$

(with the matrices integrated entry by entry). It follows from (5.55) that q is a polynomial of degree 2 without terms of degree 0 or 1. Moreover, it follows from (5.25) that

$$q(e^{As}x) = \int_0^\infty \left\| e^{At} e^{As} x \right\|^2 dt = \int_0^\infty \left\| e^{A(t+s)} x \right\|^2 dt.$$

Making the change of variables $t + s = \tau$, we obtain

$$F(s) = q(e^{As}x) = \int_s^\infty \left\| e^{A\tau} x \right\|^2 d\tau,$$

and hence, $F'(s) = -\|e^{As}x\|^2$.

Problem 5.29

Verify that the equation $x' = |x| + 1$ has no periodic solutions.

Solution

Let $x = x(t)$ be a solution. Since $|x(t)| + 1 > 0$, we have $x'(t) > 0$ for every t (in the maximal interval of x). Hence, each solution is strictly increasing, and thus it cannot be periodic.

Problem 5.30

Show that all solutions of the equation

$$\begin{cases} x' = y \cos x, \\ y' = -x \cos x, \end{cases}$$

are bounded.

Solution

If $(x, y) = (x(t), y(t))$ is a solution, then

$$(x^2 + y^2)' = 2xx' + 2yy'$$
$$= 2x(y \cos x) + 2y(-x \cos x) = 0.$$

Therefore, there exists $r \geq 0$ such that

$$x(t)^2 + y(t)^2 = r$$

for every t (in the maximal interval of the solution). This shows that the image of each solution is either contained in a circle or is the origin, and thus, in particular, it is bounded. In fact, one can show that the image of each solution is either a circle or is the origin.

Problem 5.31

Show that all solutions of the equation

$$\begin{cases} x' = y - x, \\ y' = -x - y^3 \end{cases}$$

are bounded for $t > 0$.

Solution

If $(x, y) = (x(t), y(t))$ is a solution, then

$$\begin{aligned} \left(x^2 + y^2\right)' &= 2xx' + 2yy' \\ &= 2x(y - x) + 2y\left(-x - y^3\right) \\ &= -2x^2 - 2y^4 \le 0. \end{aligned}$$

This shows that the function $t \mapsto x(t)^2 + y(t)^2 \ge 0$ is not increasing, and thus, it is bounded for $t > 0$. Since

$$|x(t)| \le \sqrt{x(t)^2 + y(t)^2} \quad \text{and} \quad |y(t)| \le \sqrt{x(t)^2 + y(t)^2},$$

the components $x(t)$ and $y(t)$ are also bounded for $t > 0$.

Problem 5.32

Write the equation

$$\begin{cases} x' = -ay, \\ y' = ax \end{cases} \tag{5.56}$$

in polar coordinates (r, θ).

Solution

We have

$$x = r \cos \theta \quad \text{and} \quad y = r \sin \theta,$$

and

$$r = \sqrt{x^2 + y^2} \quad \text{and} \quad \theta = \begin{cases} \tan^{-1}(y/x) & \text{if } x > 0, \\ \pi/2 & \text{if } x = 0 \text{ and } y > 0, \\ \tan^{-1}(y/x) + \pi & \text{if } x < 0, \\ -\pi/2 & \text{if } x = 0 \text{ and } y < 0, \end{cases}$$

where $\tan^{-1}$ is the inverse of the tangent with values in $(-\pi/2, \pi/2)$. Therefore,

$$r' = \frac{2xx' + 2yy'}{2\sqrt{x^2 + y^2}} = \frac{xx' + yy'}{r}$$

and

$$\theta' = \frac{(y/x)'}{1 + (y/x)^2} = \frac{(y'x - x'y)/x^2}{1 + y^2/x^2}$$
$$= \frac{y'x - x'y}{x^2 + y^2} = \frac{y'x - x'y}{r^2}.$$

It follows from (5.56) that

$$r' = \frac{axy - axy}{r} = 0$$

and

$$\theta' = \frac{ax^2 + ay^2}{r^2} = a,$$

and thus, in polar coordinates the equation takes the form

$$\begin{cases} r' = 0, \\ \theta' = a. \end{cases}$$

EXERCISES

5.1. Find all solutions of the equation:
(a) $x' = -tx$;
(b) $x' - (t \sin t)x = 0$;
(c) $x' = x/(1+t) + t^2$;
(d) $x' = -\cos t + x$.

5.2. Find all solutions of the equation $x' = 1/x$.

5.3. Find an equation having e^{t^2} as solution.

5.4. Find an equation having $1/(1+t)$ as solution.

5.5. Show that if $x\colon (a, b) \to \mathbb{R}^n$ is a solution of the equation $x' = f(x)$, then for each $c \in \mathbb{R}$ the function $y\colon (a + c, b + c) \to \mathbb{R}^n$ defined by $y(t) = x(t - c)$ is also a solution of the equation.

5.6. For the equation $x' = f(x)$, show that if $\cos t$ is a solution, then $-\sin t$ is also a solution.

5.7. For the equation $x'' = f(x)$, show that if $1/(1 + t)$ is a solution, then $1/(1 - t)$ is also a solution.

5.8. Find a solution of the equation:

(a) $x' = \begin{pmatrix} 1 & -1 \\ 0 & 1 \end{pmatrix} x$ with $x(0) = \begin{pmatrix} 1 \\ 0 \end{pmatrix}$;

(b) $x' = \begin{pmatrix} 2 & 1 \\ 0 & 4 \end{pmatrix} x$ with $x(1) = \begin{pmatrix} 0 \\ 1 \end{pmatrix}$;

(c) $x' = \begin{pmatrix} 2 & 4 \\ 7 & 9 \end{pmatrix} x$ with $x(4) = \begin{pmatrix} 0 \\ 0 \end{pmatrix}$.

5.9. Find all solutions of the equation

$$x' = \begin{pmatrix} 3 & 0 \\ 0 & 4 \end{pmatrix} x + \begin{pmatrix} t \\ -t \end{pmatrix}.$$

5.10. Use the Variation of parameters formula to find the solution of the equation:

(a) $x' = \begin{pmatrix} 0 & 9 \\ -1 & 0 \end{pmatrix} x + \begin{pmatrix} 0 \\ 1 \end{pmatrix}$ with $x(0) = \begin{pmatrix} 0 \\ 1 \end{pmatrix}$;

(b) $x' = \begin{pmatrix} 0 & 4 \\ -1 & 0 \end{pmatrix} x + \begin{pmatrix} 0 \\ t \end{pmatrix}$ with $x(0) = \begin{pmatrix} 0 \\ 5 \end{pmatrix}$;

(c) $x' = \begin{pmatrix} 4 & 1 \\ 0 & 4 \end{pmatrix} x + \begin{pmatrix} e^t \\ 0 \end{pmatrix}$ with $x(0) = \begin{pmatrix} 0 \\ 1 \end{pmatrix}$.

5.11. Consider the equation

$$\begin{cases} x' = 2x + y, \\ y' = -x + a. \end{cases}$$

(a) For $a = 0$, find all solutions of the equation.

(b) For $a = 1$, find the solution of the equation with $x(0) = y(0) = 0$.

5.12. Compute the complex Jordan form of the matrix:

(a) $\begin{pmatrix} 2 & 1 \\ 1 & 2 \end{pmatrix}$;

(b) $\begin{pmatrix} 4 & 1 \\ 5 & 3 \end{pmatrix}$;

(c) $\begin{pmatrix} 2 & 1 & 0 \\ 0 & 5 & 4 \\ 0 & 0 & 3 \end{pmatrix}$.

5.13. Find the solution of the equation $x'' + x' + x = 0$ with $x(0) = 0$ and $x'(0) = 3$.

5.14. Compute e^{At} for the matrix:

(a) $A = \begin{pmatrix} 0 & 1 & 0 & 0 \\ -1 & 0 & 0 & 0 \\ 0 & 0 & 2 & 1 \\ 0 & 0 & 0 & 2 \end{pmatrix}$;

(b) $A = \begin{pmatrix} 2 & 1 & 0 & 0 \\ 0 & 1 & 0 & 0 \\ 0 & 0 & 3 & 1 \\ 0 & 0 & 0 & 3 \end{pmatrix}$;

(c) $A = \begin{pmatrix} 2 & 1 & 0 & 0 \\ 0 & 3 & 0 & 0 \\ 0 & 0 & 4 & -1 \\ 0 & 0 & 0 & 4 \end{pmatrix}$;

(d) $A = \begin{pmatrix} 1 & 4 & 0 & 0 \\ -4 & 1 & 0 & 0 \\ 0 & 0 & 2 & 0 \\ 0 & 0 & 1 & 2 \end{pmatrix}$.

5.15. For each matrix A in Exercise 5.14, find all bounded solutions of the equation $x' = Ax$.

5.16. For each matrix A in Exercise 5.14, find all solutions of the equation $x' = Ax$ that are bounded for $t > 0$.

5.17. Find whether there exists a matrix A such that
$$e^{At} = \begin{pmatrix} e^{3t} & te^{2t} \\ 0 & e^{3t} \end{pmatrix}.$$

5.18. Verify that the function f is locally Lipschitz in x:
    (a) $f(t, x) = x^2$;
    (b) $f(t, x) = x|x|$.

5.19. Show that the equation $x' - x = \sqrt[3]{t}$ has solutions.

5.20. Verify that the equation $x' = 2\sqrt{x}$ has more than one solution with $x(0) = 0$.

5.21. Identify each statement as true or false.
    (a) The equation $x' = -x^3$ has solutions with maximal interval $\mathbb{R}^+$.
    (b) The equation $x' = 1 + x^2$ has nonconstant periodic solutions.

(c) The equation $x' = 1 + x^2$ has decreasing solutions.

(d) There exists a matrix A such that

$$e^{At} = \begin{pmatrix} e^t & e^{-t} \\ e^{-t} & e^t \end{pmatrix}.$$

5.22. Show that all solutions of the equation $(x', y') = (ye^x, -xe^x)$ are bounded.

5.23. Show that all solutions of the equation

$$(x', y') = (y - x^3, -x - y^3)$$

are bounded for $t > 0$.

5.24. Write the equation

$$\begin{cases} x' = ay + bx + x^2, \\ y' = -ax + by \end{cases}$$

in polar coordinates (r, θ).

5.25. Find all nonconstant periodic solutions of the equation

$$\begin{cases} r' = r(r-1)(r-2), \\ \theta' = 1. \end{cases}$$

5.26. For a square matrix A, show that $A^m e^A = e^A A^m$ for every $m \in \mathbb{N}$.

5.27. Verify that if A and B are $n \times n$ matrices, then

$$(A + B)^2 = A^2 + AB + BA + B^2.$$

5.28. Find whether the identity

$$(A + B)^2 = A^2 + 2AB + B^2$$

between $n \times n$ matrices is always satisfied.

5.29. Defining the cosine of a square matrix A by

$$\cos A = \frac{e^{iA} + e^{-iA}}{2},$$

compute

$$\cos \begin{pmatrix} 0 & 1 & 0 \\ 0 & 0 & 1 \\ 0 & 0 & 0 \end{pmatrix}.$$

5.30. Compute

$$\limsup_{t \to +\infty} \frac{1}{t} \log \|e^{At} x\|,$$

where

$$A = \begin{pmatrix} 2 & 1 \\ 0 & 2 \end{pmatrix} \quad \text{and} \quad x = \begin{pmatrix} 0 \\ 3 \end{pmatrix}.$$

5.31. Compute

$$\limsup_{t \to +\infty} \frac{1}{t} \log \|x(t)\|$$

for each nonvanishing solution $x(t)$ of the equation $x'' + 4x = 0$.

5.32. Verify that

$$\frac{d}{dt} \det e^{At} = 0 \quad \text{for } A = \begin{pmatrix} 2 & 4 \\ 10 & -2 \end{pmatrix}.$$

5.33. Verify that

$$\lim_{n \to \infty} \begin{pmatrix} 1 & 1/n^2 \\ 0 & 1 \end{pmatrix}^n = \begin{pmatrix} 1 & 0 \\ 0 & 1 \end{pmatrix},$$

with the limit computed entry by entry.

5.34. Verify that

$$\lim_{n \to \infty} \frac{1}{n} \begin{pmatrix} 1 & 1/n \\ 0 & 1 \end{pmatrix}^{n^2} = \begin{pmatrix} 0 & 1 \\ 0 & 0 \end{pmatrix}.$$

5.35. Show that $\det e^{At} = e^{t \operatorname{tr} A}$. Hint: use Proposition 5.42.

5.36. Use Exercise 5.35 to show that

$$\frac{d}{dt} \left(\det e^{At} \right) \big|_{t=0} = \operatorname{tr} A.$$

5.37. Find a necessary and sufficient condition in terms of a square matrix A in order that:
(a) all solutions of the equation $x' = Ax$ are bounded;
(b) all solutions of the equation $x' = Ax$ converge to zero when $t \to +\infty$.

Solving Differential Equations

In this chapter we present several methods for finding solutions of certain classes of differential equations. Namely, we consider exact equations, equations that can be reduced to exact, and scalar equations of order greater than 1. We also consider equations that can be solved using the Laplace transform. We note that these are only some methods among many others in the theory. On purpose, we do not consider methods adapted to very particular classes of differential equations.

6.1 Exact Equations

In this section we consider equations in $\mathbb{R}$ of the form

$$M(t,x) + N(t,x)x' = 0, \tag{6.1}$$

where M and N are continuous functions with $N \neq 0$.

Definition 6.1

The differential equation (6.1) is said to be *exact* in an open set $S \subset \mathbb{R}^2$ if there exists a differentiable function $\Phi \colon S \to \mathbb{R}$ such that

$$\frac{\partial \Phi}{\partial t}(t,x) = M(t,x) \quad \text{and} \quad \frac{\partial \Phi}{\partial x}(t,x) = N(t,x) \tag{6.2}$$

for every $(t,x) \in S$.

L. Barreira, C. Valls, *Complex Analysis and Differential Equations*,
Springer Undergraduate Mathematics Series,
DOI 10.1007/978-1-4471-4008-5_6, © Springer-Verlag London 2012

Example 6.2

Let us consider the equation

$$-4t + \left(5x^4 + 3\right)x' = 0. \tag{6.3}$$

We look for a differentiable function Φ such that

$$\frac{\partial \Phi}{\partial t} = -4t \quad \text{and} \quad \frac{\partial \Phi}{\partial x} = 5x^4 + 3.$$

It follows from the first equation that

$$\Phi(t, x) = -2t^2 + C(x)$$

for some differentiable function C, and hence,

$$\frac{\partial \Phi}{\partial x} = C'(x) = 5x^4 + 3.$$

Thus, one can take $C(x) = x^5 + 3x$, and

$$\Phi(t, x) = -2t^2 + x^5 + 3x. \tag{6.4}$$

In particular, equation (6.3) is exact.

The importance of exact equations stems from the following property.

Proposition 6.3

If equation (6.1) is exact, then each of its solutions $x(t)$ satisfies

$$\Phi\big(t, x(t)\big) = c \tag{6.5}$$

for some constant $c \in \mathbb{R}$.

Proof

Since equation (6.1) is exact, taking derivatives of $\Phi(t, x(t))$ with respect to t, we obtain

$$\frac{d}{dt}\Phi\big(t, x(t)\big) = \frac{\partial \Phi}{\partial t}\big(t, x(t)\big) + \frac{\partial \Phi}{\partial x}\big(t, x(t)\big)x'(t)$$

$$= M\big(t, x(t)\big) + N\big(t, x(t)\big)x'(t) = 0.$$

This shows that (6.5) holds for some constant $c \in \mathbb{R}$. □

Example 6.4

By Proposition 6.3, it follows from (6.4) that each solution of equation (6.3) is given implicitly by

$$-2t^2 + x(t)^5 + 3x(t) = c$$

for some constant $c \in \mathbb{R}$.

Now we describe a necessary and sufficient condition for the exactness of equation (6.1) in an open rectangle.

Theorem 6.5

Let M and N be functions of class C^1 in an open rectangle $S = (a,b) \times (c,d)$. Then equation (6.1) is exact in S if and only if

$$\frac{\partial M}{\partial x} = \frac{\partial N}{\partial t} \quad \text{in } S. \tag{6.6}$$

Proof

If equation (6.1) is exact, then there exists a differentiable function Φ satisfying (6.2) in S. Since M and N are of class C^1, the function Φ is of class C^2. Therefore,

$$\frac{\partial M}{\partial x} = \frac{\partial}{\partial x}\left(\frac{\partial \Phi}{\partial t}\right) = \frac{\partial}{\partial t}\left(\frac{\partial \Phi}{\partial x}\right) = \frac{\partial N}{\partial t}$$

in S, which establishes property (6.6).

Now we assume that property (6.6) holds. Integrating over t, we obtain

$$\int_{t_0}^{t} \frac{\partial M}{\partial x}(s,x)\,ds = N(t,x) - N(t_0,x). \tag{6.7}$$

Let us also consider the function

$$\Phi(t,x) = \int_{x_0}^{x} N(t_0,y)\,dy + \int_{t_0}^{t} M(s,x)\,ds.$$

We have

$$\frac{\partial \Phi}{\partial t}(t,x) = M(t,x),$$

and it follows from (6.7) that

$$\frac{\partial \Phi}{\partial x}(t,x) = N(t_0,x) + \int_{t_0}^{t} \frac{\partial M}{\partial x}(s,x)\,ds = N(t,x).$$

This shows that equation (6.1) is exact in S. $\qquad \square$

We give some examples.

Example 6.6

Let us consider the equation

$$3x + \sin t + \left(3t + 2e^{2x}\right)x' = 0. \tag{6.8}$$

Since

$$M(t,x) = 3x + \sin t \quad \text{and} \quad N(t,x) = 3t + 2e^{2x},$$

we have

$$\frac{\partial M}{\partial x} = 3 \quad \text{and} \quad \frac{\partial N}{\partial t} = 3.$$

Hence, by Theorem 6.5, equation (6.8) is exact in $\mathbb{R}^2$, and there exists a differentiable function Φ such that

$$\frac{\partial \Phi}{\partial t} = 3x + \sin t \quad \text{and} \quad \frac{\partial \Phi}{\partial x} = 3t + 2e^{2x}.$$

It follows from the first equation that

$$\Phi(t,x) = 3xt - \cos t + C(x)$$

for some differentiable function C. Thus,

$$\frac{\partial \Phi}{\partial x} = 3t + C'(x) = 3t + 2e^{2x},$$

which yields $C'(x) = 2e^{2x}$. Hence, one can take $C(x) = e^{2x}$ and

$$\Phi(t,x) = 3xt - \cos t + e^{2x}.$$

By Proposition 6.3, each solution of equation (6.8) satisfies $\Phi(t, x(t)) = c$ for some constant $c \in \mathbb{R}$.

Example 6.7

Now we consider the equation

$$x' = \frac{g(t)}{f(x)},$$

which is called a *separable equation*. It can be written in the form

$$M(t,x) + N(t,x)x' = 0,$$

where

$$M(t,x) = g(t) \quad \text{and} \quad N(t,x) = -f(x).$$

Since

$$\frac{\partial M}{\partial x} = \frac{\partial N}{\partial t} = 0,$$

the equation is exact and there exists a differentiable function Φ such that

$$\frac{\partial \Phi}{\partial t} = g(t) \quad \text{and} \quad \frac{\partial \Phi}{\partial x} = -f(x).$$

Indeed, one can take

$$\Phi(t,x) = \int_{t_0}^{t} g(s)\,ds - \int_{x_0}^{x} f(y)\,dy.$$

6.2 Equations Reducible to Exact

In this section we consider equations that are not exact, and we try to find a function $\mu(t,x)$ such that equation (6.1) becomes exact when multiplied by this function.

Definition 6.8

We say that equation (6.1) is *reducible to exact* in an open rectangle $S = (a,b) \times (c,d)$ if there exists a differentiable function $\mu\colon S \to \mathbb{R}$ such that the equation

$$\mu(t,x)M(t,x) + \mu(t,x)N(t,x)x' = 0 \tag{6.9}$$

is exact and has the same solutions as equation (6.1). In this case, the function μ is called an *integrating factor* of equation (6.1).

By Theorem 6.5, when the functions M, N and μ are of class C^1, equation (6.9) is exact in S if and only if

$$\frac{\partial(\mu M)}{\partial x} = \frac{\partial(\mu N)}{\partial t} \tag{6.10}$$

in S. But since this is a partial differential equation (because it contains derivatives with respect to more than one variable), we shall only consider particular cases.

Proposition 6.9

Equation (6.10) has a nonzero solution of the form:
1. $\mu(t, x) = \mu(t)$ if

$$\frac{1}{N}\left(\frac{\partial M}{\partial x} - \frac{\partial N}{\partial t}\right) \tag{6.11}$$

does not depend on x, in which case μ satisfies the equation

$$\mu' = \frac{1}{N}\left(\frac{\partial M}{\partial x} - \frac{\partial N}{\partial t}\right)\mu; \tag{6.12}$$

2. $\mu(t, x) = \mu(x)$ if

$$-\frac{1}{M}\left(\frac{\partial M}{\partial x} - \frac{\partial N}{\partial t}\right)$$

does not depend on t, in which case μ satisfies the equation

$$\mu' = -\frac{1}{M}\left(\frac{\partial M}{\partial x} - \frac{\partial N}{\partial t}\right)\mu.$$

Proof

When $\mu(t, x) = \mu(t)$, it follows from (6.10) that

$$\mu\frac{\partial M}{\partial x} = \mu'N + \mu\frac{\partial N}{\partial t},$$

that is,

$$\mu' = \frac{1}{N}\left(\frac{\partial M}{\partial x} - \frac{\partial N}{\partial t}\right)\mu.$$

Since μ and μ' do not depend on x, this equation has nonzero solutions if and only if the expression in (6.11) does not depend on x. The second property can be obtained in an analogous manner. $\qquad\square$

Example 6.10

Let us consider the equation

$$x^2 e^{-t} - 4xe^{-2t}\sin t + \left(2xe^{-t} + 4e^{-2t}\cos t\right)x' = 0. \tag{6.13}$$

We have

$$M(t, x) = x^2 e^{-t} - 4xe^{-2t}\sin t \quad \text{and} \quad N(t, x) = 2xe^{-t} + 4e^{-2t}\cos t.$$

In particular,

$$\frac{1}{N}\left(\frac{\partial M}{\partial x} - \frac{\partial N}{\partial t}\right) = 2$$

does not depend on x, and by Proposition 6.9 there exists an integrating factor of the form $\mu(t)$. Namely, since equation (6.12) takes the form $\mu' = 2\mu$, a nonzero solution is $\mu(t) = e^{2t}$. Multiplying (6.13) by e^{2t}, we obtain the exact equation

$$\left(x^2 e^t - 4x\sin t\right) + \left(2xe^t + 4\cos t\right)x' = 0.$$

It follows from

$$\frac{\partial \Phi}{\partial t} = x^2 e^t - 4x\sin t$$

that

$$\Phi(t,x) = x^2 e^t + 4x\cos t + C(x)$$

for some differentiable function C. Therefore,

$$\frac{\partial \Phi}{\partial x} = 2xe^t + 4\cos t + C'(x) = 2xe^t + 4\cos t,$$

and we obtain $C'(x) = 0$. Hence, one can take $C(x) = 0$, and each solution $x(t)$ of equation (6.13) satisfies

$$\Phi\big(t, x(t)\big) = x(t)^2 e^t + 4x(t)\cos t = c$$

for some constant $c \in \mathbb{R}$.

6.3 Scalar Equations of Order Greater than 1

In this section we consider equations in $\mathbb{R}$ of the form

$$x^{(n)} + a_{n-1}x^{(n-1)} + \cdots + a_2 x'' + a_1 x' + a_0 x = 0, \tag{6.14}$$

with $a_0, a_1, \ldots, a_{n-1} \in \mathbb{R}$.

Example 6.11

Let us consider the equation $x'' + x = 0$. Taking $y = x'$, one can write it in the matrix form

$$\begin{pmatrix} x \\ y \end{pmatrix}' = \begin{pmatrix} 0 & 1 \\ -1 & 0 \end{pmatrix}\begin{pmatrix} x \\ y \end{pmatrix},$$

which thus has the solutions

$$\begin{pmatrix} x(t) \\ y(t) \end{pmatrix} = e^{\left(\begin{smallmatrix} 0 & 1 \\ -1 & 0 \end{smallmatrix} \right)(t-t_0)} \begin{pmatrix} x(t_0) \\ y(t_0) \end{pmatrix}.$$

Nevertheless, we will show in this section that equations of the form (6.14) can be solved using an alternative method that is frequently more practical.

We first write equation (6.14) in another form. We introduce the notation

$$Dx = x'.$$

For example,

$$D^2 x = D(x') = x''.$$

Equation (6.14) can then be written in the form

$$\left(D^n + a_{n-1} D^{n-1} + \cdots + a_2 D^2 + a_1 D + a_0 \right) x = 0.$$

Now we consider the *characteristic polynomial*

$$p(\lambda) = \lambda^n + a_{n-1} \lambda^{n-1} + \cdots + a_2 \lambda^2 + a_1 \lambda + a_0.$$

It follows from the Fundamental theorem of algebra (Theorem 4.19) that there exist constants $\lambda_1, \ldots, \lambda_N \in \mathbb{C}$ and $m_1, \ldots, m_N \in \mathbb{N}$ such that

$$p(\lambda) = (\lambda - \lambda_1)^{m_1} (\lambda - \lambda_2)^{m_2} \cdots (\lambda - \lambda_N)^{m_N}, \qquad (6.15)$$

with $m_1 + \cdots + m_N = n$.

Proposition 6.12

Given constants $\lambda_1, \ldots, \lambda_N \in \mathbb{C}$ and $m_1, \ldots, m_N \in \mathbb{N}$ satisfying (6.15), equation (6.14) is equivalent to

$$(D - \lambda_1)^{m_1} (D - \lambda_2)^{m_2} \cdots (D - \lambda_N)^{m_N} x = 0. \qquad (6.16)$$

Proof

It is sufficient to note that

$$(\lambda - a_1) \cdots (\lambda - a_n) = \sum_{j=0}^{n} (-1)^j \sum_{i_1 < \cdots < i_j} a_{i_1} \cdots a_{i_j} \lambda^{n-j}$$

and

$$(D - a_1) \cdots (D - a_n) = \sum_{j=0}^{n} (-1)^j \sum_{i_1 < \cdots < i_j} a_{i_1} \cdots a_{i_j} D^{n-j}.$$

For example, for $n = 2$ we have

$$(\lambda - a)(\lambda - b) = \lambda^2 - (a + b)\lambda + ab,$$

and

$$\begin{aligned}
(D - a)(D - b)x &= (D - a)(x' - bx) \\
&= x'' - bx - ax' + abx \\
&= x'' - (a + b)x' + abx \\
&= (D^2 - (a + b)D + ab)x,
\end{aligned} \tag{6.17}$$

that is,

$$(D - a)(D - b) = D^2 - (a + b)D + ab. \tag{6.18}$$

This yields the desired result.   □

It is thus sufficient to consider equations (6.14) that are already written in the form (6.16). We first consider two particular types of equations.

Proposition 6.13

Given $\lambda \in \mathbb{R}$ and $m \in \mathbb{N}$, the solutions of the equation

$$(D - \lambda)^m x = 0 \tag{6.19}$$

are given by

$$x(t) = \sum_{k=0}^{m-1} c_k t^k e^{\lambda t} = c_0 e^{\lambda t} + \cdots + c_{m-1} t^{m-1} e^{\lambda t} \tag{6.20}$$

for $t \in \mathbb{R}$, with $c_0, \ldots, c_{m-1} \in \mathbb{R}$.

Proof

We use induction on m. For $m = 1$ the equation takes the form

$$(D - \lambda)x = x' - \lambda x = 0,$$

and thus it has the solutions $x(t) = c_0 e^{\lambda t}$ with $c_0 \in \mathbb{R}$. Now we assume that the result holds for a particular m. Since

$$(D - \lambda)^{m+1}x = (D - \lambda)^m (D - \lambda)x = 0,$$

we have

$$(D - \lambda)x = c_0 e^{\lambda t} + c_1 t e^{\lambda t} + \cdots + c_{m-1} t^{m-1} e^{\lambda t},$$

with $c_0, \ldots, c_{m-1} \in \mathbb{R}$. Therefore,

$$\begin{aligned}
\left(e^{-\lambda t} x\right)' &= e^{-\lambda t}(Dx - \lambda x) \\
&= e^{-\lambda t}(D - \lambda)x \\
&= c_0 + c_1 t + \cdots + c_{m-1} t^{m-1},
\end{aligned}$$

and hence,

$$e^{-\lambda t} x(t) = c + c_0 t + \frac{c_1}{2} t^2 + \cdots + \frac{c_{m-1}}{m} t^{m-1}$$

for some constant $c \in \mathbb{R}$. This shows that the result holds for $m + 1$, and the solution $x(t)$ is given by (6.20). $\qquad \square$

Example 6.14

Let us consider the equation

$$x'' - 8x' + 16x = 0. \tag{6.21}$$

The characteristic polynomial is

$$\lambda^2 - 8\lambda + 16 = (\lambda - 4)^2,$$

and by Proposition 6.13 the solutions of equation (6.21) are given by

$$x(t) = a e^{4t} + b t e^{4t} \quad \text{for } t \in \mathbb{R},$$

with $a, b \in \mathbb{R}$.

Example 6.15

Now we consider the equation

$$(D - a)(D - b)x = 0 \tag{6.22}$$

for some $a, b \in \mathbb{R}$ with $a \neq b$. By (6.18), we have

$$(D - a)(D - b) = (D - b)(D - a). \tag{6.23}$$

We observe that:
1. if $(D - b)x = 0$, then it follows from (6.22) that

$$(D - a)(D - b)x = (D - a)0 = 0,$$

and x is a solution of equation (6.22);
2. if $(D - a)x = 0$, then it follows from (6.23) that

$$(D - a)(D - b)x = (D - b)(D - a)x = (D - b)0 = 0,$$

and x is also a solution of equation (6.22).
Moreover, one can easily verify that any linear combination of solutions of equation (6.22) is still a solution of this equation. Hence,

$$x(t) = c_1 e^{at} + c_2 e^{bt} \tag{6.24}$$

is a solution for each $c_1, c_2 \in \mathbb{R}$, since $c_1 e^{at}$ and $c_2 e^{bt}$ are respectively the solutions of the equations $(D - a)x = 0$ and $(D - b)x = 0$.

Now we show that there are no other solutions. We write equation (6.22) in the form

$$x'' - (a + b)x' + abx = 0$$

(see (6.17)), which is the same as

$$\begin{pmatrix} x \\ y \end{pmatrix}' = \begin{pmatrix} 0 & 1 \\ -ab & a+b \end{pmatrix} \begin{pmatrix} x \\ y \end{pmatrix}. \tag{6.25}$$

By Proposition 5.32, the linear space of the solutions of equation (6.25) has dimension 2. Hence, the linear space of the solutions of equation (6.22) has at most dimension 2, since it is obtained from the first component of the solutions of equation (6.25). But since the functions in (6.24) already generate a space of dimension 2, we conclude that there are no other solutions.

Now we consider a second type of equations.

Proposition 6.16

Given $\lambda = a + ib$ with $b \neq 0$ and $m \in \mathbb{N}$, the solutions of the equation

$$(D - \lambda)^m (D - \bar{\lambda})^m x = 0 \tag{6.26}$$

are given by

$$x(t) = \sum_{k=0}^{m-1} \left[c_k t^k e^{at} \cos(bt) + d_k t^k e^{at} \sin(bt) \right]$$

for $t \in \mathbb{R}$, with $c_0, d_0, \ldots, c_{m-1}, d_{m-1} \in \mathbb{R}$.

Proof

It follows from the proof of Proposition 6.13, that the complex-valued solutions of equation (6.19), also for $\lambda \in \mathbb{C}$, are given by

$$x(t) = \sum_{k=0}^{m-1} a_k t^k e^{\lambda t}$$

for $t \in \mathbb{R}$, with $a_0, \ldots, a_{m-1} \in \mathbb{C}$. Similarly, the complex-valued solutions of the equation

$$(D - \overline{\lambda})^m x = 0$$

are given by

$$x(t) = \sum_{k=0}^{m-1} b_k t^k e^{\overline{\lambda} t}$$

for $t \in \mathbb{R}$, with $b_0, \ldots, b_{m-1} \in \mathbb{C}$. Moreover, identity (6.17) also holds for $a, b \in \mathbb{C}$, that is,

$$(D - a)(D - b) = (D - b)(D - a)$$

for every $a, b \in \mathbb{C}$. This implies that

$$(D - \lambda)^m (D - \overline{\lambda})^m = (D - \overline{\lambda})^m (D - \lambda)^m.$$

On the other hand, proceeding as in the proof of Proposition 5.32, one can show that the linear subspace of $\mathbb{C}^n = \mathbb{R}^{2n}$ formed by the $\mathbb{C}^n$-valued solutions of the equation $x' = Ax$ has real dimension $2n$. One can then proceed as in Example 6.15 to show that the complex-valued solutions of equation (6.26) are given by

$$x(t) = \sum_{k=0}^{m-1} \left(a_k t^k e^{\lambda t} + b_k t^k e^{\overline{\lambda} t} \right)$$

for $t \in \mathbb{R}$, with $a_0, b_0, \ldots, a_{m-1}, b_{m-1} \in \mathbb{C}$. Taking

$$2a_k = c_k - id_k \quad \text{and} \quad 2b_k = c_k + id_k$$

for $k = 0, \dots, m-1$, we obtain

$$x(t) = \frac{1}{2} \sum_{k=0}^{m-1} \left[(c_k - id_k) t^k e^{(a+ib)t} + (c_k + id_k) t^k e^{(a-ib)t} \right]$$

$$= \sum_{k=0}^{m-1} \left[c_k t^k e^{at} \cos(bt) + d_k t^k e^{at} \sin(bt) \right].$$

This yields the desired result. □

Example 6.17

Let us consider the equation

$$x'' + 5x = 0. \tag{6.27}$$

The characteristic polynomial is

$$\lambda^2 + 5 = (\lambda - i\sqrt{5})(\lambda + i\sqrt{5}),$$

and by Proposition 6.16 the solutions of equation (6.27) are given by

$$x(t) = a\cos(\sqrt{5}t) + b\sin(\sqrt{5}t) \quad \text{for } t \in \mathbb{R},$$

with $a, b \in \mathbb{R}$.

Example 6.18

Now we consider the equation

$$(D+2)(D-3i)^2(D+3i)^2 x = 0. \tag{6.28}$$

In an analogous manner to that in Example 6.15, we consider separately the equations

$$(D+2)x = 0$$

and

$$(D-3i)^2(D+3i)^2 x = 0.$$

These have spaces of solutions generated respectively by the functions e^{-2t} and

$$\cos(3t), \quad t\cos(3t), \quad \sin(3t) \quad \text{and} \quad t\sin(3t).$$

In a similar manner to that in Example 6.15, one can show that equation (6.28) gives rise to an equation $x' = Ax$ in $\mathbb{R}^5$, which thus has a space of solutions of dimension 5. Since the functions obtained previously already generate a space of dimension 5, the solutions of equation (6.28) are given by

$$x(t) = c_1 e^{-2t} + c_2 \cos(3t) + c_3 t \cos(3t) + c_4 \sin(3t) + c_5 t \sin(3t)$$

for $t \in \mathbb{R}$, with $c_1, c_2, c_3, c_4, c_5 \in \mathbb{R}$.

We also consider the *nonhomogeneous* case, that is, we consider equations of the form

$$x^{(n)} + a_{n-1} x^{(n-1)} + \cdots + a_2 x'' + a_1 x' + a_0 x = h(t),$$

with $a_0, a_1, \ldots, a_{n-1} \in \mathbb{R}$, for some nonzero function h.

Example 6.19

Let us consider the equation

$$x'' + 2x' - x = t. \tag{6.29}$$

One can write it in the form

$$\begin{cases} x' = y, \\ y' = x - 2y + t, \end{cases}$$

or equivalently,

$$\begin{pmatrix} x \\ y \end{pmatrix}' = \begin{pmatrix} 0 & 1 \\ 1 & -2 \end{pmatrix} \begin{pmatrix} x \\ y \end{pmatrix} + \begin{pmatrix} 0 \\ t \end{pmatrix}.$$

It then follows from the Variation of parameters formula (Theorem 5.31) that

$$\begin{pmatrix} x(t) \\ y(t) \end{pmatrix} = e^{\begin{pmatrix} 0 & 1 \\ 1 & -2 \end{pmatrix}(t-t_0)} \begin{pmatrix} x(t_0) \\ y(t_0) \end{pmatrix} + \int_{t_0}^t e^{\begin{pmatrix} 0 & 1 \\ 1 & -2 \end{pmatrix}(t-s)} \begin{pmatrix} 0 \\ s \end{pmatrix} \, ds.$$

However, one can also obtain the solutions of equation (6.29) using an alternative method, which is more automatic. We first write equation (6.29) in the form

$$(D^2 + 2D - 1)x = t. \tag{6.30}$$

Since $D^2 t = 0$, we obtain

$$D^2(D^2 + 2D - 1)x = 0,$$

that is,

$$D^2(D+1-\sqrt{2})(D+1+\sqrt{2})x = 0. \tag{6.31}$$

We note that all solutions of equation (6.29) are solutions of this equation. Proceeding as in Example 6.18, one can show that the solutions of equation (6.31) are given by

$$x(t) = a + bt + ce^{(-1+\sqrt{2})t} + de^{(-1-\sqrt{2})t} \quad \text{for } t \in \mathbb{R},$$

with $a, b, c, d \in \mathbb{R}$. However, not all of these functions are necessarily solutions of equation (6.29). Substituting $x(t)$ in (6.30), we obtain

$$\left(D^2 + 2D - 1\right)x = \left(D^2 + 2D - 1\right)(a + bt) = t,$$

since $ce^{(-1+\sqrt{2})t} + de^{(-1-\sqrt{2})t}$ is a solution of the equation $(D^2 + 2D - 1)x = 0$. We also have

$$\left(D^2 + 2D - 1\right)(a + bt) = 2b - a - bt = t,$$

and thus $-b = 1$ and $2b - a = 0$, that is,

$$a = -2 \quad \text{and} \quad b = -1.$$

Therefore, the solutions of equation (6.30) are given by

$$x(t) = -2 - t + ce^{(-1+\sqrt{2})t} + de^{(-1-\sqrt{2})t} \quad \text{for } t \in \mathbb{R},$$

with $c, d \in \mathbb{R}$.

Example 6.20

Now we consider the equation

$$x' - 2x = e^t \cos(2t), \tag{6.32}$$

that is,

$$(D - 2)x = e^t \cos(2t). \tag{6.33}$$

Any solution of equation (6.33) satisfies

$$(D - 1 - 2i)(D - 1 + 2i)(D - 2)x = 0.$$

Moreover, the solutions of this last equation are given by

$$x(t) = ae^{2t} + be^t \cos(2t) + ce^t \sin(2t) \quad \text{for } t \in \mathbb{R},$$

with $a, b, c \in \mathbb{R}$. Substituting $x(t)$ in (6.33), we obtain

$$(D - 2)\big(be^t \cos(2t) + ce^t \sin(2t)\big) = e^t \cos(2t),$$

that is,

$$be^t \cos(2t) - 2be^t \sin(2t) + ce^t \sin(2t) + 2ce^t \cos(2t) = e^t \cos(2t).$$

Therefore,

$$b + 2c = 1 \quad \text{and} \quad -2b + c = 0,$$

and hence $b = 1/5$ and $c = 2/5$. We conclude that the solutions of equation (6.32) are given by

$$x(t) = ae^{2t} + \frac{1}{5}e^t \cos(2t) + \frac{2}{5}e^t \sin(2t) \quad \text{for } t \in \mathbb{R},$$

with $a \in \mathbb{R}$.

6.4 Laplace Transform

In this section we describe another method for finding the solutions of a class of differential equations in $\mathbb{R}$. The method is based on the Laplace transform, and we consider equations of the form

$$x^{(n)} + a_{n-1}x^{(n-1)} + \cdots + a_2 x'' + a_1 x' + a_0 x = f(t), \tag{6.34}$$

with $a_0, a_1, \ldots, a_{n-1} \in \mathbb{R}$, for some nonzero function f in $\mathbb{R}_0^+ = [0, +\infty)$.

We first introduce the notion of Laplace transform. To this effect, we consider the family $\mathcal{F}$ of all functions $f \colon \mathbb{R}_0^+ \to \mathbb{R}$ such that:

1. f has at most finitely many discontinuities;
2. f has left-sided and right-sided limits at all points of $\mathbb{R}_0^+$.

Definition 6.21

Given a function $f \in \mathcal{F}$, we define the *Laplace transform* F of f by

$$F(z) = \int_0^\infty e^{-tz} f(t)\, dt \tag{6.35}$$

for each $z \in \mathbb{C}$ such that the integral is well defined. We also write $F = Lf$.

The improper integral in (6.35) is defined by

$$\int_0^\infty e^{-tz} f(t)\, dt = \lim_{R \to \infty} \int_0^R e^{-tz} f(t)\, dt, \tag{6.36}$$

whenever the limit exists. We observe that since f has at most finitely many discontinuities, for each $R > 0$ and $z \in \mathbb{C}$ the function $t \mapsto e^{-tz} f(t)$ is Riemann-integrable in the interval $[0, R]$, and hence, each integral on the right-hand side of (6.36) is well defined.

We note that

$$L(af + bg) = aLf + bLg \tag{6.37}$$

for every $a, b \in \mathbb{R}$, in the set where both Laplace transforms Lf and Lg are defined. One can also consider functions $f = u + iv \colon \mathbb{R}_0^+ \to \mathbb{C}$, and define

$$Lf = Lu + iLv.$$

Then identity (6.37) holds for every $a, b \in \mathbb{C}$.

Example 6.22

For $f = 1$ we have

$$(Lf)(z) = \int_0^\infty e^{-tz}\, dt = \lim_{R \to \infty} \int_0^R e^{-tz}\, dt$$

$$= \lim_{R \to \infty} \frac{e^{-tz}}{-z} \Big|_{t=0}^{t=R} = \lim_{R \to \infty} \frac{1 - e^{-Rz}}{z},$$

and thus, $(Lf)(z) = 1/z$ for $\operatorname{Re} z > 0$.

Example 6.23

For

$$f(t) = \begin{cases} 0 & \text{if } t \in [0, 1], \\ 1 & \text{if } t > 1, \end{cases}$$

we have

$$(Lf)(z) = \int_1^\infty e^{-tz}\, dt = \lim_{R \to \infty} \int_1^R e^{-tz}\, dt$$

$$= \lim_{R \to \infty} \frac{e^{-tz}}{-z} \Big|_{t=1}^{t=R} = \lim_{R \to \infty} \frac{e^{-z} - e^{-Rz}}{z},$$

and thus, $(Lf)(z) = e^{-z}/z$ for $\operatorname{Re} z > 0$.

Example 6.24

For $f(t) = e^{at}$ with $a \in \mathbb{C}$, we have

$$(Lf)(z) = \lim_{R \to \infty} \int_0^R e^{-tz} e^{at}\, dt$$

$$= \lim_{R \to \infty} \frac{e^{R(a-z)} - 1}{a - z},$$

and thus, $(Lf)(z) = 1/(z - a)$ for $\operatorname{Re} z > \operatorname{Re} a$.

Example 6.25

For $f(t) = \cos(at)$ and $g(t) = \sin(at)$ with $a \in \mathbb{R}$, we have

$$(Lf)(z) = \int_0^\infty e^{-tz} \cos(at)\, dt$$

and

$$(Lg)(z) = \int_0^\infty e^{-tz} \sin(at)\, dt.$$

It follows from Example 6.24 that

$$\int_0^\infty e^{-tz} e^{iat}\, dt = \frac{1}{z - ia}$$

and

$$\int_0^\infty e^{-tz} e^{-iat}\, dt = \frac{1}{z + ia}$$

for $\operatorname{Re} z > 0$. Therefore,

$$(Lf)(z) = \int_0^\infty e^{-tz} \cos(at)\, dt$$

$$= \int_0^\infty e^{-tz} \frac{e^{iat} + e^{-iat}}{2}\, dt$$

$$= \frac{1}{2}\left(\frac{1}{z - ia} + \frac{1}{z + ia} \right) = \frac{z}{z^2 + a^2} \tag{6.38}$$

and

$$(Lg)(z) = \int_0^\infty e^{-tz} \sin(at)\, dt$$

$$= \int_0^\infty e^{-tz} \frac{e^{iat} - e^{-iat}}{2i}\, dt$$

$$= \frac{1}{2i}\left(\frac{1}{z-ia} - \frac{1}{z+ia}\right) = \frac{a}{z^2+a^2}, \qquad (6.39)$$

also for $\operatorname{Re} z > 0$.

The following result gives conditions for the existence of the Laplace transform. Given $c > 0$, we denote by $\mathcal{F}_c$ the family of functions $f \in \mathcal{F}$ for which there exists $D > 0$ such that

$$|f(t)| \le De^{ct}, \quad t > 0.$$

Theorem 6.26

If $f \in \mathcal{F}_c$, then the Laplace transform of f is defined and is holomorphic for $\operatorname{Re} z > c$, with

$$(Lf)'(z) = -\int_0^\infty te^{-tz}f(t)\,dt. \qquad (6.40)$$

Proof

For $\operatorname{Re} z > c$ and $R > r$, we have

$$\int_r^R \left|e^{-tz}f(t)\right|dt \le \int_r^R e^{-t\operatorname{Re} z}De^{ct}\,dt$$

$$= D\frac{e^{t(c-\operatorname{Re} z)}}{c-\operatorname{Re} z}\bigg|_{t=r}^{t=R}$$

$$= D\frac{e^{R(c-\operatorname{Re} z)} - e^{r(c-\operatorname{Re} z)}}{c-\operatorname{Re} z} \to \frac{De^{r(c-\operatorname{Re} z)}}{\operatorname{Re} z - c}$$

when $R \to +\infty$. Therefore, $\int_0^{r_n} e^{-tz}f(t)\,dt$ is a Cauchy sequence whenever $r_n \nearrow +\infty$, and thus, the Laplace transform of f is defined for $\operatorname{Re} z > c$.

In order to show that $F = Lf$ is holomorphic, we consider the sequence of functions

$$F_n(z) = \int_0^n e^{-tz}f(t)\,dt.$$

We have

$$|F_n(z) - F(z)| = \left|\int_n^\infty e^{-tz}f(t)\,dt\right|$$

$$\le \frac{De^{t(c-z)}}{c-z}\bigg|_{t=n}^{t=\infty} = \frac{De^{n(c-z)}}{z-c},$$

and thus, $(F_n)_n$ is uniformly convergent to F on the set $\{z \in \mathbb{C} : \operatorname{Re} z > a\}$, for each $a > c$. Now we show that each function F_n is holomorphic. We first observe that there exists $C \geq 1$ such that

$$\left|e^{-z} - 1 + z\right| \leq |z|^2 \left(e^{-\operatorname{Re} z} + |z| + C\right) \tag{6.41}$$

for every $z \in \mathbb{C}$. Indeed, it follows from

$$\lim_{z \to 0} \frac{e^{-z} - 1 + z}{z^2} = \frac{1}{2}$$

that there exists $C \geq 1$ such that

$$\left|e^{-z} - 1 + z\right| \leq C|z|^2 \leq |z|^2 \left(e^{-\operatorname{Re} z} + |z| + C\right)$$

for $|z| \leq 1$. On the other hand, for $|z| \geq 1$, we have

$$\left|e^{-z} - 1 + z\right| \leq e^{-\operatorname{Re} z} + |z| + 1 \leq |z|^2 \left(e^{-\operatorname{Re} z} + |z| + C\right).$$

Given $\epsilon > 0$, it follows from (6.41) that

$$\left|\frac{e^{-ht} - 1}{h} + t\right| \leq |h|t^2 \left(e^{-(\operatorname{Re} h)t} + |ht| + C\right)$$

$$\leq |h|t^2 \left(e^{\epsilon t} + \epsilon t + C\right)$$

for $|h| < \epsilon$. Hence,

$$\left|\frac{F_n(z+h) - F_n(z)}{h} + \int_0^n ze^{-tz} f(t)\, dt\right| = \left|\int_0^n \left(\frac{e^{-th} - 1}{h} + t\right) e^{-tz} f(t)\, dt\right|$$

$$\leq |h| \int_0^n t^2 \left(e^{\epsilon t} + \epsilon t + C\right)\left|e^{-tz} f(t)\right| dt \tag{6.42}$$

for $|h| < \epsilon$. Letting $h \to 0$ we obtain

$$F_n'(z) = -\int_0^n te^{-tz} f(t)\, dt,$$

and in particular the function F_n is holomorphic. It then follows from Theorem 4.20 that F is holomorphic for $\operatorname{Re} z > a$, for each $a > c$, and hence, also for $\operatorname{Re} z > c$. Moreover, letting $n \to \infty$ and then $h \to 0$ in (6.42), since ϵ is arbitrary, we obtain identity (6.40) for $\operatorname{Re} z > c$. $\qquad\square$

Now we show that under sufficiently general conditions, it is possible to recover the function originating a given Laplace transform. In other words, there exists the inverse of the Laplace transform.

Theorem 6.27

If $f \in \mathcal{F}_c$ is of class C^1, then for each $t \in \mathbb{R}^+$ we have

$$f(t) = \lim_{R \to \infty} \frac{1}{2\pi i} \int_{\gamma_R} e^{tz} (Lf)(z) \, dz, \qquad (6.43)$$

where the path $\gamma_R \colon [-R, R] \to \mathbb{C}$ is given by $\gamma_R(y) = x + iy$ for any $x > c$.

More generally, we have the following result.

Theorem 6.28

Given $f \in \mathcal{F}_c$ and $t \in \mathbb{R}^+$, if there exist $C, \delta > 0$ such that

$$\bigl| f(u) - f(t) \bigr| \leq C|u - t| \qquad (6.44)$$

for $u \in (t - \delta, t + \delta) \cap \mathbb{R}^+$, then identity (6.43) holds.

Proof

For each $x > c$, we have

$$(Lf)(x + iy) = \int_0^\infty e^{-u(x+iy)} f(u) \, du.$$

Since $f \in \mathcal{F}_c$, we also have

$$\bigl| e^{(t-u)(x+iy)} f(u) \bigr| \leq e^{(t-u)x} D e^{cu} = D e^{tx} e^{(c-x)u}.$$

Hence,

$$\int_{-R}^R \int_0^\infty \bigl| e^{(t-u)(x+iy)} f(u) \bigr| \, du \, dy < \infty$$

for each $x > c$ and $R > 0$. By Fubini's theorem, we then obtain

$$\int_{-R}^R e^{t(x+iy)} (Lf)(x + iy) \, dy = \int_{-R}^R \int_0^\infty e^{(t-u)(x+iy)} f(u) \, du \, dy$$

$$= e^{tx} \int_0^\infty e^{-ux} f(u) \int_{-R}^R e^{i(t-u)y} \, dy \, du$$

$$= e^{tx} \int_0^\infty e^{-ux} f(u) \frac{e^{i(t-u)R} - e^{-i(t-u)R}}{i(t-u)} \, du$$

$$= 2e^{tx} \int_0^\infty e^{-ux} f(u) \frac{\sin[(t-u)R]}{t-u} \, du. \qquad (6.45)$$

Now we establish an auxiliary result, which is of independent interest. A function $h\colon I \to \mathbb{R}$ in a bounded or unbounded interval $I \subset \mathbb{R}$ is called *absolutely Riemann-integrable* (in I) if h is Riemann-integrable in I and $\int_I |h| < \infty$. We note that in bounded intervals any Riemann-integrable function is also absolutely Riemann-integrable.

Lemma 6.29 (Riemann–Lebesgue lemma)

If $h\colon I \to \mathbb{R}$ is an absolutely Riemann-integrable function in an interval I, then

$$\lim_{t \to +\infty} \int_I h(u) \cos(tu) \, du = \lim_{t \to +\infty} \int_I h(u) \sin(tu) \, du = 0. \qquad (6.46)$$

Proof of the lemma

We only consider the sine function, since the argument for the cosine is entirely analogous. We first assume that h is a step function in a bounded interval, that is,

$$h = \sum_{j=1}^N c_j \chi_{I_j}$$

for some constants $c_j \in \mathbb{R}$, where the intervals I_j form a partition of the interval I, and where

$$\chi_{I_j}(u) = \begin{cases} 1 & \text{if } u \in I_j, \\ 0 & \text{if } u \notin I_j. \end{cases}$$

Setting $I_j = [a_j, b_j]$, we obtain

$$\int_I h(u) \sin(tu) \, du = \sum_{j=1}^N c_j \int_{a_j}^{b_j} \sin(tu) \, du$$

$$= \sum_{j=1}^N c_j \frac{\cos(ta_j) - \cos(tb_j)}{t} \to 0 \qquad (6.47)$$

when $t \to +\infty$. Now let h be an absolutely Riemann-integrable function in a bounded interval I. Given $\delta > 0$, there exists a step function $g\colon I \to \mathbb{R}$ such that

$$\int_I |h - g| < \delta.$$

We have

$$\left| \int_I h(u) \sin(tu)\, du \right| \leq \left| \int_I [h(u) - g(u)] \sin(tu)\, du \right| + \left| \int_I g(u) \sin(tu)\, du \right|$$

$$\leq \int_I |h - g| + \left| \int_I g(u) \sin(tu)\, du \right|.$$

Since g is a step function, it follows from (6.47) that

$$\lim_{t \to +\infty} \int_I g(u) \sin(tu)\, du = 0,$$

and thus,

$$\left| \int_I h(u) \sin(tu)\, du \right| < 2\delta$$

for any is sufficiently large t. Since δ is arbitrary, we conclude that

$$\lim_{t \to +\infty} \int_I h(u) \sin(tu)\, du = 0. \tag{6.48}$$

Now we assume that I is unbounded. Since h is absolutely Riemann-integrable, given $\delta > 0$, there exists a bounded interval $J \subset I$ such that $\int_{I \setminus J} |h| < \delta$. Therefore,

$$\left| \int_I h(u) \sin(tu)\, du \right| \leq \left| \int_J h(u) \sin(tu)\, du \right| + \int_{I \setminus J} |h|$$

$$\leq \left| \int_J h(u) \sin(tu)\, du \right| + \delta.$$

On the other hand, it follows from (6.48) with I replaced by J that

$$\lim_{t \to +\infty} \int_J h(u) \sin(tu)\, du = 0,$$

and since δ is arbitrary, we conclude that identity (6.48) holds. $\qquad\square$

Now we divide the last integral in (6.45) into three integrals, namely over the intervals $[0, t - \delta]$, $[t - \delta, t + \delta]$ and $[t + \delta, +\infty)$. We observe that the function

$$h(u) = \frac{e^{-ux} f(u)}{t - u}$$

is absolutely Riemann-integrable in $[0, t - \delta]$ and $[t + \delta, +\infty)$, because it is continuous and

$$|h(u)| \leq \frac{De^{(c-x)u}}{|t - u|}.$$

Moreover, since

$$\sin\big[(t - u)R\big] = \sin(tR)\cos(uR) - \cos(tR)\sin(uR),$$

it follows from (6.46) that

$$\lim_{R \to \infty} \int_0^{t-\delta} h(u)\sin\big[(t - u)R\big]\, du = \lim_{R \to \infty} \int_{t+\delta}^{\infty} h(u)\sin\big[(t - u)R\big]\, du = 0. \quad (6.49)$$

On the other hand, by (6.44), we have

$$\big|e^{-ux}f(u) - e^{-tx}f(t)\big| \leq e^{-ux}\big|f(u) - f(t)\big| + \big|e^{-ux} - e^{-tx}\big| \cdot \big|f(t)\big|$$
$$\leq e^{(-t+\delta)x}C|u - t| + xe^{(-t+\delta)x}\big|f(t)\big| \cdot |u - t|$$
$$\leq \big(C + x|f(t)|\big)|u - t|$$

for each $u \in (t - \delta, t + \delta)$, with $t - \delta \geq 0$. Writing $K = C + x|f(t)|$, we obtain

$$\left|\int_{t-\delta}^{t+\delta} \big[e^{-ux}f(u) - e^{-tx}f(t)\big] \frac{\sin[(t - u)R]}{t - u}\, du\right|$$
$$\leq K \int_{t-\delta}^{t+\delta} \big|\sin\big[(t - u)R\big]\big|\, du \leq 2K\delta.$$

Hence, making the change of variables $v = (t - u)R$ yields

$$\left|\int_{t-\delta}^{t+\delta} h(u)\sin\big[(t - u)R\big]\, du - e^{-tx}f(t)\int_{-\delta R}^{\delta R} \frac{\sin v}{v}\, dv\right| \leq 2K\delta.$$

By Example 4.51, we have

$$\int_{-\infty}^{\infty} \frac{\sin x}{x}\, dx = \pi,$$

and thus,

$$\left|\int_{t-\delta}^{t+\delta} h(u)\sin\big[(t - u)R\big]\, du - e^{-tx}f(t)\pi\right| \leq 3K\delta \qquad (6.50)$$

for any sufficiently large R. On the other hand, given $\delta > 0$, it follows from (6.45) and (6.49) that

$$\left| \int_{-R}^{R} e^{t(x+iy)}(Lf)(x+iy)\,dy - 2e^{tx} \int_{t-\delta}^{t+\delta} h(u)\sin\big[(t-u)R\big]\,du \right|$$

$$= 2e^{tx} \left| \int_{0}^{t-\delta} h(u)\sin\big[(t-u)R\big]\,du + \int_{t+\delta}^{\infty} h(u)\sin\big[(t-u)R\big]\,du \right| < \delta$$

for any sufficiently large R. By (6.50), we obtain

$$\left| \int_{-R}^{R} e^{t(x+iy)}(Lf)(x+iy)\,dy - 2\pi f(t) \right| \leq \delta + 6K\delta e^{tx},$$

again for any sufficiently large R. Letting $R \to +\infty$ and then $\delta \to 0$, we conclude that

$$\int_{-\infty}^{\infty} e^{t(x+iy)}(Lf)(x+iy)\,dy = 2\pi f(t),$$

which is equivalent to

$$\lim_{R\to\infty} \int_{-R}^{R} e^{t\gamma_R(y)} F\big(\gamma_R(y)\big)\gamma_R'(y)\,dy = 2\pi i f(t).$$

This yields the desired statement. $\qquad\square$

Sometimes it is possible to use the Residue theorem to recover a function f from its Laplace transform. We establish a particular result which is sufficient for the applications to differential equations of the form (6.34).

Theorem 6.30

Given a function $f \in \mathcal{F}_c$ of class C^1, we assume that:
1. Lf is meromorphic and has a finite number of poles;
2. there exist constants $K, b, R > 0$ such that

$$\big|(Lf)(z)\big| \leq K/|z|^b \quad \text{for } |z| > R.$$

Then, for each $t \in \mathbb{R}^+$, we have

$$f(t) = \sum_{p\in A} \text{Res}(G_t, p), \tag{6.51}$$

where $G_t(z) = e^{tz}(Lf)(z)$, and where A is the set of poles of Lf.

Proof

By Theorem 6.27, there exists the inverse of the Laplace transform. It is thus sufficient to show that if $g\colon \mathbb{R}_0^+ \to \mathbb{R}$ is the function given by the right-hand side of (6.51), then $Lg = Lf$ for $\operatorname{Re} z > c$.

Given $a > c$, let γ_r^+ and γ_r^- be closed piecewise regular paths looping once in the positive direction respectively on the boundaries of the sets

$$B_r(0) \cap \{z \in \mathbb{C} : \operatorname{Re} z > a\}$$

and

$$B_r(0) \cap \{z \in \mathbb{C} : \operatorname{Re} z < a\}$$

(see Figure 6.1). By Theorem 4.48, for any sufficiently large r, we have

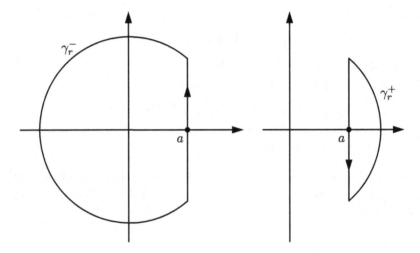

Figure 6.1 Paths γ_r^+ and γ_r^-

$$g(t) = \sum_{p \in A} \operatorname{Res}(G_t, p) = \frac{1}{2\pi i} \int_{\gamma_r^-} e^{tw} F(w) \, dw,$$

where $F = Lf$. On the other hand, by Proposition 3.34, for $\operatorname{Re} z > a$ we have

$$(Lg)(z) = \frac{1}{2\pi i} \lim_{R \to \infty} \int_0^R e^{-tz} \int_{\gamma_r^-} e^{tw} F(w) \, dw \, dt$$

$$= \frac{1}{2\pi i} \lim_{R \to \infty} \int_{\gamma_r^-} F(w) \int_0^R e^{t(w-z)} \, dt \, dw$$

$$= \frac{1}{2\pi i} \lim_{R \to \infty} \int_{\gamma_r^-} F(w) \frac{e^{R(w-z)} - 1}{w - z} \, dw$$

$$= -\frac{1}{2\pi i} \int_{\gamma_r^-} \frac{F(w)}{w - z} \, dw,$$

since

$$\left| e^{R(w-z)} \right| = e^{R(\operatorname{Re} w - \operatorname{Re} z)} \leq e^{R(a - \operatorname{Re} z)} \to 0$$

when $R \to +\infty$. It then follows from Theorem 2.60 that

$$(Lg)(z) = -\frac{1}{2\pi i} \left(\int_{\gamma_r^-} \frac{F(w)}{w - z} \, dw + \int_{\gamma_r^+} \frac{F(w)}{w - z} \, dw \right) + F(z)$$

$$= -\frac{1}{2\pi i} \int_{\gamma_r} \frac{F(w)}{w - z} \, dw + F(z), \tag{6.52}$$

where the path $\gamma_r \colon [0, 2\pi] \to \mathbb{C}$ is given by $\gamma_r(\theta) = re^{i\theta}$. Indeed, the integrals along the segments of γ_r^- and γ_r^+ in the straight line $\operatorname{Re} z = a$ cancel out. On the other hand,

$$\left| \int_{\gamma_r} \frac{F(w)}{w - z} \, dw \right| \leq \frac{2\pi r K}{r^b (r - |z|)} \to 0$$

when $r \to +\infty$, and thus, it follows from (6.52) that $Lg = Lf$ for $\operatorname{Re} z > a$. Letting $a \to c$ we obtain the desired result.   □

Example 6.31

Let

$$F(z) = \frac{1}{(z - 2)(z - 4)(z - 5)}$$

be the Laplace transform of a function f of class C^1. By Theorem 6.30, we have

$$f(t) = \sum_{p=1}^{3} \operatorname{Res}(G_t, p)$$

$$= \frac{e^{tz}}{(z - 4)(z - 5)} \bigg|_{z=2} + \frac{e^{tz}}{(z - 2)(z - 5)} \bigg|_{z=4} + \frac{e^{tz}}{(z - 2)(z - 4)} \bigg|_{z=5}$$

$$= \frac{1}{6} e^{2t} - \frac{1}{2} e^{4t} + \frac{1}{3} e^{5t}.$$

Now we start studying the relation between the Laplace transform and differential equations of the form (6.34).

Proposition 6.32

If $f' \in \mathcal{F}_c$, then

$$L(f')(z) = z(Lf)(z) - f(0^+), \quad \operatorname{Re} z > c. \qquad (6.53)$$

Proof

Since $f' \in \mathcal{F}_c$, we have

$$\left| f(t) \right| \leq \left| f(0^+) + \int_0^t f(u) \, du \right|$$

$$\leq \left| f(0^+) \right| + \int_0^t D e^{cu} \, du$$

$$= \left| f(0^+) \right| + \frac{D}{c}(e^{ct} - 1) \leq \left(\left| f(0^+) \right| + \frac{D}{c} \right) e^{ct},$$

and thus $f \in \mathcal{F}_c$. We then obtain

$$L(f')(z) = \lim_{R \to \infty} \int_0^R e^{-tz} f'(t) \, dt$$

$$= \lim_{R \to \infty} \left(e^{-tz} f(t) \Big|_{t=0}^{t=R} + z \int_0^R e^{-tz} f(t) \, dt \right)$$

$$= \lim_{R \to \infty} \left(e^{-Rz} f(R) - f(0^+) \right) + z \lim_{R \to \infty} \int_0^R e^{-tz} f(t) \, dt$$

$$= -f(0^+) + z(Lf)(z)$$

for $\operatorname{Re} z > c$. This yields the desired identity. $\qquad \square$

Example 6.33

Now we consider the second derivative. If $f'' \in \mathcal{F}_c$, then one can use Proposition 6.32 twice to obtain

$$L(f'')(z) = z^2(Lf)(z) - zf(0^+) - f'(0^+), \quad \operatorname{Re} z > c. \qquad (6.54)$$

Indeed, taking $g = f'$ we obtain

$$L(f'')(z) = L(g')(z) = zL(g)(z) - g(0^+)$$
$$= zL(f')(z) - f'(0^+)$$
$$= z[z(Lf)(z) - f(0^+)] - f'(0^+)$$
$$= z^2(Lf)(z) - zf(0^+) - f'(0^+).$$

Example 6.34

Let us consider the equation

$$x'' - 6x' + 8x = e^{5t}$$

with the conditions $x(0) = 0$ and $x'(0) = 0$. It follows from (6.53) and (6.54) that if $X = Lx$, then

$$L(x')(z) = zX(z) - x(0) = zX(z),$$

and

$$L(x'')(z) = z^2 X(z) - zx(0) - x'(0) = z^2 X(z).$$

Therefore,

$$L(x'' - 6x' + 8x)(z) = z^2 X(z) - 6zX(z) + 8X(z)$$
$$= (z^2 - 6z + 8) X(z).$$

By Example 6.24, we have

$$L(e^{5t})(z) = \frac{1}{z - 5},$$

and hence,

$$X(z) = \frac{1}{(z - 5)(z^2 - 6z + 8)}$$
$$= \frac{1}{(z - 2)(z - 4)(z - 5)}.$$

Since the solution $x(t)$ is of class C^1, it follows from Example 6.31 that

$$x(t) = \frac{1}{6}e^{2t} - \frac{1}{2}e^{4t} + \frac{1}{3}e^{5t}.$$

Alternatively, one could note that

$$X(z) = \frac{1}{6}\frac{1}{z-2} - \frac{1}{2}\frac{1}{z-4} + \frac{1}{3}\frac{1}{z-5}$$

$$= L\left(\frac{1}{6}e^{2t} - \frac{1}{2}e^{4t} + \frac{1}{3}e^{5t}\right)(z).$$

We also describe some additional properties that are useful in computing the Laplace transform.

Proposition 6.35

If $f \in \mathcal{F}_c$, then:

1. for the function $g(t) = e^{at}f(t)$, with $a \in \mathbb{R}$, we have

$$(Lg)(z) = (Lf)(z-a), \quad \mathrm{Re}\, z > a + c;$$

2. for the function $g(t) = -tf(t)$, we have

$$(Lg)(z) = (Lf)'(z), \quad \mathrm{Re}\, z > c.$$

Proof

In the first property, we have $g \in \mathcal{F}_{a+c}$ and

$$(Lg)(z) = \int_0^\infty e^{-tz}e^{at}f(t)\,dt$$

$$= \int_0^\infty e^{-t(z-a)}f(t)\,dt = (Lf)(z-a).$$

The second property follows immediately from (6.40). $\qquad\square$

Example 6.36

For $f(t) = e^{3t}$ and $g(t) = -te^{3t}$, we have

$$(Lg)(z) = (Lf)'(z) = \left(\frac{1}{z-3}\right)' = -\frac{1}{(z-3)^2}.$$

For $f(t) = \sin t$ and $g(t) = -t\sin t$, we have

$$(Lg)(z) = (Lf)'(z) = \left(\frac{1}{z^2+1}\right)' = -\frac{2z}{(z^2+1)^2}.$$

Example 6.37

For $f(t) = \cos(7t)$ and $g(t) = e^{5t}\cos(7t)$, we have

$$(Lg)(z) = (Lf)(z-5) = \frac{z-5}{(z-5)^2 + 49}.$$

Example 6.38

Now we consider the equation

$$x'' + x = \cos t \tag{6.55}$$

with the conditions $x(0) = 0$ and $x'(0) = 1$. Taking $X = Lx$, we obtain

$$L(x'')(z) = z^2 X(z) - zx(0) - x'(0) = z^2 X(z) - 1$$

and

$$L(x'' + x)(z) = (z^2 + 1)X(z) - 1.$$

It follows from (6.55) that

$$(z^2 + 1)X(z) - 1 = \frac{z}{z^2 + 1},$$

and hence,

$$X(z) = \frac{1}{z^2 + 1} + \frac{z}{(z^2 + 1)^2}$$

$$= \frac{1}{z^2 + 1} - \frac{1}{2}\left(\frac{1}{z^2 + 1}\right)' = L\left(\sin t + \frac{1}{2}t\sin t\right)(z).$$

Thus, the solution of the problem is

$$x(t) = \sin t + \frac{1}{2}t\sin t.$$

6.5 Solved Problems and Exercises

Problem 6.1

Find a differentiable function $\Phi = \Phi(t, x)$ such that

$$\frac{\partial \Phi}{\partial t} = 3x^2 + (t+1)e^t, \quad \text{and} \quad \frac{\partial \Phi}{\partial x} = 6xt.$$

Solution

It follows from the second equation that

$$\Phi(t,x) = 3x^2 t + C(t)$$

for some differentiable function C. Hence,

$$\frac{\partial \Phi}{\partial t} = 3x^2 + C'(t) = 3x^2 + (t+1)e^t,$$

and we obtain

$$C'(t) = (t+1)e^t.$$

Thus, one can take $C(t) = te^t$, and hence, $\Phi(t,x) = 3x^2 t + te^t$.

Problem 6.2

Find the solution of the equation

$$x' = (1+x^2)\cos t$$

with $x(0) = 0$.

Solution

We write the equation in the form

$$\frac{x'}{1+x^2} = \cos t.$$

Since $(\tan^{-1} x)' = x'/(1+x^2)$, integrating on both sides we obtain

$$\tan^{-1} x(t) = \sin t + c$$

for some constant $c \in \mathbb{R}$. Thus, the solutions of the equation are given by

$$x(t) = \tan(\sin t + c) \quad \text{for } t \in \mathbb{R},$$

with $c \in \mathbb{R}$. Taking $t = 0$, we obtain $x(0) = \tan c = 0$. Hence, $c = 0$, and the solution of the problem is $x(t) = \tan \sin t$ for $t \in \mathbb{R}$.

Problem 6.3

Verify that if

$$\Phi(t,x) = \int_{t_0}^{t} g(s)\,ds - \int_{x_0}^{x} f(y)\,dy,$$

where f and g are continuous functions, and $x(t)$ is a solution of the equation $x' = g(t)/f(x)$, then

$$\frac{d}{dt}\Phi\big(t,x(t)\big) = 0.$$

Solution

Using the chain rule, we obtain

$$\frac{d}{dt}\Phi\big(t,x(t)\big) = g(t) - f\big(x(t)\big)x'(t)$$

$$= g(t) - f\big(x(t)\big)g(t)/f\big(x(t)\big) = 0,$$

since $x(t)$ is a solution of the equation $x' = g(t)/f(x)$.

Problem 6.4

Find all solutions of the equation

$$-2t + \big(x^3 + 2x\big)x' = 0. \tag{6.56}$$

Solution

By Theorem 6.5, since the functions $-2t$ and $x^3 + 2x$ are of class C^1, and

$$\frac{\partial}{\partial x}(-2t) = \frac{\partial}{\partial t}\left(x^3 + 2x\right) = 0,$$

the equation is exact. Hence, there exists a differentiable function $\Phi = \Phi(t,x)$ such that

$$\frac{\partial \Phi}{\partial t} = -2t \quad \text{and} \quad \frac{\partial \Phi}{\partial x} = x^3 + 2x.$$

By the first equation, we have $\Phi(t,x) = -t^2 + C(x)$ for some differentiable function C, and thus,

$$\frac{\partial \Phi}{\partial x} = C'(x) = x^3 + 2x.$$

Hence, one can take $C(x) = x^4/4 + x^2$, and

$$\Phi(t, x) = -t^2 + \frac{1}{4}x^4 + x^2.$$

Therefore, by Proposition 6.3, each solution of equation (6.56) satisfies

$$-t^2 + \frac{1}{4}x^4 + x^2 = c$$

for some constant $c \in \mathbb{R}$. Letting $x^2 = y$, we obtain

$$y^2 + 4y - 4(t^2 + c) = 0,$$

which yields

$$y = \frac{-4 \pm \sqrt{16 + 16(t^2 + c)}}{2} = -2 \pm 2\sqrt{1 + c + t^2}.$$

Since $y = x^2$, one must take the $+$ sign, and thus,

$$x(t) = \pm\sqrt{-2 + 2\sqrt{1 + c + t^2}}.$$

Each solution is defined whenever

$$1 + c + t^2 > 0 \quad \text{and} \quad \sqrt{1 + c + t^2} > 1,$$

and thus, when $t^2 > -c$. Hence, the maximal interval is $\mathbb{R}$ for $c > 0$, it is $\mathbb{R}^+$ or $\mathbb{R}^-$ for $c = 0$, and, finally, it is $(-\infty, -\sqrt{-c})$ or $(\sqrt{-c}, +\infty)$ for $c < 0$.

Problem 6.5

Verify that the equation

$$-2tx + (x^4 + 2x^3)x' = 0 \tag{6.57}$$

is reducible to exact and find an integrating factor.

Solution

Let

$$M(t, x) = -2tx \quad \text{and} \quad N(t, x) = x^4 + 2x^3.$$

Since

$$\frac{\partial M}{\partial x} - \frac{\partial N}{\partial t} = -2t - 0 = -2t \neq 0,$$

the equation is not exact. However, by Proposition 6.9, since

$$-\frac{1}{M}\left(\frac{\partial M}{\partial x} - \frac{\partial N}{\partial t}\right) = \frac{1}{x}$$

does not depend on t, there exists an integrating factor $\mu = \mu(x)$. It satisfies the equation

$$\mu' = -\frac{1}{M}\left(\frac{\partial M}{\partial x} - \frac{\partial N}{\partial t}\right)\mu, \tag{6.58}$$

that is, $\mu' = -(1/x)\mu$. Since

$$\left(\log|\mu|\right)' = \frac{\mu'}{\mu} = -\frac{1}{x},$$

we obtain

$$\log|\mu(x)| = -\log|x| + c$$

for some constant $c \in \mathbb{R}$. Hence, a solution of equation (6.58) is for example $\mu(x) = 1/x$ (this corresponds to taking $c = 0$). Multiplying equation (6.57) by $\mu(x)$, we obtain the exact equation

$$-2t + \left(x^3 + 2x^2\right)x' = 0.$$

Problem 6.6

Solve the equation

$$x'' - 6x' + 9x = 0. \tag{6.59}$$

Solution

Since $\lambda^2 - 6\lambda + 9 = (\lambda - 3)^2$, equation (6.59) can be written in the form $(D - 3)^2 x = 0$. By Proposition 6.13, the solutions of equation (6.59) are given by

$$x(t) = c_1 e^{3t} + c_2 t e^{3t} \quad \text{for } t \in \mathbb{R},$$

with $c_1, c_2 \in \mathbb{R}$.

Problem 6.7

Solve the equation

$$x'' + 2x = 0. \tag{6.60}$$

Solution

We write the equation in the form

$$\left(D^2 + 2\right)x = \left(D - i\sqrt{2}\right)\left(D + i\sqrt{2}\right)x = 0.$$

By Proposition 6.16, the solutions of equation (6.60) are given by

$$x(t) = c_1 \cos(\sqrt{2}t) + c_2 \sin(\sqrt{2}t) \quad \text{for } t \in \mathbb{R},$$

with $c_1, c_2 \in \mathbb{R}$.

Problem 6.8

Solve the equation

$$(D + 1)\left(D^2 + 1\right)x = 0. \tag{6.61}$$

Solution

We first consider separately the equations

$$(D + 1)x = 0 \quad \text{and} \quad \left(D^2 + 1\right)x = 0,$$

which have spaces of solutions generated respectively by e^{-t} and by $\cos t$ and $\sin t$. Proceeding as in Example 6.18, we find that the space of solutions of equation (6.61) is generated by these three functions, and thus the solutions are given by

$$x(t) = c_1 e^{-t} + c_2 \cos t + c_3 \sin t \quad \text{for } t \in \mathbb{R},$$

with $c_1, c_2, c_3 \in \mathbb{R}$.

Problem 6.9

Solve the equation

$$x'' + 4x = e^t. \tag{6.62}$$

Solution

We write equation (6.62) in the form

$$\left(D^2 + 4\right)x = e^t. \tag{6.63}$$

Since

$$(D-1)e^t = De^t - e^t = 0,$$

it follows from (6.63) that

$$(D-1)(D^2+4)x = 0,$$

that is,

$$(D-1)(D-2i)(D+2i)x = 0.$$

In an analogous manner to that in Problem 6.8, the solutions of this equation are given by

$$x(t) = c_1 e^t + c_2 \cos(2t) + c_3 \sin(2t) \quad \text{for } t \in \mathbb{R},$$

with $c_1, c_2, c_3 \in \mathbb{R}$. However, not all of these functions are necessarily solutions of equation (6.62). Substituting $x(t)$ in (6.62), we obtain

$$(D^2+4)x = (D^2+4)(c_1 e^t) = e^t,$$

since $c_2 \cos(2t) + c_3 \sin(2t)$ is a solution of the equation $(D^2+4)x = 0$. Hence,

$$(D^2+4)(c_1 e^t) = c_1 e^t + 4c_1 e^t = e^t,$$

and thus, $5c_1 = 1$, that is, $c_1 = 1/5$. Therefore, the solutions of equation (6.62) are given by

$$x(t) = \frac{1}{5}e^t + c_2 \cos(2t) + c_3 \sin(2t) \quad \text{for } t \in \mathbb{R},$$

with $c_2, c_3 \in \mathbb{R}$.

Problem 6.10

Solve the equation

$$x'' - 9x = \cos t. \tag{6.64}$$

Solution

We write equation (6.64) in the form $(D^2-9)x = \cos t$, that is,

$$(D-3)(D+3)x = \cos t. \tag{6.65}$$

Now we observe that

$$(D-i)(D+i)\cos t = (D^2+1)\cos t = -\cos t + \cos t = 0,$$

and thus, it follows from (6.65) that

$$(D-i)(D+i)(D-3)(D+3)x = 0.$$

Proceeding as in Problem 6.8, we find that the solutions of this equation are given by

$$x(t) = c_1 e^{3t} + c_2 e^{-3t} + c_3 \cos t + c_4 \sin t \quad \text{for } t \in \mathbb{R},$$

with $c_1, c_2, c_3, c_4 \in \mathbb{R}$. Substituting $x(t)$ in (6.64), we obtain

$$(D^2-9)x = (D^2-9)(c_3 \cos t + c_4 \sin t),$$

since $c_1 e^{3t} + c_2 e^{-3t}$ is a solution of the equation $(D^2-9)x = 0$. Hence,

$$(D^2-9)(c_3 \cos t + c_4 \sin t) = -10c_3 \cos t - 10c_4 \sin t = \cos t,$$

and thus, $c_3 = -1/10$ and $c_4 = 0$. Therefore, the solutions of equation (6.64) are given by

$$x(t) = c_1 e^{3t} + c_2 e^{-3t} - \frac{1}{10}\cos t \quad \text{for } t \in \mathbb{R},$$

with $c_1, c_2 \in \mathbb{R}$.

Problem 6.11

Find all nonconstant periodic solutions of the equation

$$x'' - 5x' + 6 = 0. \tag{6.66}$$

Solution

We write the equation in the form $(D^2 - 5D)x = -6$. Since $D(-6) = 0$, we obtain

$$D(D^2-5D)x = 0,$$

or equivalently, $D^2(D-5)x = 0$. The solutions of this equation are given by

$$x(t) = c_1 + c_2 t + c_3 e^{5t} \quad \text{for } t \in \mathbb{R},$$

with $c_1, c_2, c_3 \in \mathbb{R}$. Substituting $x(t)$ in (6.66), we obtain

$$\left(D^2 - 5D\right)x = 25c_3 e^{5t} - 5c_2 - 25c_3 e^{5t} = -5c_2 = -6,$$

and thus, $c_2 = 6/5$. Therefore, the solutions of equation (6.66) are given by

$$x(t) = c_1 + \frac{6}{5}t + c_3 e^{5t} \quad \text{for } t \in \mathbb{R},$$

with $c_1, c_3 \in \mathbb{R}$. In particular, all periodic solutions are constant.

Problem 6.12

Consider the equation

$$(D - 1)\left(D^2 - 1\right)x = h(t). \tag{6.67}$$

1. Find all solutions for $h(t) = 0$.
2. Find all solutions for $h(t) = e^{-t}$ such that $x(0) = x'(0) = 0$.
3. Find all solutions for $h(t) = 0$ such that $x(t) = x(-t)$ for every $t \in \mathbb{R}$.

Solution

1. We have

   $$(D - 1)\left(D^2 - 1\right)x = (D - 1)^2(D + 1)x,$$

   and thus, the solutions of equation (6.67) for $h(t) = 0$ are given by

   $$x(t) = ae^t + bte^t + ce^{-t} \quad \text{for } t \in \mathbb{R}, \tag{6.68}$$

   with $a, b, c \in \mathbb{R}$.
2. Applying $D + 1$ to the equation

   $$(D - 1)^2(D + 1)x = e^{-t} \tag{6.69}$$

   yields

   $$(D - 1)^2(D + 1)^2 x = 0,$$

   which has the solutions

   $$x(t) = ae^t + bte^t + ce^{-t} + dte^{-t} \quad \text{for } t \in \mathbb{R},$$

   with $a, b, c, d \in \mathbb{R}$. Substituting $x(t)$ in (6.69), we obtain

   $$(D - 1)^2(D + 1)\left(dte^{-t}\right) = e^{-t},$$

that is, $4de^{-t} = e^{-t}$. Hence, $d = 1/4$. Moreover, since $x(0) = x'(0) = 0$, we must have

$$a + c = 0 \quad \text{and} \quad a + b - c + \frac{1}{4} = 0.$$

We thus obtain $b = -2a - 1/4$ and $c = -a$, and the solutions of equation (6.67) for $h(t) = e^{-t}$ are given by

$$x(t) = ae^t - \left(2a + \frac{1}{4}\right)te^t - ae^{-t} + \frac{1}{4}te^{-t} \quad \text{for } t \in \mathbb{R},$$

with $a \in \mathbb{R}$.

3. By (6.68), the condition $x(t) = x(-t)$ is equivalent to

$$ae^t + bte^t + ce^{-t} = ae^{-t} - bte^{-t} + ce^t,$$

which yields

$$(a - c)e^t + bte^t - bte^{-t} + (c - a)e^{-t} = 0.$$

Hence, $a = c$ and $b = 0$. The desired solutions are thus

$$x(t) = a\left(e^t + e^{-t}\right)$$

for $t \in \mathbb{R}$, with $a \in \mathbb{R}$.

Problem 6.13

Consider the equation

$$\left(D^2 + 1\right)(D + 1)x = h(t). \tag{6.70}$$

1. Find all solutions for $h(t) = 0$.
2. Find all solutions for $h(t) = 0$ having a limit when $t \to +\infty$.
3. Find all solutions for $h(t) = t$ such that $x'(0) = 0$.

Solution

1. The roots of the polynomial $(\lambda^2 + 1)(\lambda + 1)$ are i, $-i$ and -1, and thus, the solutions of equation (6.70) for $h(t) = 0$ are given by

$$x(t) = a\cos t + b\sin t + ce^{-t} \quad \text{for } t \in \mathbb{R}, \tag{6.71}$$

with $a, b, c \in \mathbb{R}$.

2. Since the component ce^{-t} in (6.71) has a limit when $t \to +\infty$, in order to determine whether $x(t)$ has a limit when $t \to +\infty$, it remains to study the component $a \cos t + b \sin t$. Since this is a nonconstant periodic function for $a \neq 0$ or $b \neq 0$, we must take $a = b = 0$. Hence, the solutions of the equation for $h(t) = 0$ having a limit when $t \to +\infty$ are the functions ce^{-t}, with $c \in \mathbb{R}$.

3. Applying D^2 to equation (6.70) for $h(t) = t$, we obtain

$$D^2(D^2 + 1)(D + 1)x = 0,$$

which has the solutions

$$x(t) = a \cos t + b \sin t + ce^{-t} + dt + e \quad \text{for } t \in \mathbb{R},$$

with $a, b, c, d, e \in \mathbb{R}$. Substituting $x(t)$ in equation (6.70) for $h(t) = t$, we obtain

$$(D^2 + 1)(D + 1)(dt + e) = (D^2 + 1)(d + dt + e)$$
$$= dt + (d + e) = t$$

and thus, $d = 1$ and $e = -1$. Hence,

$$x(t) = a \cos t + b \sin t + ce^{-t} + t - 1 \quad \text{for } t \in \mathbb{R},$$

with $a, b, c \in \mathbb{R}$. Since

$$x'(t) = -a \sin t + b \cos t - ce^{-t} + 1,$$

we have $x'(0) = b - c + 1$, and hence, $c = b + 1$. The desired solutions are thus

$$x(t) = a \cos t + b \sin t + (b + 1)e^{-t} + t - 1 \quad \text{for } t \in \mathbb{R},$$

with $a, b \in \mathbb{R}$.

Problem 6.14

Find the Laplace transform of the function

$$f(t) = \cos(3t) + e^t \sin(4t).$$

Solution

Let $f_1(t) = \cos(3t)$ and $f_2(t) = \sin(4t)$. It follows from (6.38) and (6.39) that

$$(Lf_1)(z) = \frac{z}{z^2 + 9} \quad \text{and} \quad (Lf_2)(z) = \frac{4}{z^2 + 16}$$

for $\operatorname{Re} z > 0$. Hence, for the function $f_3(t) = e^t \sin(4t)$, we have

$$\begin{aligned}
(Lf_3)(z) &= \int_0^\infty e^{-tz} e^t \sin(4t)\, dt \\
&= \int_0^\infty e^{-t(z-1)} \sin(4t)\, dt \\
&= (Lf_2)(z-1) = \frac{4}{(z-1)^2 + 16}
\end{aligned}$$

for $\operatorname{Re} z > 1$. We then obtain

$$\begin{aligned}
(Lf)(z) &= (Lf_1)(z) + (Lf_3)(z) \\
&= \frac{z}{z^2 + 9} + \frac{4}{(z-1)^2 + 16}
\end{aligned}$$

for $\operatorname{Re} z > 1$.

Problem 6.15

For each $a, b \in \mathbb{R}$, find the Laplace transform of the function

$$f(t) = te^{-t} + e^{at} \cos^2(bt).$$

Solution

Let $f_1(t) = e^{-t}$. By Example 6.24, we have $(Lf_1)(z) = 1/(z+1)$ for $\operatorname{Re} z > -1$. Hence, for the function $f_2(t) = te^{-t}$, it follows from Proposition 6.35 that

$$(Lf_2)(z) = -(Lf_1)'(z) = \frac{1}{(z+1)^2}$$

for $\operatorname{Re} z > -1$. On the other hand, since

$$\cos^2(bt) = \frac{1 + \cos(2bt)}{2},$$

by Example 6.22 and (6.38), for the function $f_3(t) = \cos^2(bt)$ we have

$$(Lf_3)(z) = \frac{1}{2z} + \frac{z}{2(z^2 + 4b^2)}$$

for $\operatorname{Re} z > 0$. Hence, for the function $f_4(t) = e^{at} \cos^2(bt)$, it follows from Proposition 6.35 that

$$(Lf_4)(z) = (Lf_3)(z-a)$$

$$= \frac{1}{2(z-a)} + \frac{z-a}{2[(z-a)^2 + 4b^2]}$$

for $\operatorname{Re} z > a$. We then obtain

$$(Lf)(z) = (Lf_2)(z) + (Lf_4)(z)$$

$$= \frac{1}{(z+1)^2} + \frac{1}{2(z-a)} + \frac{z-a}{2[(z-a)^2 + 4b^2]}$$

for $\operatorname{Re} z > \max\{-1, a\}$.

Problem 6.16

Find the Laplace transform of the function

$$f(t) = \begin{cases} t & \text{if } t \in [0, 1], \\ e^{2t} & \text{if } t > 1. \end{cases}$$

Solution

We have

$$(Lf)(z) = \int_0^1 t e^{-tz}\, dt + \int_1^\infty e^{2t} e^{-tz}\, dt$$

$$= -\frac{t}{z} e^{-tz}\Big|_{t=0}^{t=1} + \frac{1}{z}\int_0^1 e^{-tz}\, dt + \lim_{R\to\infty}\int_1^R e^{(2-z)t}\, dt$$

$$= -\frac{e^{-z}}{z} - \frac{1}{z^2} e^{-tz}\Big|_{t=0}^{t=1} + \lim_{R\to\infty}\frac{e^{(2-z)t}}{2-z}\Big|_{t=1}^{t=R}$$

$$= -\frac{e^{-z}}{z} - \frac{1}{z^2}(e^{-z} - 1) + \lim_{R\to\infty}\frac{e^{(2-z)R} - e^{2-z}}{z-2},$$

and thus,

$$(Lf)(z) = -\frac{e^{-z}}{z} - \frac{1}{z^2}(e^{-z} - 1) - \frac{e^{2-z}}{z-2}$$

for $\operatorname{Re} z > 2$.

Problem 6.17

Find the function f of class C^1 whose Laplace transform is

$$F(z) = \frac{1}{z(z-2)(z+4)}.$$

Solution

We note that the function F is meromorphic and has a finite number of poles. Moreover, given $R > 4$, for each $|z| > R$ we have

$$|z(z-2)(z+4)| \geq |z|(|z|-2)(|z|-4)$$
$$> |z|(R-2)(R-4),$$

and thus,

$$|F(z)| < \frac{1}{(R-2)(R-4)|z|}.$$

By Theorem 6.30, we obtain

$$f(t) = \sum_{p \in A} \text{Res}(G_t, p),$$

where $G_t(z) = e^{tz} F(z)$, and where A is the set of poles of F. Therefore,

$$f(t) = \text{Res}(G_t, 0) + \text{Res}(G_t, 2) + \text{Res}(G_t, -4)$$

$$= \frac{e^{tz}}{(z-2)(z+4)}\bigg|_{z=0} + \frac{e^{tz}}{z(z+4)}\bigg|_{z=2} + \frac{e^{tz}}{z(z-2)}\bigg|_{z=-4}$$

$$= -\frac{1}{8} + \frac{1}{12}e^{2t} + \frac{1}{24}e^{-4t}.$$

Problem 6.18

Use the Laplace transform to find the solution of the equation

$$x'' - 5x' + 6x = e^t \quad \text{with } x(0) = x'(0) = 0. \tag{6.72}$$

Solution

Letting $X = Lx$, we obtain

$$L(x')(z) = zX(z) - x(0) = zX(z)$$

and

$$L(x'')(z) = z^2 X(z) - zx(0) - x'(0) = z^2 X(z).$$

Hence,

$$L(x'' - 5x' + 6x)(z) = z^2 X(z) - 5zX(z) + 6X(z)$$
$$= (z^2 - 5z + 6) X(z).$$

By Example 6.24, it follows from (6.72) that

$$(z^2 - 5z + 6) X(z) = \frac{1}{z - 1},$$

that is,

$$X(z) = \frac{1}{(z - 1)(z^2 - 5z + 6)} = \frac{1}{(z - 1)(z - 2)(z - 3)}.$$

Proceeding as in Problem 6.17, we then obtain

$$x(t) = \frac{e^{tz}}{(z - 2)(z - 3)}\Big|_{z=1} + \frac{e^{tz}}{(z - 1)(z - 3)}\Big|_{z=2} + \frac{e^{tz}}{(z - 1)(z - 2)}\Big|_{z=3}$$
$$= \frac{1}{2}e^t - e^{2t} + \frac{1}{2}e^{3t}.$$

Alternatively, note that

$$X(z) = \frac{1}{2}\frac{1}{z - 1} - \frac{1}{z - 2} + \frac{1}{2}\frac{1}{z - 3}$$
$$= L\left(\frac{1}{2}e^t - e^{2t} + \frac{1}{2}e^{3t}\right)(z),$$

and thus again

$$x(t) = \frac{1}{2}e^t - e^{2t} + \frac{1}{2}e^{3t}.$$

Problem 6.19

Use the Laplace transform to find the solution of the equation

$$x'' - x = \sin t \quad \text{with } x(0) = 0 \text{ and } x'(0) = 1. \tag{6.73}$$

Solution

Letting $X = Lx$, we obtain

$$L(x'')(z) = z^2 X(z) - zx(0) - x'(0) = z^2 X(z) - 1,$$

and hence,

$$L(x'' - x)(z) = (z^2 - 1)X(z) - 1.$$

By (6.39), we have

$$L(\sin t)(z) = \frac{1}{z^2 + 1},$$

and thus, it follows from (6.73) that

$$(z^2 - 1)X(z) - 1 = \frac{1}{z^2 + 1}.$$

Hence,

$$X(z) = \frac{1}{z^2 - 1} + \frac{1}{(z^2 + 1)(z^2 - 1)}$$

$$= \frac{1}{z - 1} \cdot \frac{1}{z + 1} + \frac{1}{z - 1} \cdot \frac{1}{z + 1} \cdot \frac{1}{z - i} \cdot \frac{1}{z + i}.$$

Proceeding as in Problem 6.17, we then obtain

$$x(t) = \frac{e^{tz}}{z + 1}\bigg|_{z=1} + \frac{e^{tz}}{z - 1}\bigg|_{z=-1}$$

$$+ \frac{e^{tz}}{(z + 1)(z - i)(z + i)}\bigg|_{z=1} + \frac{e^{tz}}{(z - 1)(z - i)(z + i)}\bigg|_{z=-1}$$

$$+ \frac{e^{tz}}{(z + 1)(z - 1)(z + i)}\bigg|_{z=i} + \frac{e^{tz}}{(z - 1)(z + 1)(z - i)}\bigg|_{z=-i}$$

$$= \frac{3}{4}e^t - \frac{3}{4}e^{-t} - \frac{1}{2}\left(\frac{e^{it} - e^{-it}}{2i}\right)$$

$$= \frac{3}{4}e^t - \frac{3}{4}e^{-t} - \frac{\sin t}{2}.$$

Problem 6.20

Given continuous functions $f, g \colon \mathbb{R}_0^+ \to \mathbb{R}$, we define their convolution by

$$(f * g)(t) = \int_0^t f(u)g(t - u)\, du.$$

Show that if

$$|f(t)|, |g(t)| \le ce^{at}, \quad t \in \mathbb{R}_0^+ \tag{6.74}$$

for some constants $a, c > 0$, then

$$L(f * g)(z) = (Lf)(z)(Lg)(z), \quad \operatorname{Re} z > a.$$

Solution

We have

$$
(Lf)(z)(Lg)(z) = \int_0^\infty e^{-tz} f(t)\, dt \int_0^\infty e^{-sz} g(s)\, ds
$$
$$
= \int_0^\infty \int_0^\infty e^{-(t+s)z} f(t) g(s)\, dt\, ds
$$

for $\operatorname{Re} z > a$. Letting $t + s = \tau$, we obtain

$$
(Lf)(z)(Lg)(z) = \int_0^\infty \int_t^\infty e^{-\tau z} f(t) g(\tau - t)\, d\tau\, dt.
$$

By Fubini's theorem, since the functions f and g are continuous and satisfy (6.74), one can interchange the order of integration in the last integral (for $\operatorname{Re} z > a$). Therefore,

$$
(Lf)(z)(Lg)(z) = \int_0^\infty \int_0^\tau e^{-\tau z} f(t) g(\tau - t)\, dt\, d\tau
$$
$$
= \int_0^\infty e^{-\tau z} \left(\int_0^\tau f(t) g(\tau - t)\, dt \right) d\tau
$$
$$
= \int_0^\infty e^{-\tau z} (f * g)(\tau)\, d\tau
$$
$$
= L(f * g)(z).
$$

Problem 6.21

Verify that the function $1/t^2$ does not have a Laplace transform.

Solution

For $f(t) = 1/t^2$ we have

$$
(Lf)(z) = \int_0^\infty e^{-tz} \frac{1}{t^2}\, dt
$$

whenever the integral is well defined. Moreover,

$$\left| e^{-tz} \frac{1}{t^2} \right| = \frac{e^{-t\operatorname{Re}z}}{t^2},$$

and thus,

$$\int_1^\infty \left| e^{-tz} \frac{1}{t^2} \right| dt \le \int_1^\infty \frac{1}{t^2} dt = 1$$

for $\operatorname{Re}z \ge 0$. Hence, the integral $\int_1^\infty e^{-tz}/t^2 \, dt$ is well defined for $\operatorname{Re}z \ge 0$. On the other hand, for $x \in \mathbb{R}$ we have

$$\int_0^1 e^{-tx} \frac{1}{t^2} dt \ge \min\{1, e^{-x}\} \int_0^1 \frac{1}{t^2} dt = +\infty,$$

and thus, the integral $\int_0^1 e^{-tx}/t^2 \, dt$ is not defined for $\operatorname{Re}z > c$, for any c. This shows that the Laplace transform of $1/t^2$ does not exist.

EXERCISES

6.1. Find all solutions of the equation:
   (a) $x'' + x' - x = 0$;
   (b) $x^{(5)} - x' = 0$.

6.2. Solve the equation:
   (a) $x'' - 2x' + x = 0$ with $x(0) = 0$ and $x'(0) = 2$;
   (b) $x^{(3)} - x' = 0$ with $x(0) = x'(0) = 0$ and $x''(0) = 3$.

6.3. Find all solutions of the equation:
   (a) $(D+1)(D-2)x = e^t$;
   (b) $(D^2 + 1)x = \cos t$;
   (c) $(D^2 + 4)x = t$;
   (d) $(D^2 + 2D + 1)x = 2t + \cos t$;
   (e) $(D^3 + 2D^2)x = \cos t$.

6.4. Solve the equation:
   (a) $x'' + x = \cos t$ with $x(0) = x'(0) = 1$;
   (b) $x'' - 4x = t$ with $x(0) = 1$ and $x'(0) = 2$;
   (c) $x'' - x = te^t$ with $x(0) = x'(0) = 0$;
   (d) $x'' + x = e^t - \sin t$ with $x(0) = x'(0) = 1$.

6.5. Consider the equation

$$(3t^2 x^2 + t) + (t^3 + 1)xx' = 0.$$

   (a) Verify that it is not exact.
   (b) Find an integrating factor.
   (c) Find as explicitly as possible the solution with $x(0) = 1$.
6.6. Solve the equation:
   (a) $\cos x - t \sin x x' = 0$;
   (b) $x + x x' = 0$;
   (c) $1 - 2t \sin x - t^2 \cos x x' = 0$ with $x(2) = \pi/6$;
   (d) $(2 - x)x' = 2\cos(4t)$ with $x(0) = 0$;
   (e) $(1 - x)x' = \sin(2t)$ with $x(0) = 0$;
   (f) $2t x x' = t + x^2$ with $x(1) = 1$;
   (g) $t^2 x^2 x' = t^2 + 1$ with $x(1) = 1$.
6.7. Show that the equation

$$x^2 + tx \log x + (t^2 + tx \log t)x' = 0$$

   has an integrating factor of the form $\mu(tx)$.
6.8. Find all solutions of the equation $x'' - 2x' + x = 0$ satisfying the
   condition $x(0) + x(1) = 0$.
6.9. Find all bounded solutions of the equation:
   (a) $x^{(4)} + x' = 0$;
   (b) $x'' - x' - x = 0$.
6.10. Consider the equation $(D^2 + 4)(D + 2)x = h(t)$.
   (a) Find all solutions for $h(t) = 0$.
   (b) Find all solutions for $h(t) = 1$ such that $x'(0) = 4$.
   (c) Find all bounded solutions for $h(t) = 0$.
6.11. Find all solutions of the equation $x'' - x' - 6x = 0$ that are bounded
   for $t > 0$.
6.12. Identify each statement as true or false.
   (a) The equation $x'' + x' = 0$ has nonconstant periodic solutions.
   (b) The equation $x' = 6t\sqrt[3]{x^2}$ has more than one solution with
   $x(0) = 0$.
6.13. Find whether the equation $x'''x = x''x'$ has solutions for which x''/x
   is bounded. Hint: note that $(x''/x)' = (x'''x - x''x')/x^2$.
6.14. Consider the equation

$$x' = p(t)x^2 + q(t)x + r(t),$$

   where p, q and r are continuous functions with $p \neq 0$.
   (a) Verify that if $x(t)$ is a solution, then the function $y(t) = p(t)x(t)$
   satisfies the equation

$$y' = y^2 + Q(t)y + R(t),$$

where

$$Q(t) = q(t) + \frac{p'(t)}{p(t)} \quad \text{and} \quad R(t) = p(t)r(t).$$

(b) Verify that the function $z(t) = e^{-Y(t)}$, where $Y(t)$ is a primitive of $y(t)$, satisfies the equation

$$z'' - Q(t)z' + R(t)z = 0.$$

6.15. Find the Laplace transform of the function:
  (a) $t^2 e^t$;
  (b) $t^2 \cos t$;
  (c) $t^4 \sin t$;
  (d) $t + e^{-t} \cos t$;
  (e) $e^t \sin^2 t$;
  (f) $te^{2t} \sin(3t)$;
  (g) $\cos(3t) \sin(2t)$;
  (h) $\cos^3 t$.

6.16. Find the function with Laplace transform:
  (a) $z/(z^2 - 1)$;
  (b) $(z + 1)/(z^3 - 1)$;
  (c) $(3z^2 - 1)/(z^2 + 1)^3$;
  (d) $(3z^2 + 1)/[z^2(z^2 + 1)^2]$.

6.17. Find the Laplace transform of the solution of the equation

$$x'' + 3x' + x = \cos t \quad \text{with } x(0) = 0, \; x'(0) = 1.$$

6.18. Given $f \in \mathcal{F}_c$ and $n \in \mathbb{N}$, show that

$$(Lf)^{(n)}(z) = (-1)^n \int_0^\infty t^n e^{-tz} f(t) \, dt, \quad \operatorname{Re} z > c.$$

6.19. Under the assumptions of Problem 6.20, show that:
  (a) $f * g = g * f$.
  (b) $f * (g + h) = f * g + f * h$.
  (c) $(f * g) * h = f * (g * h)$.

6.20. For the function $g(t) = \int_0^t f(s) \, ds$, show that

$$(Lg)(z) = \frac{(Lf)(z)}{z}.$$

6.21. Show that if f is a T-periodic function, then

$$(Lf)(z) = \frac{1}{1 - e^{-Tz}} \int_0^T e^{-tz} f(t) \, dt.$$

6.22. Show that if $g(t) = f(t)\cosh(at)$, then

$$(Lg)(z) = \frac{1}{2}\big[(Lf)(z - a) + (Lf)(z + a)\big].$$

6.23. Use the Laplace transform to find a function f such that

$$f(t) + \int_0^t f(s)\,ds = 1.$$

6.24. Find all constant solutions of the equation

$$x'' = x(x - 1)(x - 2).$$

6.25. Verify that the equation $x' = x(x - 1)(x - 2)$ has bounded solutions that are not constant.

6.26. Verify that if $(x, y) = (x(t), y(t))$ is a solution of the equation

$$\begin{cases} x' = y, \\ y' = x^2 + x, \end{cases}$$

then

$$\frac{d}{dt}\left(\frac{1}{2}y(t)^2 - \frac{1}{3}x(t)^3 - \frac{1}{2}x(t)^2\right) = 0.$$

6.27. Given a function $H\colon \mathbb{R}^2 \to \mathbb{R}$ of class C^1, verify that if

$$(x, y) = (x(t), y(t))$$

is a solution of the equation

$$\begin{cases} x' = \partial H/\partial y, \\ y' = -\partial H/\partial x, \end{cases} \tag{6.75}$$

then

$$\frac{d}{dt}H\big(x(t), y(t)\big) = 0.$$

6.28. For the equation

$$\begin{cases} x' = x, \\ y' = -y - 1, \end{cases}$$

find a function $H(x, y)$ satisfying (6.75).

6.29. For the equation

$$\begin{cases} x' = x^2 - 2x^2y, \\ y' = -2xy + 2xy^2, \end{cases}$$

find a function $H(x,y)$ satisfying (6.75).

6.30. Verify that the image of each solution of the equation

$$\begin{cases} x' = y, \\ y' = -x \end{cases}$$

is a circle or a point.

6.31. Verify that the image of each solution of the equation

$$\begin{cases} x' = y, \\ y' = x \end{cases}$$

is a branch of a hyperbola, a ray, or a point.

7

Fourier Series

In this chapter we introduce the notion of Fourier series of a given function. In particular, we study the convergence as well as the uniform convergence of Fourier series. We also show how to expand a sufficiently regular function as a series of cosines and as a series of sines. As a by-product of the theory, we obtain several identities expressing π and other numbers as series of real numbers.

7.1 An Example

As a motivation for the study of Fourier series, we consider the *heat equation*

$$\frac{\partial u}{\partial t} = \kappa \frac{\partial^2 u}{\partial x^2}, \tag{7.1}$$

for $t \geq 0$ and $x \in [0, l]$, with $\kappa, l > 0$. This equation models the evolution of the temperature $u(t, x)$ at time t and at each point x of a bar of length l. We assume that

$$u(t, 0) = u(t, l) = 0, \quad t \geq 0. \tag{7.2}$$

L. Barreira, C. Valls, *Complex Analysis and Differential Equations*,
Springer Undergraduate Mathematics Series,
DOI 10.1007/978-1-4471-4008-5_7, © Springer-Verlag London 2012

This means that the endpoints of the bar are kept permanently at zero temperature. We first look for solutions of the form

$$u(t,x) = T(t)X(x).$$

This is the first step of the method of separation of variables.

Proposition 7.1

If TX does not vanish, then there exists $\lambda \in \mathbb{R}$ such that

$$T' = -\lambda \kappa T \quad \text{and} \quad X'' + \lambda X = 0. \tag{7.3}$$

Proof

Substituting $u(t,x) = T(t)X(x)$ in (7.1), we obtain

$$T'X = \kappa T X'',$$

and thus,

$$\frac{T'}{\kappa T} = \frac{X''}{X}, \tag{7.4}$$

since TX does not vanish. We note that the left-hand side of (7.4) does not depend on x and that the right-hand side does not depend on t. Hence, there exists a constant $\lambda \in \mathbb{R}$ such that

$$\frac{T'}{\kappa T} = \frac{X''}{X} = -\lambda$$

for every t and x. This yields the desired identities. $\square$

The solutions of the first equation in (7.3) are given by

$$T(t) = ce^{-\lambda \kappa t}, \tag{7.5}$$

with $c \neq 0$ (so that $T \neq 0$). On the other hand, it follows from (7.2) that

$$T(t)X(0) = T(t)X(l) = 0 \quad \text{for } t \geq 0,$$

which in view of (7.5) is equivalent to

$$X(0) = X(l) = 0.$$

Hence, it remains to solve the problem

$$X'' + \lambda X = 0, \qquad X(0) = X(l) = 0. \tag{7.6}$$

Proposition 7.2

Problem (7.6) has nonzero solutions if and only if $\lambda = n^2\pi^2/l^2$ for some $n \in \mathbb{N}$, in which case

$$X(x) = b\sin\left(\frac{n\pi x}{l}\right), \qquad \text{with } b \neq 0. \tag{7.7}$$

Proof

We consider three cases.

1. When $\lambda = 0$, we have $X'' = 0$, and thus,

$$X(x) = a + bx, \tag{7.8}$$

with $a, b \in \mathbb{R}$. It follows from $X(0) = 0$ that $a = 0$. Thus, $X(l) = bl = 0$, which yields $b = 0$ and $X(x) = 0$.

2. When $\lambda < 0$, the solutions of the equation

$$X'' + \lambda X = \left(D^2 + \lambda\right)X = \left(D - \sqrt{|\lambda|}\right)\left(D + \sqrt{|\lambda|}\right)X = 0$$

are given by

$$X(x) = ae^{\sqrt{|\lambda|}x} + be^{-\sqrt{|\lambda|}x}, \tag{7.9}$$

with $a, b \in \mathbb{R}$. It follows from $X(0) = 0$ that $a + b = 0$. Hence,

$$X(l) = a\left(e^{\sqrt{|\lambda|}l} - e^{-\sqrt{|\lambda|}l}\right) = 0. \tag{7.10}$$

Since $\sqrt{|\lambda|}l > 0$, we have $e^{\sqrt{|\lambda|}l} > 1$ and $e^{-\sqrt{|\lambda|}l} < 1$. Thus, it follows from (7.10) that $a = 0$, which yields $b = 0$ and $X(x) = 0$.

3. When $\lambda > 0$, the solutions of the equation

$$X'' + \lambda X = \left(D^2 + \lambda\right)X = 0$$

are given by

$$X(x) = a\cos(\sqrt{\lambda}x) + b\sin(\sqrt{\lambda}x), \tag{7.11}$$

with $a, b \in \mathbb{R}$. It follows from $X(0) = 0$ that $a = 0$. Hence, $X(l) = b\sin(\sqrt{\lambda}l) = 0$. Therefore, $b = 0$ (which would give $X = 0$), or

$$\sin(\sqrt{\lambda}l) = 0. \tag{7.12}$$

It follows from (7.12) that $\sqrt{\lambda}l = n\pi$ for some $n \in \mathbb{N}$ (we recall that $\lambda > 0$), that is,

$$\lambda = \frac{n^2\pi^2}{l^2}, \quad n \in \mathbb{N}.$$

We thus obtain the solutions in (7.7).

This completes the proof of the proposition. $\square$

Combining (7.5) and (7.7), we obtain the solutions of the heat equation (7.1) given by

$$u(t,x) = c_n e^{-n^2\pi^2\kappa t/l^2} \sin\left(\frac{n\pi x}{l}\right), \tag{7.13}$$

with $n \in \mathbb{N}$ and $c_n \in \mathbb{R}$. We note that all of them satisfy condition (7.2).

The solutions of the heat equation have the following property.

Proposition 7.3

If u_1 and u_2 are solutions of equation (7.1), then $c_1 u_1 + c_2 u_2$ is also a solution of equation (7.1) for each $c_1, c_2 \in \mathbb{R}$. Moreover, if u_1 and u_2 satisfy condition (7.2), then $c_1 u_1 + c_2 u_2$ also satisfies condition (7.2) for each $c_1, c_2 \in \mathbb{R}$.

Proof

We have

$$\frac{\partial}{\partial t}(c_1 u_1 + c_2 u_2) = c_1 \frac{\partial u_1}{\partial t} + c_2 \frac{\partial u_2}{\partial t}$$

$$= c_1 \kappa \frac{\partial^2 u_1}{\partial x^2} + c_2 \kappa \frac{\partial^2 u_2}{\partial x^2}$$

$$= \kappa \frac{\partial^2}{\partial x^2}(c_1 u_1 + c_2 u_2),$$

which establishes the first property. The second property is immediate. $\square$

The following is an immediate consequence of (7.13) and Proposition 7.3.

Proposition 7.4

For each $N \in \mathbb{N}$ and $c_1, \ldots, c_N \in \mathbb{R}$, the function

$$u(t,x) = \sum_{n=1}^{N} c_n e^{-n^2 \pi^2 \kappa t / l^2} \sin\left(\frac{n\pi x}{l}\right) \tag{7.14}$$

is a solution of equation (7.1) satisfying condition (7.2).

Now we make the additional assumption that

$$u(0,x) = f(x), \quad x \in [0,l], \tag{7.15}$$

for a given function $f \colon [0,l] \to \mathbb{R}$. For the solution in (7.14), it follows from (7.15) that

$$u(0,x) = \sum_{n=1}^{N} c_n \sin\left(\frac{n\pi x}{l}\right) = f(x), \tag{7.16}$$

and thus, we would like to find constants c_n such that the second identity in (7.16) holds. For example, for

$$f(x) = \sin\left(\frac{\pi x}{l}\right) + 2\sin\left(\frac{3\pi x}{l}\right),$$

one can take

$$c_n = \begin{cases} 1 & \text{if } n = 1, \\ 2 & \text{if } n = 2, \\ 0 & \text{otherwise}, \end{cases}$$

and thus, the solution in (7.14) takes the form

$$u(t,x) = e^{-\pi^2 \kappa t / l^2} \sin\left(\frac{\pi x}{l}\right) + 2e^{-9\pi^2 \kappa t / l^2} \sin\left(\frac{3\pi x}{l}\right).$$

Unfortunately, in general (that is, for an arbitrary function f), it is not possible to find constants c_n such that identity (7.16) holds. However, we verify in the following sections that it is possible to find constants c_n for a large class of functions f, provided that one can take $N = \infty$, that is, provided that one can consider solutions given by series of the form

$$u(t,x) = \sum_{n=1}^{\infty} c_n e^{-n^2 x^2 \kappa t / l^2} \sin\left(\frac{n\pi x}{l}\right).$$

In particular, we will have to discuss the convergence of the series

$$u(0,x) = \sum_{n=1}^{\infty} c_n \sin\left(\frac{n\pi x}{l}\right).$$

(7.17)

7.2 Fourier Series

In this section we discuss the convergence of the series in (7.17). To this effect, we first introduce a class of functions in the interval $[-l, l]$.

Definition 7.5

Let $\mathcal{D}_l$ be the family of functions $f \colon [-l, l] \to \mathbb{R}$ such that:
1. f has at most finitely many discontinuities;
2. f has left-sided and right-sided limits at all points of $[-l, l]$;
3. for each interval $(a, b) \subset [-l, l]$ where the function f is continuous, the (continuous) function $g \colon [a, b] \to \mathbb{R}$ defined by

$$g(x) = \begin{cases} f(a^+) & \text{if } x = a, \\ f(x) & \text{if } x \in (a, b), \\ f(b^-) & \text{if } x = b \end{cases}$$

(7.18)

has left-sided and right-sided (finite) derivatives at all points of $[a, b]$.

Now we define the Fourier coefficients.

Definition 7.6

Given a function $f \in \mathcal{D}_l$, we define its *Fourier coefficients* by

$$a_n = \frac{1}{l} \int_{-l}^{l} f(x) \cos\left(\frac{n\pi x}{l}\right) dx, \quad n \in \mathbb{N} \cup \{0\}$$

and

$$b_n = \frac{1}{l} \int_{-l}^{l} f(x) \sin\left(\frac{n\pi x}{l}\right) dx, \quad n \in \mathbb{N}.$$

We note that each function $f \in \mathcal{D}_l$ is Riemann-integrable, and thus, the coefficients a_n and b_n are well defined.

Definition 7.7

We define the *Fourier series* of a function $f \in \mathcal{D}_l$ by

$$F(x) = \frac{a_0}{2} + \sum_{n=1}^{\infty} \left[a_n \cos\left(\frac{n\pi x}{l}\right) + b_n \sin\left(\frac{n\pi x}{l}\right) \right],$$

whenever it converges.

The following result establishes the convergence of Fourier series.

Theorem 7.8

For each function $f \in \mathcal{D}_l$, we have

$$F(x) = \frac{f(x^+) + f(x^-)}{2} \quad \text{for } x \in (-l, l), \tag{7.19}$$

and

$$F(x) = \frac{f(-l^+) + f(l^-)}{2} \quad \text{for } x \in \{-l, l\}. \tag{7.20}$$

Proof

The partial sums

$$S_n(x) = \frac{a_0}{2} + \sum_{k=1}^{n} \left[a_k \cos\left(\frac{k\pi x}{l}\right) + b_k \sin\left(\frac{k\pi x}{l}\right) \right] \tag{7.21}$$

satisfy

$$S_n(x) - \frac{1}{2l} \int_{-l}^{l} f(y)\, dy$$

$$= \frac{1}{l} \int_{-l}^{l} f(y) \sum_{k=1}^{n} \left(\cos\left(\frac{k\pi y}{l}\right) \cos\left(\frac{k\pi x}{l}\right) + \sin\left(\frac{k\pi y}{l}\right) \sin\left(\frac{k\pi x}{l}\right) \right) dy$$

$$= \frac{1}{l} \int_{-l}^{l} f(y) \sum_{k=1}^{n} \cos\left(\frac{k\pi(y - x)}{l}\right) dy,$$

where

$$\sum_{k=1}^{n} \cos\left(\frac{k\pi(y-x)}{l}\right) = \mathrm{Re}\sum_{k=1}^{n} e^{ik\pi(y-x)/l}$$

$$= \mathrm{Re}\left(\frac{1 - e^{in\pi(y-x)/l}}{1 - e^{i\pi(y-x)/l}} e^{i\pi(y-x)/l}\right)$$

$$= \mathrm{Re}\left(\frac{e^{i\pi(y-x)/(2l)} - e^{i(n+1/2)\pi(y-x)/l}}{e^{-i\pi(y-x)/(2l)} - e^{i\pi(y-x)/(2l)}}\right)$$

$$= \mathrm{Re}\left(\frac{e^{i\pi(y-x)/(2l)} - e^{i(n+1/2)\pi(y-x)/l}}{-2i\sin(\pi(y-x)/(2l))}\right)$$

$$= -\frac{1}{2} + \frac{\sin((n+1/2)\pi(y-x)/(2l))}{\sin(\pi(y-x)/(2l))}.$$

Letting

$$D_n(y) = \frac{\sin((n+1/2)\pi y/(2l))}{\sin(\pi y/(2l))},$$

one can then write

$$S_n(x) = \frac{1}{l}\int_{-l}^{l} f(y) D_n(y-x)\,dy$$

$$= \frac{1}{l}\int_{-l}^{l} g(y+x) D_n(y)\,dy, \tag{7.22}$$

where the function $g\colon \mathbb{R}\to\mathbb{R}$ is given by

$$g(x) = f(x - 2kl) \tag{7.23}$$

when $x - 2kl \in [-l, l)$ for some $k \in \mathbb{Z}$. On the other hand, if $f = 1$, then $a_n = b_n = 0$ for every $n \in \mathbb{N}$, and it follows from (7.21) and (7.22) that

$$1 = \frac{1}{l}\int_{-l}^{l} D_n(y)\,dy, \quad n \in \mathbb{N}.$$

Therefore,

$$S_n(x) - \frac{g(x^+) + g(x^-)}{2} = \frac{1}{l}\int_{-l}^{l}\left(g(y+x) - \frac{g(x^+) + g(x^-)}{2}\right) D_n(y)\,dy$$

$$= \frac{1}{l}\int_{-l}^{0}\left(g(y+x) - g(x^-)\right) D_n(y)\,dy$$

$$+ \frac{1}{l}\int_{0}^{l}\left(g(y+x) - g(x^+)\right) D_n(y)\,dy. \tag{7.24}$$

Now we use the Riemann–Lebesgue lemma (Lemma 6.29) to show that the integrals in (7.24) converge to zero when $n \to \infty$. For each $y \in [-l, 0)$, we have

$$\left(g(y+x) - g(x^-)\right) D_n(y) = \frac{g(y+x) - g(x^-)}{y} \cdot \frac{\sin((n+1/2)\pi y/(2l))}{\sin(\pi y/(2l))/y}.$$

We note that the function $\chi \colon [-l, l] \to \mathbb{R}^+$ given by

$$\chi(y) = \begin{cases} \sin(\pi y/(2l))/y & \text{if } y \neq 0, \\ \pi/(2l) & \text{if } y = 0 \end{cases}$$

is continuous. On the other hand, since $f \in \mathcal{D}_l$, the function $\psi \colon [-l, 0] \to \mathbb{R}$ given by

$$\psi(y) = \begin{cases} (g(y+x) - g(x^-))/y & \text{if } y \neq 0, \\ g'(x^-) & \text{if } y = 0 \end{cases}$$

has at most finitely many discontinuities, and has left-sided and right-sided limits at all points of $[-l, 0]$. Hence, ψ is Riemann-integrable in $[-l, 0]$ (it is also absolutely Riemann-integrable, since the interval is bounded). The same happens with the function $h = \psi/\chi$, and thus, it follows from Lemma 6.29 that

$$\int_{-l}^{0} \left(g(y+x) - g(x^-)\right) D_n(y) \, dy$$

$$= \int_{-l}^{0} h(y) \sin\left(\frac{(n+1/2)\pi y}{2l}\right) dy$$

$$= \int_{-l}^{0} h(y) \sin\left(\frac{\pi y}{4l}\right) \cos\left(\frac{n\pi y}{2l}\right) dy + \int_{-l}^{0} h(y) \cos\left(\frac{\pi y}{4l}\right) \sin\left(\frac{n\pi y}{2l}\right) dy \to 0$$

when $n \to \infty$. One can show in an analogous manner that

$$\lim_{n \to \infty} \int_{0}^{l} \left(g(y+x) - g(x^+)\right) D_n(y) \, dy = 0.$$

It then follows from (7.24) that

$$S_n(x) \to \frac{g(x^+) + g(x^-)}{2} \tag{7.25}$$

when $n \to \infty$. Identity (7.19) follows immediately from (7.25), since $g = f$ in $(-l, l)$. For $x \in \{-l, l\}$, we note that

$$g(l^-) = g(-l^-) = f(l^-) \quad \text{and} \quad g(l^+) = g(-l^+) = f(-l^+).$$

This establishes identity (7.20). □

The following property is an immediate consequence of Theorem 7.8.

Proposition 7.9

If a function $f \in \mathcal{D}_l$ is continuous at a point $x \in (-l, l)$, then $F(x) = f(x)$.

Now we give some examples.

Example 7.10

Let us consider the function

$$f(x) = \begin{cases} 0 & \text{if } -1 \le x < 0, \\ 2 & \text{if } 0 \le x \le 1. \end{cases}$$

Clearly, $f \in \mathcal{D}_1$. Taking $l = 1$, we have

$$a_0 = \int_{-1}^{1} f(x)\, dx = 2$$

and

$$a_n = \int_{-1}^{1} f(x) \cos(n\pi x)\, dx = 2 \int_{0}^{1} \cos(n\pi x)\, dx = 0$$

for $n \in \mathbb{N}$. We also have

$$b_n = \int_{-1}^{1} f(x) \sin(n\pi x)\, dx = 2 \int_{0}^{1} \sin(n\pi x)\, dx$$
$$= \frac{2}{n\pi} \left(1 - \cos(n\pi)\right) = \frac{2(1 - (-1)^n)}{n\pi}$$

for $n \in \mathbb{N}$. Hence, the Fourier series of f is

$$F(x) = 1 + \sum_{m=1}^{\infty} \frac{4}{(2m-1)\pi} \sin\big((2m-1)\pi x\big).$$

On the other hand, by Theorem 7.8, we have

$$F(x) = \begin{cases} f(x) & \text{if } x \in (-1, 0) \cup (0, 1), \\ 1 & \text{if } x \in \{-1, 0, 1\}. \end{cases}$$

In particular, taking $x = 1/2$, we obtain

$$2 = 1 + \sum_{m=1}^{\infty} \frac{4(-1)^{m-1}}{(2m-1)\pi} = 1 + \frac{4}{\pi} - \frac{4}{3\pi} + \frac{4}{5\pi} - \cdots,$$

that is,

$$\frac{\pi}{4} = \sum_{m=1}^{\infty} \frac{(-1)^{m-1}}{2m-1} = 1 - \frac{1}{3} + \frac{1}{5} - \frac{1}{7} + \cdots.$$

Example 7.11

Now we consider the function $f: [-1,1] \to \mathbb{R}$ given by $f(x) = |x|$. Clearly, $f \in \mathcal{D}_1$. Taking $l = 1$, we have

$$a_0 = \int_{-1}^{1} |x| \, dx = 1$$

and

$$a_n = \int_{-1}^{1} |x| \cos(n\pi x) \, dx = 2 \int_{0}^{1} x \cos(n\pi x) \, dx$$

$$= 2 \left(x \frac{\sin(n\pi x)}{n\pi} + \frac{\cos(n\pi x)}{(n\pi)^2} \right) \Big|_{x=0}^{x=1}$$

$$= 2 \frac{\cos(n\pi) - 1}{(n\pi)^2} = 2 \frac{(-1)^n - 1}{(n\pi)^2}$$

for $n \in \mathbb{N}$. We also have

$$b_n = \int_{-1}^{1} |x| \sin(n\pi x) \, dx = 0$$

for $n \in \mathbb{N}$. Hence, the Fourier series of f is

$$F(x) = \frac{1}{2} + \sum_{m=1}^{\infty} \frac{-4}{(2m-1)^2 \pi^2} \cos\big((2m-1)\pi x\big). \tag{7.26}$$

By Theorem 7.8, we have $F(x) = |x|$ for every $x \in [-1, 1]$. In particular, taking $x = 0$, we obtain

$$0 = \frac{1}{2} - \frac{4}{\pi^2} \sum_{m=1}^{\infty} \frac{1}{(2m-1)^2}$$

$$= \frac{1}{2} - \frac{4}{\pi^2} \left(1 + \frac{1}{3^2} + \frac{1}{5^2} + \cdots \right),$$

that is,

$$\frac{\pi^2}{8} = \sum_{m=1}^{\infty} \frac{1}{(2m-1)^2} = 1 + \frac{1}{3^2} + \frac{1}{5^2} + \cdots.$$

Example 7.12

Let us consider the function

$$f(x) = \begin{cases} -x & \text{if } -2 \le x < 0, \\ 1 & \text{if } 0 \le x \le 2. \end{cases}$$

Clearly, $f \in \mathcal{D}_2$. Taking $l = 2$, we have

$$a_0 = \frac{1}{2} \int_{-2}^{2} f(x)\,dx = 2$$

and

$$a_n = \frac{1}{2} \int_{-2}^{2} f(x) \cos\left(\frac{n\pi x}{2}\right) dx$$

$$= -\frac{1}{2} \int_{-2}^{0} x \cos\left(\frac{n\pi x}{2}\right) dx + \frac{1}{2} \int_{0}^{2} \cos\left(\frac{n\pi x}{2}\right) dx$$

$$= \left. \left(-\frac{2}{n^2\pi^2} \cos\left(\frac{n\pi x}{2}\right) - \frac{x}{n\pi} \sin\left(\frac{n\pi x}{2}\right) \right) \right|_{x=-2}^{x=0}$$

$$= \frac{2}{(n\pi)^2} \left(\cos(n\pi) - 1 \right) = \frac{2[(-1)^n - 1]}{(n\pi)^2}$$

for $n \in \mathbb{N}$. We also have

$$b_n = \frac{1}{2} \int_{-2}^{2} f(x) \sin\left(\frac{n\pi x}{2}\right) dx$$

$$= -\frac{1}{2} \int_{-2}^{0} x \sin\left(\frac{n\pi x}{2}\right) dx + \frac{1}{2} \int_{0}^{2} \sin\left(\frac{n\pi x}{2}\right) dx$$

$$= \left. \left(\frac{x}{n\pi} \cos\left(\frac{n\pi x}{2}\right) - \frac{2}{n^2\pi^2} \sin\left(\frac{n\pi x}{2}\right) \right) \right|_{x=-2}^{x=0} - \left. \frac{1}{n\pi} \cos\left(\frac{n\pi x}{2}\right) \right|_{x=0}^{x=2}$$

$$= \frac{1}{n\pi} (1 + \cos(n\pi)) = \frac{1 + (-1)^n}{n\pi}$$

for $n \in \mathbb{N}$. Therefore,

$$a_n = \begin{cases} 0 & \text{if } n \text{ is even}, \\ -4/(n^2\pi^2) & \text{if } n \text{ is odd}, \end{cases}$$

and

$$b_n = \begin{cases} 0 & \text{if } n \text{ is odd}, \\ 2/(n\pi) & \text{if } n \text{ is even}. \end{cases}$$

Hence, the Fourier series of f is

$$F(x) = 1 + \sum_{m=1}^{\infty} \frac{-4}{(2m-1)^2\pi^2} \cos\left(\frac{(2m-1)\pi x}{2}\right) + \sum_{m=1}^{\infty} \frac{1}{m\pi} \sin(m\pi x).$$

By Theorem 7.8, we have

$$F(x) = \begin{cases} 3/2 & \text{if } x = -2 \text{ or } x = 2, \\ -x & \text{if } -2 < x < 0, \\ 1/2 & \text{if } x = 0, \\ 1 & \text{if } 0 < x < 2. \end{cases}$$

We also give a condition for the uniform convergence of a Fourier series.

Theorem 7.13

If $f \colon [-l, l] \to \mathbb{R}$ is the restriction to the interval $[-l, l]$ of a $2l$-periodic function $g \colon \mathbb{R} \to \mathbb{R}$ of class C^1, then the Fourier series of f converges uniformly to f on $[-l, l]$.

More generally, we have the following result.

Theorem 7.14

Let $f \colon [-l, l] \to \mathbb{R}$ be a continuous function with $f(-l) = f(l)$. If there exist points $-l = x_0 < x_1 < \cdots < x_m = l$ such that the restriction of f to (x_i, x_{i+1}) has an extension of class C^1 to some open interval containing $[x_i, x_{i+1}]$, for $i = 0, \ldots, m-1$, then the Fourier series of f converges uniformly to f on the interval $[-l, l]$.

Proof

We note that $f \in \mathcal{D}_l$, and we use the same notation as in the proof of Theorem 7.8. Given $\varepsilon > 0$ and $x \in [-l, l]$, we define a function $h_x \colon [-l, l] \setminus [-\varepsilon, \varepsilon] \to \mathbb{R}$ by

$$h_x(y) = \frac{g(y+x) - g(x)}{\sin(\pi y/(2l))},$$

with g as in (7.23). Writing $k_n = (n+1/2)\pi/(2l)$, we have

$$\int_{-l}^{-\varepsilon} \left[g(y+x) - g(x)\right] D_n(y) \, dy$$

$$= \int_{-l}^{-\varepsilon} h_x(y) \sin(k_n y) \, dy$$

$$= -\frac{h_x(y)\cos(k_n y)}{k_n}\bigg|_{y=-l}^{y=-\varepsilon} + \int_{-l}^{-\varepsilon} \frac{h_x'(y)\cos(k_n y)}{k_n} \, dy \to 0 \qquad (7.27)$$

when $n \to \infty$, uniformly in x on the interval $(-l, -\varepsilon)$. One can also establish an analogous result in the interval (ε, l). On the other hand, it follows from the hypotheses on f that there exists $C > 0$ such that

$$\left|g(y+x) - g(x)\right| \le C|y| \quad \text{for } y \in [-\varepsilon, \varepsilon].$$

Hence,

$$\left|\int_{-\varepsilon}^{\varepsilon} \left[g(y+x) - g(x)\right] D_n(y) \, dy\right| \le \int_{-\varepsilon}^{\varepsilon} \frac{C|y|}{|\sin(\pi y/(2l))|} \, dy \le \frac{8lC\varepsilon}{\pi} \qquad (7.28)$$

for any sufficiently small ε, independently of x and n, since

$$\lim_{y \to 0} \frac{y}{\sin(\pi y/(2l))} = \frac{2l}{\pi}.$$

Given $\delta > 0$, let us take $\varepsilon > 0$ such that $8lC\varepsilon/\pi < \delta$. By (7.27) and the analogous result in the interval (ε, l), there exists $p \in \mathbb{N}$ such that

$$\alpha := \left|\int_{[-l,l] \setminus [-\varepsilon, \varepsilon]} \left[g(y+x) - g(x)\right] D_n(y) \, dy\right| < \delta$$

for $n \ge p$. It then follows from (7.28) that

$$\left|\int_{-l}^{l} \left[g(y+x) - g(x)\right] D_n(y) \, dy\right| \le \left|\int_{-\varepsilon}^{\varepsilon} \left[g(y+x) - g(x)\right] D_n(y) \, dy\right| + \alpha < 2\delta,$$

also for $n \ge p$. Finally, since $f(-l) = f(l)$ and

$$g(x^+) = g(x^-) = f(x)$$

for every $x \in [-l, l]$, it follows from (7.24) that the convergence in (7.25) is uniform on the interval $[-l, l]$. $\qquad\square$

Example 7.15

It follows from Theorem 7.14 that the Fourier series in (7.26) converges uniformly to $|x|$ on the interval $[-1, 1]$.

7.3 Uniqueness and Orthogonality

In this section we establish some additional properties of Fourier series. In particular, we show that the Fourier coefficients uniquely determine the function defining them.

We start with an auxiliary result.

Theorem 7.16

Given a function $f \in \mathcal{D}_l$, if all of its Fourier coefficients are zero, that is, if

$$a_n = 0 \quad \text{for } n \geq 0 \quad \text{and} \quad b_n = 0 \quad \text{for } n \geq 1, \qquad (7.29)$$

then $f = 0$.

Proof

If f is not identically zero, then there exist $x_0 \in (-l, l)$ and $\varepsilon, \delta > 0$ such that

$$\left| f(x) \right| > \varepsilon \quad \text{for } x \in J := (x_0 - \delta, x_0 + \delta).$$

Without loss of generality, we always assume that f has a single sign in J. Now we consider the functions $g_n : [-l, l] \to \mathbb{R}$ given by

$$g_n(x) = \left(1 + \cos\left(\frac{\pi(x - x_0)}{l} \right) - \cos\left(\frac{\pi \delta}{l} \right) \right)^n$$

for $n \in \mathbb{N}$. One can show that each g_n is a linear combination of the functions

$$1, \quad \cos\left(\frac{m\pi x}{l} \right), \quad \sin\left(\frac{m\pi x}{l} \right), \quad m \in \mathbb{N}.$$

It then follows from (7.29) that

$$\frac{1}{l} \int_{-l}^{l} f(x) g_n(x) \, dx = 0, \quad n \in \mathbb{N}.$$

On the other hand, $g_n(x) > 1$ for $x \in J$, and

$$|g_n(x)| \leq 1 \quad \text{for } x \notin J. \tag{7.30}$$

Moreover, in the interval $K = [x_0 - \delta/2, x_0 + \delta/2]$ we have

$$\inf_{x \in K} g_n(x) = g_n(x_0 \pm \delta/2) = \left(1 + \cos\left(\frac{\pi\delta}{2l}\right) - \cos\left(\frac{\pi\delta}{l}\right)\right)^n \to +\infty$$

when $n \to \infty$, since $\cos(\pi\delta/2l) > \cos(\pi\delta/l)$. Now we observe that

$$0 = \int_{-l}^{l} f(x)g_n(x)\,dx$$

$$= \int_{[-l,l] \setminus J} f(x)g_n(x)\,dx + \int_{J} f(x)g_n(x)\,dx.$$

By (7.30), we have

$$\left|\int_{[-l,l] \setminus J} f(x)g_n(x)\,dx\right| \leq \int_{-l}^{l} |f(x)|\,dx < +\infty.$$

On the other hand,

$$\left|\int_{J} f(x)g_n(x)\,dx\right| \geq \left|\int_{K} f(x)g_n(x)\,dx\right|$$

$$\geq \delta\varepsilon \inf_{x \in K} g_n(x) \to +\infty$$

when $n \to \infty$ (because f has a single sign in J). Therefore,

$$0 = \int_{-l}^{l} f(x)g_n(x)\,dx \to +\infty$$

when $n \to \infty$. This contradiction shows that $f = 0$. $\qquad\square$

Theorem 7.16 implies that the Fourier coefficients uniquely determine the function defining them.

Theorem 7.17

Given functions $f, g \in \mathcal{D}_l$, if f and g have the same Fourier coefficients, then $f = g$.

Proof

We note that all Fourier coefficients of $f - g$ are zero. Indeed,

$$\frac{1}{l} \int_{-l}^{l} [f(x) - g(x)] \cos\left(\frac{n\pi x}{l}\right) dx$$

$$= \frac{1}{l} \int_{-l}^{l} f(x) \cos\left(\frac{n\pi x}{l}\right) dx - \frac{1}{l} \int_{-l}^{l} g(x) \cos\left(\frac{n\pi x}{l}\right) dx = 0$$

for $n \geq 0$, and

$$\frac{1}{l} \int_{-l}^{l} [f(x) - g(x)] \sin\left(\frac{n\pi x}{l}\right) dx$$

$$= \frac{1}{l} \int_{-l}^{l} f(x) \sin\left(\frac{n\pi x}{l}\right) dx - \frac{1}{l} \int_{-l}^{l} g(x) \sin\left(\frac{n\pi x}{l}\right) dx = 0$$

for $n \geq 1$. It then follows from Theorem 7.16 that $f - g = 0$. $\square$

Now we briefly consider a different point of view about Fourier series. We introduce an inner product in $\mathcal{D}_l$ by

$$\langle f, g \rangle = \frac{1}{l} \int_{-l}^{l} f(x) g(x) \, dx$$

for each $f, g \in \mathcal{D}_l$. It is easy to verify that this is indeed an inner product:
1. the functions $f \mapsto \langle f, g \rangle$ and $g \mapsto \langle f, g \rangle$ are linear;
2. $\langle f, f \rangle \geq 0$, and $\langle f, f \rangle = 0$ if and only if $f = 0$;
3. $\langle f, g \rangle = \langle g, f \rangle$.

In particular, the Fourier series of a function $f \in \mathcal{D}_l$ can be written in the form

$$\langle f, 1 \rangle \frac{1}{2} + \sum_{n=1}^{\infty} \langle f, u_n \rangle \cos\left(\frac{n\pi x}{l}\right) + \langle f, v_n \rangle \sin\left(\frac{n\pi x}{l}\right),$$

where

$$u_n = \cos\left(\frac{n\pi x}{l}\right) \quad \text{and} \quad v_n = \sin\left(\frac{n\pi x}{l}\right).$$

Moreover, the norm of a function $f \in \mathcal{D}_l$ is defined by

$$\|f\| = \langle f, f \rangle^{1/2} = \left(\frac{1}{l} \int_{-l}^{l} f(x)^2 \, dx\right)^{1/2}.$$

Now we recall the notion of orthogonality.

Definition 7.18

Two functions $f, g \in \mathcal{D}_l$ are said to be *orthogonal* if $\langle f, g \rangle = 0$.

Proposition 7.19

The functions

$$\frac{1}{\sqrt{2}}, \quad u_n, \quad v_n, \quad n \in \mathbb{N} \tag{7.31}$$

have norm 1 and are pairwise orthogonal.

Proof

For each $n, m \in \mathbb{N} \cup \{0\}$, we have the identities

$$\cos\left(\frac{n\pi x}{l}\right) \cos\left(\frac{m\pi x}{l}\right) = \frac{1}{2} \cos\left(\frac{(n-m)\pi x}{l}\right) + \frac{1}{2} \cos\left(\frac{(n+m)\pi x}{l}\right),$$

$$\sin\left(\frac{n\pi x}{l}\right) \sin\left(\frac{m\pi x}{l}\right) = \frac{1}{2} \cos\left(\frac{(n-m)\pi x}{l}\right) - \frac{1}{2} \cos\left(\frac{(n+m)\pi x}{l}\right), \tag{7.32}$$

$$\cos\left(\frac{n\pi x}{l}\right) \sin\left(\frac{m\pi x}{l}\right) = \frac{1}{2} \sin\left(\frac{(n+m)\pi x}{l}\right) - \frac{1}{2} \sin\left(\frac{(n-m)\pi x}{l}\right).$$

Hence, integrating over x yields

$$\int_{-l}^{l} \cos\left(\frac{n\pi x}{l}\right) \cos\left(\frac{m\pi x}{l}\right) dx = 0,$$

$$\int_{-l}^{l} \sin\left(\frac{n\pi x}{l}\right) \sin\left(\frac{m\pi x}{l}\right) dx = 0,$$

$$\int_{-l}^{l} \cos\left(\frac{n\pi x}{l}\right) \sin\left(\frac{m\pi x}{l}\right) dx = 0$$

for every $n, m \in \mathbb{N} \cup \{0\}$ with $n \neq m$. This shows that the functions in (7.31) are pairwise orthogonal. Moreover,

$$\left\| \frac{1}{\sqrt{2}} \right\|^2 = \frac{1}{l} \int_{-l}^{l} \frac{1}{2} dx = 1.$$

It also follows from (7.32) with $n = m \in \mathbb{N}$ that

$$\cos^2\left(\frac{n\pi x}{l}\right) = \frac{1}{2} + \frac{1}{2}\cos\left(\frac{2n\pi x}{l}\right)$$

and

$$\sin^2\left(\frac{n\pi x}{l}\right) = \frac{1}{2} - \frac{1}{2}\cos\left(\frac{2n\pi x}{l}\right).$$

Therefore,

$$\left\|\frac{1}{\sqrt{2}}\right\|^2 = \frac{1}{l}\int_{-l}^{l}\frac{1}{2}\,dx = 1,$$

$$\|u_n\|^2 = \frac{1}{l}\int_{-l}^{l}\cos^2\left(\frac{n\pi x}{l}\right)\,dx$$

$$= \left(\frac{x}{2l} + \frac{1}{4n\pi}\sin\left(\frac{2n\pi x}{l}\right)\right)\Big|_{x=-l}^{x=l} = 1,$$

and

$$\|v_n\|^2 = \frac{1}{l}\int_{-l}^{l}\sin^2\left(\frac{n\pi x}{l}\right)\,dx$$

$$= \left(\frac{x}{2l} - \frac{1}{4n\pi}\sin\left(\frac{2n\pi x}{l}\right)\right)\Big|_{x=-l}^{x=l} = 1.$$

This completes the proof of the proposition. $\qquad\qquad\qquad\qquad\qquad\square$

Theorem 7.16 can now be reformulated as follows.

Theorem 7.20

If $f \in \mathcal{D}_l$ is orthogonal to all functions in (7.31), then $f = 0$.

We conclude this section with a discussion about the norm of a function.

Theorem 7.21 (Bessel's inequality)

For each $f \in \mathcal{D}_l$, we have

$$\frac{a_0^2}{2} + \sum_{n=1}^{\infty}(a_n^2 + b_n^2) \leq \frac{1}{l}\int_{-l}^{l}f(x)^2\,dx. \qquad\qquad (7.33)$$

Proof

We consider the sequence of partial sums

$$S_m(x) = \frac{a_0}{2} + \sum_{n=1}^{m} \left[a_n \cos\left(\frac{n\pi x}{l}\right) + b_n \sin\left(\frac{n\pi x}{l}\right) \right]$$

$$= \left\langle f, \frac{1}{\sqrt{2}} \right\rangle \frac{1}{\sqrt{2}} + \sum_{n=1}^{m} \left[\langle f, u_n \rangle \cos\left(\frac{n\pi x}{l}\right) + \langle f, v_n \rangle \sin\left(\frac{n\pi x}{l}\right) \right]. \quad (7.34)$$

It follows easily from Proposition 7.19 that

$$\|S_m\|^2 = \frac{a_0^2}{2} + \sum_{n=1}^{m} (a_n^2 + b_n^2). \quad (7.35)$$

Now we show that $f - S_m$ is orthogonal to the functions

$$\frac{1}{\sqrt{2}}, \quad u_n, \quad v_n, \quad n \le m.$$

Indeed, if g is any of these functions, then by Proposition 7.19 and (7.34), we obtain

$$\langle f - S_m, g \rangle = \langle f, g \rangle - \langle S_m, g \rangle$$
$$= \langle f, g \rangle - \langle f, g \rangle = 0.$$

It then follows from (7.35) that

$$\|f\|^2 = \langle f - S_m + S_m, f - S_m + S_m \rangle$$
$$= \|f - S_m\|^2 + 2\langle f - S_m, S_m \rangle + \|S_m\|^2$$
$$= \|f - S_m\|^2 + \|S_m\|^2$$
$$= \|f - S_m\|^2 + \frac{a_0^2}{2} + \sum_{n=1}^{m} (a_n^2 + b_n^2)$$
$$\ge \frac{a_0^2}{2} + \sum_{n=1}^{m} (a_n^2 + b_n^2). \quad (7.36)$$

Finally, letting $m \to +\infty$ in (7.36), we obtain inequality (7.33). $\square$

In fact, Bessel's inequality is an identity. Here we consider only a particular class of functions.

Theorem 7.22 (Parseval's formula)

Let $f: [-l, l] \to \mathbb{R}$ be a continuous function with $f(-l) = f(l)$. If there exist points $-l = x_0 < x_1 < \cdots < x_m = l$ such that the restriction of f to (x_i, x_{i+1}) has an extension of class C^1 to some open interval containing $[x_i, x_{i+1}]$, for $i = 0, \ldots, m-1$, then

$$\frac{1}{l} \int_{-l}^{l} f(x)^2 \, dx = \frac{a_0^2}{2} + \sum_{n=1}^{\infty} (a_n^2 + b_n^2). \qquad (7.37)$$

Proof

By Theorem 7.14, the Fourier series of f converges uniformly to f on the interval $[-l, l]$; that is, if S_m are the partial sums in (7.34), then

$$\lim_{m \to \infty} \sup_{x \in [-l,l]} |f(x) - S_m(x)| = 0.$$

On the other hand,

$$\begin{aligned}
\|f - S_m\|^2 &= \frac{1}{l} \int_{-l}^{l} [f(x) - S_m(x)]^2 \, dx \\
&\leq 2 \sup_{x \in [-l,l]} \left(|f(x) - S_m(x)|^2 \right) \\
&= 2 \left(\sup_{x \in [-l,l]} |f(x) - S_m(x)| \right)^2,
\end{aligned}$$

and it follows from (7.35) and (7.36) that

$$\|f\|^2 - \|S_m\|^2 = \|f\|^2 - \frac{a_0^2}{2} - \sum_{n=1}^{m} (a_n^2 + b_n^2)$$

$$= \|f - S_m\|^2 \to 0$$

when $m \to \infty$. This establishes identity (7.37). $\qquad \square$

7.4 Even and Odd Functions

In this section we consider the particular classes of even functions and odd functions.

Definition 7.23

A function $f\colon [-l,l] \to \mathbb{R}$ is said to be *even* if

$$f(-x) = f(x) \quad \text{for every } x \in [-l,l],$$

and it is said to be *odd* if

$$f(-x) = -f(x) \quad \text{for every } x \in [-l,l].$$

Example 7.24

The functions 1 and $\cos(n\pi x/l)$ are even, while the function $\sin(n\pi x/l)$ is odd, for each $n \in \mathbb{N}$.

Example 7.25

The function $f(x) = |x|$ is even.

Example 7.26

The function $f(x) = x^3 + 3x$ is odd.

We show that Fourier series of even functions are series of cosines, and that Fourier series of odd functions are series of sines.

Proposition 7.27

For a function $f \in \mathcal{D}_l$:
 1. if f is even, then $b_n = 0$ for every $n \in \mathbb{N}$;
 2. if f is odd, then $a_n = 0$ for every $n \in \mathbb{N} \cup \{0\}$.

Proof

We have

$$b_n = \frac{1}{l} \int_{-l}^{l} f(x) \sin\left(\frac{n\pi x}{l}\right) dx$$

$$= \frac{1}{l} \left(\int_{-l}^{0} f(x) \sin\left(\frac{n\pi x}{l}\right) dx + \int_{0}^{l} f(x) \sin\left(\frac{n\pi x}{l}\right) dx \right). \qquad (7.38)$$

Now we assume that f is even. Making the change of variables $y = -x$, we obtain

$$\int_{-l}^{0} f(x) \sin\left(\frac{n\pi x}{l}\right) dx = -\int_{l}^{0} f(-y) \sin\left(-\frac{n\pi y}{l}\right) dy$$

$$= -\int_{0}^{l} f(y) \sin\left(\frac{n\pi y}{l}\right) dy.$$

Hence, it follows from (7.38) that $b_n = 0$. The second property can be obtained in a similar manner. □

Example 7.28

Let us consider the odd function $f(x) = x$ in the interval $[-l, l]$. By Proposition 7.27, we have $a_n = 0$ for $n \in \mathbb{N} \cup \{0\}$. Moreover,

$$b_n = \frac{1}{l} \int_{-l}^{l} x \sin\left(\frac{n\pi x}{l}\right) dx$$

$$= \left(-\frac{x}{n\pi} \cos\left(\frac{n\pi x}{l}\right) + \frac{l}{n^2\pi^2} \sin\left(\frac{n\pi x}{l}\right)\right)\Bigg|_{x=-l}^{x=l} = -\frac{2l}{n\pi}(-1)^n$$

for $n \in \mathbb{N}$. Hence, it follows from Theorem 7.8 that

$$x = \sum_{n=1}^{\infty} (-1)^{n+1} \frac{2l}{n\pi} \sin\left(\frac{n\pi x}{l}\right) \tag{7.39}$$

for each $x \in (-l, l)$.

7.5 Series of Cosines and Series of Sines

Now we consider a function $f \colon [0, l] \to \mathbb{R}$ satisfying the same conditions as the functions in $\mathcal{D}_l$ but with $-l$ replaced by 0 in the interval $[-l, l]$. More precisely, we assume that:

1. f has at most finitely many discontinuities;
2. f has left-sided and right-sided limits at all points of $[0, l]$;
3. for each open interval $(a, b) \subset [0, l]$ where f is continuous, the function $g \colon [a, b] \to \mathbb{R}$ defined by (7.18) has left-sided and right-sided (finite) derivatives at all points of $[a, b]$.

The following result shows that it is always possible to write f as a series of cosines and as a series of sines.

Theorem 7.29

The functions

$$G(x) = \frac{a_0}{2} + \sum_{n=1}^{\infty} a_n \cos\left(\frac{n\pi x}{l}\right) \quad \text{and} \quad H(x) = \sum_{n=1}^{\infty} b_n \sin\left(\frac{n\pi x}{l}\right), \quad (7.40)$$

where

$$a_n = \frac{2}{l} \int_0^l f(x) \cos\left(\frac{n\pi x}{l}\right) dx, \quad n \in \mathbb{N} \cup \{0\}, \quad (7.41)$$

and

$$b_n = \frac{2}{l} \int_0^l f(x) \sin\left(\frac{n\pi x}{l}\right) dx, \quad n \in \mathbb{N}, \quad (7.42)$$

satisfy

$$G(x) = H(x) = \frac{f(x^+) + f(x^-)}{2} \quad \text{for } x \in (0, l),$$

and

$$G(x) = f(x) \quad \text{and} \quad H(x) = 0 \quad \text{for } x \in \{0, l\}.$$

Proof

We consider the functions $g, h \colon [-l, l] \to \mathbb{R}$ given by

$$g(x) = \begin{cases} f(x) & \text{if } 0 \le x \le l, \\ f(-x) & \text{if } -l \le x < 0, \end{cases}$$

and

$$h(x) = \begin{cases} f(x) & \text{if } 0 < x \le l, \\ 0 & \text{if } x = 0, \\ -f(-x) & \text{if } -l \le x < 0. \end{cases}$$

One can easily verify that g is even and that h is odd. Moreover, it follows from the properties of the function f that $g, h \in \mathcal{D}_l$. By Proposition 7.27, the Fourier series of g is then the function G in (7.40), where

$$a_n = \frac{1}{l} \int_{-l}^l g(x) \cos\left(\frac{n\pi x}{l}\right) dx$$

$$= \frac{1}{l} \left(\int_{-l}^0 g(x) \cos\left(\frac{n\pi x}{l}\right) dx + \int_0^l f(x) \cos\left(\frac{n\pi x}{l}\right) dx \right).$$

Since g is even, making the change of variables $y = -x$, we obtain

$$\int_{-l}^{0} g(x) \cos\left(\frac{n\pi x}{l}\right) dx = -\int_{l}^{0} f(-y) \cos\left(-\frac{n\pi y}{l}\right) dy$$

$$= \int_{0}^{l} f(y) \cos\left(\frac{n\pi y}{l}\right) dy,$$

and thus a_n is given by (7.41).

It also follows from Proposition 7.27 that the Fourier series of h is the function H in (7.40), where

$$b_n = \frac{1}{l} \int_{-l}^{l} h(x) \sin\left(\frac{n\pi x}{l}\right) dx$$

$$= \frac{1}{l} \left(\int_{-l}^{0} h(x) \sin\left(\frac{n\pi x}{l}\right) dx + \int_{0}^{l} f(x) \sin\left(\frac{n\pi x}{l}\right) dx \right).$$

Since h is odd, making the change of variables $y = -x$, we conclude that b_n is given by (7.42). The remaining properties follow easily from Theorem 7.8.   □

Definition 7.30

The series in (7.40) are called respectively the *series of cosines* and the *series of sines* of the function f.

We give some examples.

Example 7.31

Let us consider the function $f: [0, \pi] \to \mathbb{R}$ given by $f(x) = 1$. It follows from (7.42) that

$$b_n = \frac{2}{\pi} \int_{0}^{\pi} \sin(nx)\, dx = \frac{2}{n\pi} \left(1 - (-1)^n\right)$$

for $n \in \mathbb{N}$. Therefore,

$$1 = \sum_{n=1}^{\infty} b_n \sin(nx) = \frac{4}{\pi} \left(\sin x + \frac{1}{3} \sin(3x) + \frac{1}{5} \sin(5x) + \cdots \right) \qquad (7.43)$$

for each $x \in (0, \pi)$. We note that this identity does not hold for $x = 0$ or $x = \pi$ (this also follows from Theorem 7.29). Taking $x = \pi/4$ in (7.43), we obtain

$$1 = \frac{4}{\pi} \left(\frac{1}{\sqrt{2}} + \frac{1}{3} \cdot \frac{1}{\sqrt{2}} - \frac{1}{5} \cdot \frac{1}{\sqrt{2}} - \frac{1}{7} \cdot \frac{1}{\sqrt{2}} + \cdots \right),$$

that is,

$$\frac{\pi}{2\sqrt{2}} = 1 + \frac{1}{3} - \frac{1}{5} - \frac{1}{7} + \frac{1}{9} + \frac{1}{11} - \cdots.$$

On the other hand, taking $x = \pi/3$ in (7.43), we obtain

$$1 = \frac{4}{\pi}\left(\frac{\sqrt{3}}{2} + \frac{1}{3} \cdot 0 - \frac{1}{5} \cdot \frac{\sqrt{3}}{2} + \frac{1}{7} \cdot \frac{\sqrt{3}}{2} + \cdots\right),$$

that is,

$$\frac{\pi}{2\sqrt{3}} = 1 - \frac{1}{5} + \frac{1}{7} - \frac{1}{11} + \frac{1}{13} - \frac{1}{17} + \cdots.$$

Example 7.32

Now we consider the function $f\colon [0,1] \to \mathbb{R}$ given by $f(x) = x$. It follows from (7.41) that

$$a_0 = 2\int_0^1 x\,dx = 1$$

and

$$a_n = 2\int_0^1 x\cos(n\pi x)\,dx = 2\frac{(-1)^n - 1}{(n\pi)^2}$$

for $n \in \mathbb{N}$. We also have

$$b_n = 2\int_0^1 x\sin(n\pi x)\,dx = -\frac{2(-1)^n}{n\pi}$$

for $n \in \mathbb{N}$. Therefore,

$$x = \frac{1}{2} + \sum_{m=1}^{\infty} \frac{-4}{(2m-1)^2\pi^2}\cos\bigl((2m-1)\pi x\bigr)$$

and

$$x = \sum_{n=1}^{\infty} -\frac{2(-1)^n}{n\pi}\sin(n\pi x)$$

for each $x \in (0,1)$. For example, taking $x = 1/4$ in the first series, we obtain

$$\frac{1}{4} = \frac{1}{2} - \frac{4}{\pi^2}\left(\frac{1}{\sqrt{2}} - \frac{1}{3^2}\cdot\frac{1}{\sqrt{2}} - \frac{1}{5^2}\cdot\frac{1}{\sqrt{2}} + \cdots\right),$$

that is,

$$\frac{\pi^2}{8\sqrt{2}} = 1 - \frac{1}{3^2} - \frac{1}{5^2} + \frac{1}{7^2} + \frac{1}{9^2} - \cdots .$$

7.6   Integration and Differentiation Term by Term

In this section we show that Fourier series can be integrated and differentiated term by term. In other words, the integral and the derivative of a given function (in some appropriate class) can be computed by taking respectively integrals and derivatives term by term in its Fourier series.

We start with integration.

Theorem 7.33

Let $f\colon [-l, l] \to \mathbb{R}$ be a function having at most finitely many discontinuities, and with left-sided and right-sided limits at all points of $[-l, l]$. For each interval $[a, b] \subset [-l, l]$, we have

$$\int_a^b f(x)\, dx = \frac{a_0}{2}(b - a) + \sum_{n=1}^{\infty} \left[a_n \sin\left(\frac{n\pi x}{l}\right) - b_n \cos\left(\frac{n\pi x}{l}\right) \right]\Big|_{x=a}^{x=b}. \quad (7.44)$$

Proof

Let us consider the continuous function $g\colon [-l, l] \to \mathbb{R}$ given by

$$g(x) = \int_{-l}^{x} \left(f(y) - \frac{a_0}{2} \right) dy$$

(we note that f is Riemann-integrable). One can easily verify that $g \in \mathcal{D}_l$. Moreover, $g(-l) = 0$ and

$$g(l) = \int_{-l}^{l} \left(f(y) - \frac{a_0}{2} \right) dy = \int_{-l}^{l} f(y)\, dy - a_0 l = 0.$$

By Theorem 7.8, we then have

$$g(x) = \frac{A_0}{2} + \sum_{n=1}^{\infty} \left[A_n \cos\left(\frac{n\pi x}{l}\right) + B_n \sin\left(\frac{n\pi x}{l}\right) \right]$$

for every $x \in [-l, l]$, where A_n and B_n are the Fourier coefficients of the function g. For each $n \in \mathbb{N}$, we have

$$
\begin{aligned}
A_n &= \frac{1}{l} \int_{-l}^{l} g(x) \cos\left(\frac{n\pi x}{l}\right) dx \\
&= \frac{1}{n\pi} g(x) \sin\left(\frac{n\pi x}{l}\right)\Big|_{x=-l}^{x=l} \\
&\quad - \frac{1}{n\pi} \int_{-l}^{l} \left(f(x) - \frac{a_0}{2}\right) \sin\left(\frac{n\pi x}{l}\right) dx = -\frac{b_n}{n\pi/l} \quad (7.45)
\end{aligned}
$$

and

$$
\begin{aligned}
B_n &= \frac{1}{l} \int_{-l}^{l} g(x) \sin\left(\frac{n\pi x}{l}\right) dx \\
&= -\frac{1}{n\pi} g(x) \cos\left(\frac{n\pi x}{l}\right)\Big|_{x=-l}^{x=l} \\
&\quad + \frac{1}{n\pi} \int_{-l}^{l} \left(f(x) - \frac{a_0}{2}\right) \cos\left(\frac{n\pi x}{l}\right) dx = \frac{a_n}{n\pi/l}. \quad (7.46)
\end{aligned}
$$

Therefore,

$$
g(x) = \frac{A_0}{2} + \sum_{n=1}^{\infty} \frac{a_n \sin(n\pi x/l) - b_n \cos(n\pi x/l)}{n\pi/l},
$$

that is,

$$
\int_{-l}^{x} f(y)\, dy = \frac{a_0(x+l)}{2} + \frac{A_0}{2} + \sum_{n=1}^{\infty} \frac{a_n \sin(n\pi x/l) - b_n \cos(n\pi x/l)}{n\pi/l}, \quad (7.47)
$$

for every $x \in [-l, l]$. Since

$$
\int_{a}^{b} f(y)\, dy = \int_{-l}^{b} f(y)\, dy - \int_{-l}^{a} f(y)\, dy,
$$

identity (7.44) now follows immediately from (7.47). $\square$

We note that one does not assume in Theorem 7.33 that $f \in \mathcal{D}_l$. Now we consider the differentiation of Fourier series.

Theorem 7.34

If $f \colon [-l, l] \to \mathbb{R}$ is the restriction to the interval $[-l, l]$ of a $2l$-periodic function $g \colon \mathbb{R} \to \mathbb{R}$ of class C^1 such that g' has left-sided and right-sided (finite) derivatives at all points, then

$$f'(x) = \sum_{n=1}^{\infty} \left[\frac{n\pi}{l} b_n \cos\left(\frac{n\pi x}{l}\right) - \frac{n\pi}{l} a_n \sin\left(\frac{n\pi x}{l}\right) \right]$$

for every $x \in (-l, l)$.

More generally, we have the following result.

Theorem 7.35

If $f \colon [-l, l] \to \mathbb{R}$ is the restriction to the interval $[-l, l]$ of a function $g \colon \mathbb{R} \to \mathbb{R}$ of class C^1 such that g' has left-sided and right-sided (finite) derivatives at all points, then

$$f'(x) = \frac{c}{2} + \sum_{n=1}^{\infty} \left[\left(\frac{n\pi}{l} b_n + (-1)^n c \right) \cos\left(\frac{n\pi x}{l}\right) - \frac{n\pi}{l} a_n \sin\left(\frac{n\pi x}{l}\right) \right] \quad (7.48)$$

for every $x \in (-l, l)$, where $c = [f(l) - f(-l)]/l$.

Proof

Since g' is continuous and its restriction to $[-l, l]$ is in $\mathcal{D}_l$, we have

$$f'(x) = \frac{\alpha_0}{2} + \sum_{n=1}^{\infty} \left[\alpha_n \cos\left(\frac{n\pi x}{l}\right) + \beta_n \sin\left(\frac{n\pi x}{l}\right) \right]$$

for every $x \in (-l, l)$, where α_n and β_n are the Fourier coefficients of the function g'. We note that

$$\alpha_0 = \frac{1}{l} \int_{-l}^{l} f'(x)\, dx = \frac{f(l) - f(-l)}{l} = c,$$

and thus,

$$f'(x) - \frac{c}{2} = \sum_{n=1}^{\infty} \left[\alpha_n \cos\left(\frac{n\pi x}{l}\right) + \beta_n \sin\left(\frac{n\pi x}{l}\right) \right] \quad (7.49)$$

for every $x \in (-l, l)$. Now we consider the function $h \colon [-l, l] \to \mathbb{R}$ given by

$$h(x) = \int_{-l}^{x} \left(f'(y) - \frac{c}{2} \right) dy = f(x) - f(-l) - \frac{c}{2}(x + l).$$

By (7.45) and (7.46), the Fourier coefficients of h satisfy

$$A_n = -\frac{B_n}{n\pi/l} \quad \text{and} \quad B_n = \frac{\alpha_n}{n\pi/l}$$

for each $n \in \mathbb{N}$. It then follows from (7.39) that the Fourier series of f is given by

$$f(x) = f(-l) + \frac{cl}{2} + \frac{A_0}{2}$$
$$+ \sum_{n=1}^{\infty} \left[-\frac{B_n l}{n\pi} \cos\left(\frac{n\pi x}{l}\right) + \left(\frac{\alpha_n l}{n\pi} + (-1)^{n+1}\frac{cl}{n\pi}\right) \sin\left(\frac{n\pi x}{l}\right) \right],$$

where

$$A_0 = \frac{1}{l} \int_{-l}^{l} h(x)\, dx.$$

Hence,

$$a_n = -\frac{B_n l}{n\pi} \quad \text{and} \quad b_n = \frac{\alpha_n l}{n\pi} + (-1)^{n+1}\frac{cl}{n\pi},$$

that is,

$$a_n = \frac{n\pi}{l} b_n + (-1)^n c \quad \text{and} \quad \beta_n = -\frac{n\pi}{l} a_n,$$

for each $n \in \mathbb{N}$. Identity (7.48) now follows readily from (7.49).   □

7.7 Solved Problems and Exercises

Problem 7.1

Find the Fourier series of the function

$$f(x) = \begin{cases} 0 & \text{if } -1 \le x < 0, \\ 2 & \text{if } 0 \le x \le 1. \end{cases} \tag{7.50}$$

Solution

Clearly, $f \in \mathcal{D}_1$. We have

$$a_0 = \int_{-1}^{1} f(x)\, dx = \int_{0}^{1} 2\, dx = 2,$$

and

$$a_n = \int_{-1}^{1} f(x)\cos(n\pi x)\,dx = 2\int_{0}^{1}\cos(n\pi x)\,dx = 0$$

for $n \in \mathbb{N}$. We also have

$$b_n = \int_{-1}^{1} f(x)\sin(n\pi x)\,dx = 2\int_{0}^{1}\sin(n\pi x)\,dx$$

$$= \frac{2}{n\pi}(1 - \cos(n\pi)) = 2\frac{1 - (-1)^n}{n\pi} = \begin{cases} 4/(n\pi) & \text{if } n \text{ is odd,} \\ 0 & \text{if } n \text{ is even} \end{cases}$$

for $n \in \mathbb{N}$. The Fourier series of f is then

$$F(x) = 1 + \sum_{m=1}^{\infty} \frac{4}{(2m-1)\pi}\sin\big((2m-1)\pi x\big). \tag{7.51}$$

Problem 7.2

For the function f in Problem 7.1, find explicitly the values of the Fourier series at each point of the interval $[-1, 1]$.

Solution

By Theorem 7.8, the Fourier series in (7.51) takes the values

$$F(x) = \frac{f(x^+) + f(x^-)}{2} \quad \text{for } x \in (-1, 1),$$

and

$$F(x) = \frac{f(-1^+) + f(1^-)}{2} \quad \text{for } x \in \{-1, 1\}.$$

Therefore,

$$F(x) = \begin{cases} f(x) & \text{if } x \in (-1, 0) \cup (0, 1), \\ 1 & \text{if } x \in \{-1, 0, 1\}. \end{cases} \tag{7.52}$$

Problem 7.3

Use Problem 7.2 to show that

$$\frac{\pi}{4} = 1 - \frac{1}{3} + \frac{1}{5} - \frac{1}{7} + \cdots \tag{7.53}$$

and

$$\frac{\pi}{2\sqrt{2}} = 1 + \frac{1}{3} - \frac{1}{5} - \frac{1}{7} + \frac{1}{9} + \frac{1}{11} - \cdots . \tag{7.54}$$

Solution

Taking $x = 1/2$ in (7.51), it follows from (7.52) that

$$2 = 1 + \sum_{m=1}^{\infty} \frac{4}{(2m-1)\pi} \sin\left(\frac{(2m-1)\pi}{2}\right)$$

$$= 1 + \frac{4}{\pi} - \frac{4}{3\pi} + \frac{4}{5\pi} - \cdots .$$

This yields identity (7.53). On the other hand, taking $x = 1/4$ in (7.51), it follows from (7.52) that

$$2 = 1 + \sum_{m=1}^{\infty} \frac{4}{(2m-1)\pi} \sin\left(\frac{(2m-1)\pi}{4}\right)$$

$$= 1 + \frac{4}{\pi\sqrt{2}} + \frac{4}{3\pi\sqrt{2}} - \frac{4}{5\pi\sqrt{2}} - \frac{4}{7\pi\sqrt{2}} + \cdots ,$$

which yields identity (7.54).

Problem 7.4

Find the Fourier series of the function $g \colon [-1,1] \to \mathbb{R}$ given by

$$g(x) = \int_0^x y f(y) \, dy,$$

with f as in (7.50).

Solution

It follows from (7.50) that

$$g(x) = \begin{cases} 0 & \text{if } -1 \le x < 0, \\ x^2 & \text{if } 0 \le x \le 1. \end{cases} \tag{7.55}$$

Clearly, $g \in \mathcal{D}_1$. The Fourier coefficients of the function g are then

$$a_0 = \int_0^1 x^2 \, dx = \frac{1}{3}$$

and

$$a_n = \int_0^1 x^2 \cos(n\pi x)\, dx$$

$$= \frac{2n\pi x \cos(n\pi x) + (n^2\pi^2 x^2 - 2)\sin(n\pi x)}{(n\pi)^3}\Big|_{x=0}^{x=1} = \frac{2(-1)^n}{(n\pi)^2}$$

for $n \in \mathbb{N}$. We also have

$$b_n = \int_0^1 x^2 \sin(n\pi x)\, dx$$

$$= \frac{2n\pi x \sin(n\pi x) - (n^2\pi^2 x^2 - 2)\cos(n\pi x)}{(n\pi)^3}\Big|_{x=0}^{x=1}$$

$$= \frac{(2 - n^2\pi^2)(-1)^n - 2}{(n\pi)^3} = \begin{cases} (n^2\pi^2 - 4)/(n\pi)^3 & \text{if } n \text{ is odd,} \\ -1/(n\pi) & \text{if } n \text{ is even} \end{cases}$$

for $n \in \mathbb{N}$. Therefore, the Fourier series of g is

$$G(x) = \frac{1}{6} + \sum_{n=1}^{\infty} \frac{2(-1)^n}{(n\pi)^2} \cos(n\pi x)$$

$$+ \sum_{m=1}^{\infty} \left[\frac{(2m-1)^2\pi^2 - 4}{(2m-1)^3\pi^3} \sin\big((2m-1)\pi x\big) - \frac{1}{2m\pi} \sin(2m\pi x) \right]. \quad (7.56)$$

Problem 7.5

Use Problem 7.4 to show that

$$\frac{\pi^2}{12} = 1 - \frac{1}{2^2} + \frac{1}{3^2} - \frac{1}{4^2} + \cdots. \quad (7.57)$$

Solution

By Theorem 7.8, the Fourier series of the function g in (7.55) has the values

$$G(x) = \begin{cases} g(x) & \text{if } x \in (-1,1), \\ 1/2 & \text{if } x \in \{-1,1\}. \end{cases}$$

Thus, taking $x = 0$ in (7.56), we obtain

$$\frac{1}{6} + \sum_{n=1}^{\infty} \frac{2(-1)^n}{(n\pi)^2} = \frac{1}{6} - \frac{2}{\pi^2}\left(1 - \frac{1}{2^2} + \frac{1}{3^2} - \frac{1}{4^2} + \cdots\right) = 0.$$

This yields identity (7.57).

Problem 7.6

Find the Fourier series of the function $f(x) = x - |x|$ in the interval $[-1, 1]$.

Solution

We have

$$f(x) = x - |x| = \begin{cases} 2x & \text{if } -1 \le x \le 0, \\ 0 & \text{if } 0 \le x \le 1. \end{cases} \tag{7.58}$$

Clearly, $f \in \mathcal{D}_1$. Thus,

$$a_0 = \int_{-1}^{1} (x - |x|)\, dx = \int_{-1}^{0} 2x\, dx = -1$$

and

$$a_n = \int_{-1}^{1} (x - |x|)\cos(n\pi x)\, dx = 2\int_{-1}^{0} x\cos(n\pi x)\, dx$$

$$= 2\left(x\frac{\sin(n\pi x)}{n\pi} + \frac{\cos(n\pi x)}{(n\pi)^2} \right)\Bigg|_{x=-1}^{x=0} = 2\frac{1 - \cos(n\pi)}{(n\pi)^2}.$$

$$= 2\frac{1 - (-1)^n}{(n\pi)^2} = \begin{cases} 4/(n\pi)^2 & \text{if } n \text{ is odd,} \\ 0 & \text{if } n \text{ is even} \end{cases}$$

for $n \in \mathbb{N}$. We also have

$$b_n = \int_{-1}^{1} (x - |x|)\sin(n\pi x)\, dx = 2\int_{-1}^{0} x\sin(n\pi x)\, dx$$

$$= 2\left(-x\frac{\cos(n\pi x)}{n\pi} + \frac{\sin(n\pi x)}{(n\pi)^2} \right)\Bigg|_{x=-1}^{x=0} = \frac{2(-1)^{n+1}}{n\pi}$$

for $n \in \mathbb{N}$. Therefore, the Fourier series of f is

$$F(x) = -\frac{1}{2} + \sum_{m=1}^{\infty} \frac{4}{(2m-1)^2\pi^2}\cos((2m-1)\pi x)$$

$$+ \sum_{n=1}^{\infty} \frac{2(-1)^{n+1}}{n\pi}\sin(n\pi x). \tag{7.59}$$

Problem 7.7

Use Problem 7.6 to show that

$$\frac{\pi^2}{8} = 1 + \frac{1}{3^2} + \frac{1}{5^2} + \frac{1}{7^2} + \cdots. \qquad (7.60)$$

Solution

By Theorem 7.8 and (7.58), the Fourier series of the function $x - |x|$ is given by

$$F(x) = \begin{cases} x - |x| & \text{if } x \in (-1, 1), \\ -1 & \text{if } x \in \{-1, 1\}. \end{cases}$$

Thus, taking $x = 0$ in (7.59), we obtain

$$0 = -\frac{1}{2} + \sum_{m=1}^{\infty} \frac{4}{(2m-1)^2 \pi^2}$$

$$= -\frac{1}{2} + \frac{4}{\pi^2}\left(1 + \frac{1}{3^2} + \frac{1}{5^2} + \frac{1}{7^2} + \cdots\right).$$

This yields identity (7.60).

Problem 7.8

Verify that the product of odd functions is an even function.

Solution

Let $f, g \colon [-l, l] \to \mathbb{R}$ be odd functions. This means that

$$f(-x) = -f(x) \quad \text{and} \quad g(-x) = -g(x)$$

for every $x \in [-l, l]$. Then the function $h(x) = f(x)g(x)$ satisfies

$$h(-x) = f(-x)g(-x)$$
$$= \left(-f(x)\right)\left(-g(x)\right)$$
$$= f(x)g(x) = h(x)$$

for every $x \in [-l, l]$; that is, h is an even function.

Problem 7.9

Verify that the Fourier coefficients a_n of an odd function are zero.

Solution

We have

$$a_0 = \frac{1}{l} \int_{-l}^{l} f(x)\, dx = \frac{1}{l} \left(\int_{-l}^{0} f(x)\, dx + \int_{0}^{l} f(x)\, dx \right), \qquad (7.61)$$

and

$$a_n = \frac{1}{l} \int_{-l}^{l} f(x) \cos\left(\frac{n\pi x}{l} \right) dx$$

$$= \frac{1}{l} \left(\int_{-l}^{0} f(x) \cos\left(\frac{n\pi x}{l} \right) dx + \int_{0}^{l} f(x) \cos\left(\frac{n\pi x}{l} \right) dx \right) \qquad (7.62)$$

for $n \in \mathbb{N}$. Now we assume that f is odd. Making the change of variables $y = -x$, we obtain

$$\int_{-l}^{0} f(x)\, dx = -\int_{l}^{0} f(-y)\, dy$$

$$= \int_{0}^{l} f(-y)\, dy = -\int_{0}^{l} f(y)\, dy,$$

and it follows from (7.61) that $a_0 = 0$. Similarly, again making the change of variables $y = -x$, we obtain

$$\int_{-l}^{0} f(x) \cos\left(\frac{n\pi x}{l} \right) dx = -\int_{l}^{0} f(-y) \cos\left(-\frac{n\pi y}{l} \right) dy$$

$$= -\int_{0}^{l} f(y) \cos\left(\frac{n\pi y}{l} \right) dy,$$

and it follows from (7.62) that $a_n = 0$ for $n \in \mathbb{N}$.

Problem 7.10

Find the series of cosines of the function x in the interval $[0, 1]$.

Solution

By Theorem 7.29, the series of cosines of a function f in the interval $[0, l]$ is

$$G(x) = \frac{a_0}{2} + \sum_{n=1}^{\infty} a_n \cos\left(\frac{n\pi x}{l}\right),$$

where

$$a_n = \frac{2}{l} \int_0^l f(x) \cos\left(\frac{n\pi x}{l}\right) dx, \quad n \in \mathbb{N} \cup \{0\}.$$

For $f(x) = x$ and $l = 1$, we obtain

$$a_0 = 2 \int_0^1 x\, dx = 1$$

and

$$
\begin{aligned}
a_n &= 2 \int_0^1 x \cos(n\pi x)\, dx \\
&= 2\left(x\frac{\sin(n\pi x)}{n\pi} + \frac{\cos(n\pi x)}{(n\pi)^2} \right)\Bigg|_{x=0}^{x=1} = 2\frac{(-1)^n - 1}{(n\pi)^2}
\end{aligned}
$$

for $n \in \mathbb{N}$. Thus, the series of cosines of the function x is

$$G(x) = \frac{1}{2} - \sum_{m=1}^{\infty} \frac{4}{(2m-1)^2\pi^2} \cos\big((2m-1)\pi x\big).$$

Problem 7.11

Find the series of sines of the function $\cos x$ in the interval $[0, \pi]$.

Solution

By Theorem 7.29, the series of sines of a function f in the interval $[0, l]$ is

$$H(x) = \sum_{n=1}^{\infty} b_n \sin\left(\frac{n\pi x}{l}\right),$$

where

$$b_n = \frac{2}{l} \int_0^l f(x) \sin\left(\frac{n\pi x}{l}\right) dx, \quad n \in \mathbb{N}.$$

For $f(x) = \cos x$ and $l = \pi$, we obtain

$$b_1 = \frac{2}{\pi} \int_0^\pi \cos x \sin x \, dx = 0$$

and

$$b_n = \frac{2}{\pi} \int_0^\pi \cos x \sin(nx) \, dx$$

$$= \frac{2}{\pi} \left. \frac{n \cos x \cos(nx) + \sin x \sin(nx)}{1 - n^2} \right|_{x=0}^{x=\pi} = \frac{2n(1 + (-1)^n)}{(n^2 - 1)\pi}$$

for $n > 1$. Thus, the series of sines of $\cos x$ is

$$H(x) = \sum_{m=1}^\infty \frac{8m}{(4m^2 - 1)\pi} \sin(2mx).$$

Problem 7.12

Let $f \colon \mathbb{R} \to \mathbb{R}$ be a function of class C^1. For each $n \in \mathbb{N}$, show that the Fourier coefficient

$$a_n = \frac{1}{l} \int_{-l}^l f(x) \cos\left(\frac{n\pi x}{l}\right) dx$$

is also given by

$$a_n = -\frac{1}{n\pi} \int_{-l}^l f'(x) \sin\left(\frac{n\pi x}{l}\right) dx.$$

Solution

We have

$$a_n = \frac{1}{l} \int_{-l}^l f(x) \cos\left(\frac{n\pi x}{l}\right) dx$$

$$= \frac{1}{n\pi} \left. f(x) \sin\left(\frac{n\pi x}{l}\right) \right|_{x=-l}^{x=l} - \frac{1}{n\pi} \int_{-l}^l f'(x) \sin\left(\frac{n\pi x}{l}\right) dx$$

$$= \frac{1}{n\pi} [f(l)\sin(n\pi) - f(-l)\sin(-n\pi)] - \frac{1}{n\pi} \int_{-l}^l f'(x) \sin\left(\frac{n\pi x}{l}\right) dx$$

$$= -\frac{1}{n\pi} \int_{-l}^l f'(x) \sin\left(\frac{n\pi x}{l}\right) dx,$$

because $\sin(n\pi) = \sin(-n\pi) = 0$.

EXERCISES

7.1. Find the Fourier series of the function:

(a) $\begin{cases} 1 & \text{if } -1 \le x < 0, \\ 2 & \text{if } 0 \le x \le 1; \end{cases}$

(b) $\sin(x/2)$ in the interval $[-\pi, \pi]$;

(c) e^x in the interval $[-1, 1]$;

(d) e^{2x} in the interval $[-\pi, \pi]$;

(e) $(1+x)/2$ in the interval $[-1, 1]$;

(f) $|x|$ in the interval $[-3, 3]$;

(g) $x + |x|$ in the interval $[-2, 2]$.

7.2. Write the function

$$f(x) = \begin{cases} 0 & \text{if } 0 < x \le 2, \\ 1 & \text{if } 2 < x \le 3 \end{cases}$$

as a:

(a) series of sines for $0 < x < 3$;

(b) series of cosines for $0 < x < 3$.

7.3. Find the Fourier series of the even function:

(a) $\cos(2x)$ in the interval $[-\pi, \pi]$;

(b) $1 + |x|$ in the interval $[-1, 1]$.

7.4. Find the Fourier series of the odd function:

(a) x in the interval $[-\pi, \pi]$;

(b) $x - \sin x$ in the interval $[-\pi, \pi]$.

7.5. Verify that the sum of even functions is an even function.

7.6. Find all even polynomials.

7.7. Consider the function $f \colon [-\pi, \pi] \to \mathbb{R}$ given by $f(x) = x^2$.

(a) Verify that the Fourier series of f is

$$\frac{\pi^2}{3} + \sum_{n=1}^{\infty} \frac{4(-1)^n}{n^2} \cos(nx).$$

(b) Use the Fourier series to show that

$$\sum_{n=1}^{\infty} \frac{1}{n^2} = \frac{\pi^2}{6}.$$

(c) Find the series of sines of f in $[0, \pi]$.

7.8. Consider the function $f \colon [0, \pi] \to \mathbb{R}$ given by $f(x) = x(\pi - x)$.

(a) Find the series of sines of f.

(b) Use the series to show that

$$\sum_{n=0}^{\infty} \frac{(-1)^n}{(2n+1)^3} = \frac{\pi^3}{32}.$$

7.9. Verify that if $f: \mathbb{R} \to \mathbb{R}$ is a function of class C^2, then $na_n \to 0$ when $n \to \infty$.

7.10. Show that

$$|\sin x| = \frac{2}{\pi} - \frac{4}{\pi} \sum_{n=1}^{\infty} \frac{\cos(2nx)}{4n^2 - 1}$$

for every $x \in \mathbb{R}$.

7.11. For each $a \in \mathbb{R} \setminus \mathbb{Z}$, verify that:

(a) the Fourier series of $\sin(ax)$ in the interval $[-\pi, \pi]$ is

$$\frac{2\sin(ax)}{\pi} \sum_{n=1}^{\infty} (-1)^n \frac{n\sin(nx)}{a^2 - n^2};$$

(b) the Fourier series of $\cos(ax)$ in the interval $[-\pi, \pi]$ is

$$\frac{2\sin(ax)}{\pi} \left(\frac{1}{2a} + \sum_{n=1}^{\infty} (-1)^n \frac{a\cos(nx)}{a^2 - n^2} \right).$$

8

Partial Differential Equations

In this chapter we study some classes of partial differential equations, including the heat equation, the Laplace equation, and the wave equation. In particular, based on the study of Fourier series, we find solutions for several equations and several types of boundary conditions. We mainly use the method of separation of variables. In contrast to what happens in all former chapters, here not everything is proved since this would require additional techniques. One notable exception is the proof of existence and uniqueness of solutions for the heat equation under certain assumptions.

8.1 Heat Equation and Its Modifications

Again we consider the heat equation

$$\frac{\partial u}{\partial t} = \kappa \frac{\partial^2 u}{\partial x^2},$$

(8.1)

for $t \geq 0$ and $x \in [0, l]$, with $\kappa, l > 0$. By Proposition 7.4, for each $N \in \mathbb{N}$ and $c_1, \ldots, c_N \in \mathbb{R}$, the function

$$u(t, x) = \sum_{n=1}^{N} c_n e^{-n^2 \pi^2 \kappa t / l^2} \sin\left(\frac{n\pi x}{l}\right)$$

L. Barreira, C. Valls, *Complex Analysis and Differential Equations*,
Springer Undergraduate Mathematics Series,
DOI 10.1007/978-1-4471-4008-5_8, © Springer-Verlag London 2012

is a solution of equation (8.1) satisfying the condition

$$u(t,0) = u(t,l) = 0, \quad t \geq 0. \tag{8.2}$$

Now we make the additional assumption that

$$u(0,x) = f(x), \quad x \in [0,l], \tag{8.3}$$

and show that for a certain class of functions $f \colon [0,l] \to \mathbb{R}$ there exist constants c_n such that the function

$$u(t,x) = \sum_{n=1}^{\infty} c_n e^{-n^2 \pi^2 \kappa t / l^2} \sin\left(\frac{n\pi x}{l}\right) \tag{8.4}$$

is well defined and is a solution of equation (8.1) satisfying conditions (8.2) and (8.3).

Proposition 8.1

If $f \colon [0,l] \to \mathbb{R}$ is a continuous function with $f(0) = f(l) = 0$, and there exist points $0 = x_0 < x_1 < \cdots < x_m = l$ such that the restriction of f to (x_i, x_{i+1}) has an extension of class C^1 to some open interval containing $[x_i, x_{i+1}]$, for $i = 0, \ldots, m-1$, then the function $u(t,x)$ in (8.4), with

$$c_n = \frac{2}{l} \int_0^l f(x) \sin\left(\frac{n\pi x}{l}\right) dx \quad \text{for } n \in \mathbb{N}, \tag{8.5}$$

is well defined and is the unique function with the following properties:
1. u is continuous in $\mathbb{R}_0^+ \times [0,l]$;
2. u satisfies equation (8.1) in $\mathbb{R}^+ \times (0,l)$, which in particular includes the existence of the derivatives $\partial u / \partial t$ and $\partial^2 u / \partial x^2$;
3. u satisfies conditions (8.2) and (8.3).

Moreover, the function u is of class C^∞ in $\mathbb{R}^+ \times (0,l)$.

Proof

We first show that the series in (8.4) converges. Letting

$$I = \frac{2}{l} \int_0^l |f(x)| \, dx,$$

it follows from (8.5) that $|c_n| \le I$ for each $n \in \mathbb{N}$. Hence, for $t \ge \tau > 0$, we have

$$\sum_{n=m}^{\infty} |c_n| e^{-n^2\pi^2\kappa t/l^2} \le I \sum_{n=m}^{\infty} e^{-n^2\pi^2\kappa\tau/l^2}$$

$$\le I \sum_{n=m}^{\infty} e^{-n\pi^2\kappa\tau/l^2}$$

$$= I \frac{e^{-m\pi^2\kappa\tau/l^2}}{1 - e^{-\pi^2\kappa\tau/l^2}} \to 0$$

when $m \to \infty$, and the series

$$\sum_{n=1}^{\infty} |c_n| e^{-n^2\pi^2\kappa t/l^2}$$

is uniformly convergent on $[\tau, +\infty)$, for each $\tau > 0$. By Exercise 3.21, the series in (8.4) is uniformly convergent on $[\tau, +\infty) \times [0, l]$, for each $\tau > 0$, and the function u is continuous in that set. Hence, u is also continuous in $\mathbb{R}^+ \times [0, l]$.

For each $p, q \in \mathbb{N} \cup \{0\}$, let us consider the series

$$\sum_{n=1}^{\infty} |c_n| \cdot \left| \frac{\partial^{p+q}}{\partial t^p \partial x^q} \left(e^{-n^2\pi^2\kappa t/l^2} \sin\left(\frac{n\pi x}{l}\right) \right) \right|$$

$$\le I \sum_{n=1}^{\infty} \left(\frac{n^2\pi^2\kappa}{l^2} \right)^p \left(\frac{n\pi}{l} \right)^q e^{-n^2\pi^2\kappa t/l^2}. \tag{8.6}$$

Since the power series $\sum_{n=1}^{\infty} n^{2p+q} z^n$ has radius of convergence $R = 1$, it follows from Theorem 4.8 that for each $p \in \mathbb{N}$ and $\tau > 0$ the last series in (8.6) is uniformly convergent on $[\tau, +\infty)$. By Exercise 3.21, it follows by induction that u is of class C^∞ in $\mathbb{R}^+ \times (0, l)$, with

$$\frac{\partial^{p+q}u}{\partial t^p \partial x^q} = \sum_{n=1}^{\infty} c_n \left(-\frac{n^2\pi^2\kappa}{l^2} \right)^p e^{-n^2\pi^2\kappa t/l^2} \frac{\partial^q}{\partial x^q} \sin\left(\frac{n\pi x}{l}\right)$$

for each $p, q \in \mathbb{N} \cup \{0\}$. In particular,

$$\frac{\partial u}{\partial t} = \sum_{n=1}^{\infty} c_n \left(-\frac{n^2\pi^2\kappa}{l^2} \right) e^{-n^2\pi^2\kappa t/l^2} \sin\left(\frac{n\pi x}{l}\right)$$

and

$$\frac{\partial^2 u}{\partial x^2} = \sum_{n=1}^{\infty} c_n \left(\frac{n\pi}{l} \right)^2 e^{-n^2\pi^2\kappa t/l^2} \sin\left(\frac{n\pi x}{l}\right),$$

which shows that u satisfies equation (8.1) in $\mathbb{R}^+ \times (0, l)$.

In order to establish the uniqueness of the solution, we first prove the following result.

Lemma 8.2 (Weak maximum principle)

Let $v\colon \mathbb{R}_0^+ \times [0,l] \to \mathbb{R}$ be a continuous function. If v is of class C^2 in $\mathbb{R}^+ \times (0,l)$ and satisfies equation (8.1) in this set, then for each $\tau > 0$, we have

$$v(t,x) \le \sup\{v(s,y) : (s,y) \in \{0\} \times [0,l] \cup [0,\tau] \times \{0,l\}\} \qquad (8.7)$$

for $(t,x) \in (0,\tau) \times (0,l)$.

Proof of the lemma

Otherwise, the maximum of v in $[0,\tau] \times [0,l]$ would be attained at a point (s,y) in $(0,\tau) \times (0,l)$ or in $\{\tau\} \times [0,l]$. Then $v(s,y) > L$, where L is the supremum in (8.7). Given $\varepsilon > 0$, let us consider the function

$$w(t,x) = v(t,x) + \varepsilon(x-y)^2.$$

We note that

$$\sup\{w(t,x) : (t,x) \in \{0\} \times [0,l] \cup [0,\tau] \times \{0,l\}\} \le L + \varepsilon l^2 < v(s,y)$$

for any sufficiently small ε. On the other hand, $w(s,y) = v(s,y)$, and hence, the function w has a maximum greater than or equal to $v(s,y)$ at a point $p \in (0,s] \times (0,l)$. One can easily verify that if $p \in (0,s) \times (0,l)$, then

$$\frac{\partial w}{\partial t}(p) = 0 \quad \text{and} \quad \frac{\partial^2 w}{\partial x^2}(p) \le 0,$$

and that if $p \in \{s\} \times (0,l)$, then

$$\frac{\partial w}{\partial t}(p) \ge 0 \quad \text{and} \quad \frac{\partial^2 w}{\partial x^2}(p) \le 0.$$

In both cases, we have

$$\frac{\partial w}{\partial t}(p) - \kappa \frac{\partial^2 w}{\partial x^2}(p) \ge 0.$$

On the other hand, we also have

$$\frac{\partial w}{\partial t}(p) - \kappa \frac{\partial^2 w}{\partial x^2}(p) = \frac{\partial v}{\partial t}(p) - \kappa \frac{\partial^2 v}{\partial x^2}(p) - 2\varepsilon = -2\varepsilon < 0.$$

This contradiction yields the desired result.

Now let $u, v \colon \mathbb{R}_0^+ \times [0, l] \to \mathbb{R}$ be continuous functions that are of class C^∞ and satisfy equation (8.1) in $\mathbb{R}^+ \times (0, l)$, together with the conditions

$$u(t, 0) = u(t, l) = v(t, 0) = v(t, l) = 0, \quad t \geq 0,$$

and

$$u(0, x) = v(0, x) = f(x), \quad x \in [0, l].$$

The function $w = u - v$ is continuous in $\mathbb{R}_0^+ \times [0, l]$. Moreover, it is of class C^∞ and satisfies equation (8.1) in $\mathbb{R}^+ \times (0, l)$, together with the conditions

$$w(t, 0) = w(t, l) = 0, \quad t \geq 0,$$

and

$$w(0, x) = 0, \quad x \in [0, l].$$

It then follows from Lemma 8.2 that

$$\big| w(t, x) \big| \leq \sup\big\{ \big| w(s, y) \big| : s = 0 \text{ or } y \in \{0, l\} \big\} = 0,$$

and thus $w = 0$. This shows that $u = v$.

We also show that u is continuous at $t = 0$. By Theorem 7.14, the series

$$\sum_{n=1}^{\infty} c_n \sin\left(\frac{n\pi x}{l} \right)$$

is uniformly convergent on $[0, l]$ (since the Fourier series of the odd extension of f to the interval $[-l, l]$ is uniformly convergent). Now let

$$S_m(x) = \sum_{n=1}^{m} c_n \sin\left(\frac{n\pi x}{l} \right)$$

be the corresponding sequence of partial sums. Given $\varepsilon > 0$, there exists $p \in \mathbb{N}$ such that

$$\big| S_m(x) - f(x) \big| < \varepsilon \quad \text{for } x \in [0, l], \ m \geq p.$$

We also consider the functions

$$S_m(t, x) = \sum_{n=1}^{m} c_n e^{-n^2 \pi^2 \kappa t / l^2} \sin\left(\frac{n\pi x}{l} \right).$$

For each $m, n \geq p$, the function

$$v(t, x) = S_m(t, x) - S_n(t, x)$$

is of class C^∞ and satisfies equation (8.1) in $\mathbb{R}^+ \times (0, l)$. Moreover,

$$v(t, 0) = v(t, l) = 0, \quad t \geq 0,$$

and

$$|v(0, x)| = |S_m(0, x) - S_n(0, x)|$$
$$\leq |S_m(0, x) - f(x)| + |S_n(0, x) - f(x)| < 2\varepsilon$$

for $x \in [0, l]$. It then follows from Lemma 8.2 that

$$|S_m(t, x) - S_n(t, x)| \leq 2\varepsilon$$

for every $m, n \geq p$, $t \geq 0$ and $x \in [0, l]$. This shows that the series of continuous functions in (8.4) is uniformly convergent on $\mathbb{R}_0^+ \times [0, l]$. By Exercise 3.21, we conclude that u is continuous in $\mathbb{R}_0^+ \times [0, l]$.

Finally, it follows from (8.4) that

$$u(t, 0) = u(t, l) = 0,$$

that is, condition (8.2) holds. Moreover,

$$u(0, x) = \sum_{n=1}^{\infty} c_n \sin\left(\frac{n\pi x}{l}\right).$$

By Theorem 7.29, taking the constants c_n in (8.5), condition (8.3) is satisfied (the assumption $f(0) = f(l) = 0$ guarantees that (8.3) holds for $x = 0$ and $x = l$). $\square$

Example 8.3

Let us consider the equation

$$\frac{\partial u}{\partial t} = \frac{\partial^2 u}{\partial x^2}, \tag{8.8}$$

for $t \geq 0$ and $x \in [0, \pi]$. We also consider the function $f\colon [0, \pi] \to \mathbb{R}$ given by $f(x) = x(\pi - x)$. One can easily verify that

$$c_n = \frac{2}{\pi} \int_0^\pi x(1 - x) \sin(nx)\, dx$$

$$= \begin{cases} 0 & \text{if } n \text{ is even,} \\ 8/(n^3 \pi) & \text{if } n \text{ is odd} \end{cases}$$

for $n \in \mathbb{N}$. Hence, by Proposition 8.1, the solution of equation (8.8) satisfying conditions (8.2) and (8.3) is given by

$$u(t,x) = \sum_{n=1}^{\infty} c_n e^{-n^2 t} \sin(nx).$$

The following is a modification of the former example.

Example 8.4

Let us consider the heat equation (8.1) together with the conditions

$$\frac{\partial u}{\partial x}(t,0) = \frac{\partial u}{\partial x}(t,l) = 0, \quad t > 0 \tag{8.9}$$

and (8.3), for some function $f \colon [0,l] \to \mathbb{R}$ satisfying the hypotheses of Proposition 8.1.

We first obtain solutions of equation (8.1) of the form $u(t,x) = T(t)X(x)$. By Proposition 7.1, there exists $\lambda \in \mathbb{R}$ such that

$$T(t) = ce^{-\lambda \kappa t} \quad \text{and} \quad X'' + \lambda X = 0,$$

with $c \neq 0$. On the other hand, it follows from (8.9) that

$$T(t)X'(0) = T(t)X'(l) = 0, \quad t > 0,$$

which is equivalent to

$$X'(0) = X'(l) = 0.$$

Hence, we must solve the problem

$$X'' + \lambda X = 0, \quad X'(0) = X'(l) = 0.$$

We consider three cases.

1. When $\lambda = 0$, we have the solutions in (7.8). It follows from $X'(0) = 0$ that $b = 0$. Hence, $X(x) = a$ with $a \neq 0$, which also satisfies $X'(l) = 0$.
2. When $\lambda < 0$, we have the solutions in (7.9). Since

$$X'(x) = a\sqrt{|\lambda|}e^{\sqrt{|\lambda|}x} - b\sqrt{|\lambda|}e^{-\sqrt{|\lambda|}x},$$

we obtain

$$X'(0) = a\sqrt{|\lambda|} - b\sqrt{|\lambda|}.$$

It follows from $X'(0) = 0$ that $a = b$, and thus,

$$X'(l) = a\sqrt{|\lambda|}\left(e^{\sqrt{|\lambda|}l} - e^{-\sqrt{|\lambda|}l}\right) = 0.$$

Since $e^{\sqrt{|\lambda|}l} > 1$ and $e^{-\sqrt{|\lambda|}l} < 1$, we have $a = b = 0$ and $X(x) = 0$.

3. When $\lambda > 0$, we have the solutions in (7.11). Since

$$X'(x) = -a\sqrt{\lambda}\sin(\sqrt{\lambda}x) + b\sqrt{\lambda}\cos(\sqrt{\lambda}x),$$

we obtain $X'(0) = b\sqrt{\lambda}$. It follows from $X'(0) = 0$ that $b = 0$, and thus,

$$X'(l) = -a\sqrt{\lambda}\sin(\sqrt{\lambda}l).$$

Hence, $a = 0$ (which would give $X = 0$), or

$$\sin(\sqrt{\lambda}l) = 0.$$

Therefore,

$$\lambda = \frac{n^2\pi^2}{l^2}, \quad n \in \mathbb{N},$$

and we obtain the solutions

$$X(x) = a\cos\left(\frac{n\pi x}{l}\right), \quad \text{with } a \neq 0.$$

Now we look for (formal) solutions of the form

$$u(t, x) = \sum_{n=0}^{\infty} c_n e^{-n^2\pi^2\kappa t/l^2} \cos\left(\frac{n\pi x}{l}\right).$$

We note that

$$u(0, x) = \sum_{n=0}^{\infty} c_n \cos\left(\frac{n\pi x}{l}\right)$$

$$= \frac{2c_0}{2} + \sum_{n=1}^{\infty} c_n \cos\left(\frac{n\pi x}{l}\right). \tag{8.10}$$

By Theorem 7.29, in order that condition (8.3) is satisfied, we take

$$c_0 = \frac{1}{l}\int_0^l f(x)\,dx$$

and

$$c_n = \frac{2}{l}\int_0^l f(x)\cos\left(\frac{n\pi x}{l}\right)dx \quad \text{for } n \in \mathbb{N}.$$

By the hypotheses in f, it follows from Theorem 7.14 that the series in (8.10) are uniformly convergent on $[0, l]$ (since the Fourier series of the even extension of f to the interval $[-l, l]$ is uniformly convergent). Proceeding as in the proof of Proposition 8.1, one can show that u is continuous in $\mathbb{R}^+ \times [0, l]$, and that it is of class C^∞ and satisfies equation (8.1) in $\mathbb{R}^+ \times (0, l)$. One can also show that u is continuous in $\mathbb{R}_0^+ \times [0, l]$ and that it is the unique solution of the problem, although this would require additional techniques.

Example 8.5

Now we consider the equation

$$\frac{\partial u}{\partial t} = \frac{\partial^2 u}{\partial x^2} + u, \tag{8.11}$$

with the conditions

$$u(t, 0) = u(t, \pi) = 0, \quad t > 0, \tag{8.12}$$

and

$$u(0, x) = \sin(2x), \quad x \in (0, \pi). \tag{8.13}$$

It is easy to show that any linear combination of solutions of equation (8.11) satisfying condition (8.12) is still a solution of this equation and satisfies condition (8.12). Now we look for solutions of the form $u(t, x) = T(t)X(x)$. Substituting $u(t, x)$ in (8.11), we obtain

$$T'X = TX'' + TX,$$

and thus,

$$\frac{T'}{T} = \frac{X''}{X} + 1 = -\lambda$$

for some constant $\lambda \in \mathbb{R}$ (whenever $TX \neq 0$). This yields the equations

$$T' = -\lambda T \quad \text{and} \quad X'' + (\lambda + 1)X = 0.$$

The solutions of the first equation are given by $T(t) = ce^{-\lambda t}$, with $c \neq 0$ (so that $T \neq 0$). The second equation can be written in the form $X'' + \mu X = 0$, with $\mu = \lambda + 1$. By Proposition 7.2, there exist nonzero solutions $X(x)$ if and only if

$$\mu = \frac{n^2 \pi^2}{l^2} = n^2, \quad \text{that is,} \quad \lambda = n^2 - 1,$$

for some $n \in \mathbb{N}$ (since $l = \pi$). We then look for (formal) solutions of the form

$$u(t,x) = \sum_{n=1}^{\infty} c_n e^{(1-n^2)t} \sin(nx). \tag{8.14}$$

In order that condition (8.13) is satisfied, we take

$$c_n = \frac{2}{\pi} \int_0^{\pi} \sin(2x) \sin(nx)\, dx = \begin{cases} 1 & \text{if } n = 2, \\ 0 & \text{if } n \neq 2, \end{cases}$$

which yields the solution

$$u(t,x) = e^{-3t} \sin(2x).$$

Example 8.6

We present an alternative argument to obtain a solution of the problem in Example 8.5. Let us consider the function $v = ue^{-t}$, that is,

$$v(t,x) = u(t,x)e^{-t}.$$

Substituting $u = ve^t$ in (8.11), we obtain

$$\frac{\partial v}{\partial t}e^t + ve^t = \frac{\partial^2 v}{\partial x^2}e^t + ve^t,$$

or equivalently,

$$\frac{\partial v}{\partial t} = \frac{\partial^2 v}{\partial x^2};$$

that is, v satisfies the heat equation. On the other hand, it follows from (8.12) that

$$v(t,0) = v(t,\pi) = 0, \quad t > 0,$$

and hence, by Proposition 8.1, we obtain the solutions

$$v(t,x) = \sum_{n=1}^{\infty} c_n e^{-n^2 t} \sin(nx).$$

This yields

$$u(t,x) = v(t,x)e^t = \sum_{n=1}^{\infty} c_n e^{(1-n^2)t} \sin(nx),$$

which is the series already obtained in (8.14).

8.2 Laplace Equation

Now we consider the *Laplace equation*

$$\Delta u = \frac{\partial^2 u}{\partial x^2} + \frac{\partial^2 u}{\partial y^2} = 0, \tag{8.15}$$

for $(x, y) \in [0, a] \times [0, b]$. For example, the real and imaginary parts of a holomorphic function satisfy the Laplace equation (see Exercise 4.36).

Example 8.7

Given a function $f \colon [0, b] \to \mathbb{R}$, let us consider the conditions

$$u(x, 0) = u(x, b) = 0 \tag{8.16}$$

and

$$u(0, y) = 0, \qquad u(a, y) = f(y) \tag{8.17}$$

for $(x, y) \in (0, a) \times (0, b)$. Substituting $u(x, y) = X(x)Y(y)$ in (8.15), we obtain

$$X''Y + XY'' = 0$$

and hence,

$$\frac{X''}{X} = -\frac{Y''}{Y} = \lambda$$

for some constant $\lambda \in \mathbb{R}$ (whenever $XY \neq 0$). This yields the equations

$$X'' - \lambda X = 0 \quad \text{and} \quad Y'' + \lambda Y = 0. \tag{8.18}$$

On the other hand, it follows from (8.16) and (8.17) that

$$Y(0) = Y(b) = 0 \quad \text{and} \quad X(0) = 0.$$

By Proposition 7.2, the nonzero solutions of the problem

$$Y'' + \lambda Y = 0, \quad Y(0) = Y(l) = 0$$

are given by

$$Y(y) = c\sin\left(\frac{n\pi y}{b}\right), \quad \text{with } c \neq 0,$$

for $\lambda = n^2\pi^2/b^2$ with $n \in \mathbb{N}$. Now we solve the problem

$$X'' - \lambda X = 0, \quad X(0) = 0.$$

Since $\lambda = n^2\pi^2/b^2 > 0$, we have

$$X(x) = ce^{\sqrt{\lambda}x} + de^{-\sqrt{\lambda}x}, \quad \text{with } c, d \in \mathbb{R}.$$

It follows from $X(0) = 0$ that $c + d = 0$. Therefore,

$$X(x) = c(e^{\sqrt{\lambda}x} - e^{-\sqrt{\lambda}x}) = 2c\sinh(\sqrt{\lambda}x),$$

and we look for (formal) solutions of the Laplace equation of the form

$$u(x, y) = \sum_{n=1}^{\infty} c_n \sinh\left(\frac{n\pi x}{b}\right) \sin\left(\frac{n\pi y}{b}\right). \tag{8.19}$$

Taking $x = a$, we obtain

$$u(a, y) = \sum_{n=1}^{\infty} c_n \sinh\left(\frac{n\pi a}{b}\right) \sin\left(\frac{n\pi y}{b}\right) = f(y),$$

and thus,

$$c_n \sinh\left(\frac{n\pi a}{b}\right) = \frac{2}{b} \int_0^b f(y) \sin\left(\frac{n\pi y}{b}\right) dy$$

for each $n \in \mathbb{N}$. Substituting the constants c_n in (8.19), we obtain a (formal) solution of the Laplace equation satisfying conditions (8.16) and (8.17).

Example 8.8

Given functions $f, g: [0, b] \to \mathbb{R}$, we consider the conditions (8.16) and

$$u(0, y) = f(y), \qquad u(a, y) = g(y) \tag{8.20}$$

for $(x, y) \in (0, a) \times (0, b)$. Again, for a solution of the Laplace equation of the form $u(x, y) = X(x)Y(y)$, we obtain the equations in (8.18), and it follows from (8.16) that $Y(0) = Y(b) = 0$. By Proposition 7.2, the problem

$$Y'' + \lambda Y = 0, \quad Y(0) = Y(b) = 0$$

has the nonzero solutions given by

$$Y(y) = c\sin\left(\frac{n\pi y}{b}\right), \quad \text{with } c \neq 0,$$

for $\lambda = n^2\pi^2/b^2$ with $n \in \mathbb{N}$. Thus, it remains to solve the equation

$$X'' - \frac{n^2\pi^2}{b^2}X = 0.$$

Its solutions are given by

$$X(x) = c_n \sinh\left(\frac{n\pi x}{b}\right) + d_n \cosh\left(\frac{n\pi x}{b}\right).$$

Hence, we look for (formal) solutions of the form

$$u(x,y) = \sum_{n=1}^{\infty} \left[c_n \sinh\left(\frac{n\pi x}{b}\right) + d_n \cosh\left(\frac{n\pi x}{b}\right) \right] \sin\left(\frac{n\pi y}{b}\right). \qquad (8.21)$$

Taking $x = 0$ and $x = a$, we obtain respectively

$$u(0,y) = \sum_{n=1}^{\infty} d_n \sin\left(\frac{n\pi y}{b}\right)$$

and

$$u(a,y) = \sum_{n=1}^{\infty} \left[c_n \sinh\left(\frac{n\pi a}{b}\right) + d_n \cosh\left(\frac{n\pi a}{b}\right) \right] \sin\left(\frac{n\pi y}{b}\right).$$

Thus,

$$d_n = \frac{2}{b} \int_0^b f(y) \sin\left(\frac{n\pi y}{b}\right) dy$$

and

$$c_n \sinh\left(\frac{n\pi a}{b}\right) = \frac{2}{b} \int_0^b g(y) \sin\left(\frac{n\pi y}{b}\right) dy - d_n \cosh\left(\frac{n\pi a}{b}\right)$$

for each $n \in \mathbb{N}$. Substituting the constants c_n and d_n in (8.21), we obtain a (formal) solution of the Laplace equation satisfying conditions (8.16) and (8.20).

Example 8.9

Given functions $f: [0,a] \to \mathbb{R}$ and $g: [0,b] \to \mathbb{R}$, we consider the conditions

$$u(x,0) = 0, \qquad u(x,b) = f(x), \qquad (8.22)$$

and

$$u(0,y) = 0, \qquad u(a,y) = g(y), \qquad (8.23)$$

for $(x, y) \in (0, a) \times (0, b)$. We note that if v and w are solutions of the Laplace equation satisfying the conditions

$$
\begin{aligned}
v(x, 0) &= 0, & v(x, b) &= f(x), \\
v(0, y) &= v(a, y) = 0, \\
w(x, 0) &= w(x, b) = 0, \\
w(0, y) &= 0, & w(a, y) &= g(y),
\end{aligned}
\tag{8.24}
$$

then

$$
u = v + w
$$

is a solution of the Laplace equation and satisfies conditions (8.22) and (8.23). By Example 8.7, one can take

$$
v(x, y) = \sum_{n=1}^{\infty} c_n \sin\left(\frac{n\pi x}{a}\right) \sinh\left(\frac{n\pi y}{a}\right)
$$

and

$$
w(x, y) = \sum_{n=1}^{\infty} d_n \sinh\left(\frac{n\pi x}{b}\right) \sin\left(\frac{n\pi y}{b}\right),
$$

with constants c_n and d_n such that

$$
c_n \sinh\left(\frac{n\pi b}{a}\right) = \frac{2}{a} \int_0^a f(x) \sin\left(\frac{n\pi x}{a}\right) dx
$$

and

$$
d_n \sinh\left(\frac{n\pi a}{b}\right) = \frac{2}{b} \int_0^b g(y) \sin\left(\frac{n\pi y}{b}\right) dy;
$$

that is, v and w are solutions of the Laplace equation satisfying the conditions in (8.24). Then

$$
u(x, y) = v(x, y) + w(x, y)
$$

$$
= \sum_{n=1}^{\infty} \left[c_n \sin\left(\frac{n\pi x}{a}\right) \sinh\left(\frac{n\pi y}{a}\right) + d_n \sinh\left(\frac{n\pi x}{b}\right) \sin\left(\frac{n\pi y}{b}\right) \right]
$$

is a solution of the Laplace equation and satisfies conditions (8.22) and (8.23).

Example 8.10

Given a function $f\colon [0,b] \to \mathbb{R}$ such that

$$\int_0^b f(y)\,dy = 0,$$

let us consider the conditions

$$\frac{\partial u}{\partial y}(x,0) = \frac{\partial u}{\partial y}(x,b) = 0 \qquad (8.25)$$

and

$$\frac{\partial u}{\partial x}(0,y) = 0, \qquad \frac{\partial u}{\partial x}(a,y) = f(y) \qquad (8.26)$$

for $(x,y) \in (0,a) \times (0,b)$. Substituting the function $u(x,y) = X(x)Y(y)$ in equation (8.15) and in conditions (8.25) and (8.26), we obtain

$$\begin{cases} X'' - \lambda X = 0, \\ X'(0) = 0 \end{cases} \quad \text{and} \quad \begin{cases} Y'' + \lambda Y = 0, \\ Y'(0) = Y'(b) = 0, \end{cases} \qquad (8.27)$$

for some constant $\lambda \in \mathbb{R}$. It follows from Example 8.4 that

$$Y(y) = c\cos\left(\frac{n\pi y}{b}\right), \quad \text{with } c \neq 0,$$

for $\lambda = n^2\pi^2/b^2$ with $n \in \mathbb{N}$. It remains to solve the first problem in (8.27). Since $\lambda > 0$, we have

$$X(x) = ce^{\sqrt{\lambda}x} + de^{-\sqrt{\lambda}x}, \quad \text{with } c,d \in \mathbb{R},$$

and it follows from $X'(0) = 0$ that $c = d$. Therefore,

$$X(x) = 2c\frac{e^{\sqrt{\lambda}x} + e^{-\sqrt{\lambda}x}}{2} = 2c\cosh(\sqrt{\lambda}x),$$

and we look for (formal) solutions of the form

$$u(x,y) = \sum_{n=0}^{\infty} c_n \cosh\left(\frac{n\pi x}{b}\right)\cos\left(\frac{n\pi y}{b}\right).$$

Assuming that the derivative $\partial u/\partial x$ is well defined and that it can be computed term by term, we obtain

$$\frac{\partial u}{\partial x}(a,y) = \sum_{n=1}^{\infty} \frac{n\pi}{b}c_n \sinh\left(\frac{n\pi a}{b}\right)\cos\left(\frac{n\pi y}{b}\right).$$

Hence, in order that condition (8.26) is satisfied, we must take constants c_n such that

$$\frac{n\pi}{b} c_n \sinh\left(\frac{n\pi a}{b}\right) = \frac{2}{b} \int_0^b f(y) \cos\left(\frac{n\pi y}{b}\right) dy$$

for each $n \in \mathbb{N}$. We note that the constant c_0 is arbitrary.

8.3 Wave Equation

In this section we consider the *wave equation*

$$\frac{\partial^2 u}{\partial t^2} = c^2 \frac{\partial^2 u}{\partial x^2}, \tag{8.28}$$

where $c \in \mathbb{R} \setminus \{0\}$, for $t \geq 0$ and $x \in [0,l]$, with $l > 0$. We also consider the conditions

$$u(0,x) = f(x), \qquad \frac{\partial u}{\partial t}(0,x) = g(x) \tag{8.29}$$

for $x \in (0,l)$, and

$$u(t,0) = u(t,l) = 0, \quad t > 0, \tag{8.30}$$

for some functions $f, g \colon [0,l] \to \mathbb{R}$.

Example 8.11

Let us assume that f is the restriction to $[0,l]$ of a function of class C^2 in $\mathbb{R}$, and that g is the restriction to $[0,l]$ of a function of class C^1 in $\mathbb{R}$. Letting $r = x - ct$ and $s = x + ct$, we consider the function

$$v(r,s) = u(t,x).$$

When u is of class C^2, we have

$$\frac{\partial u}{\partial x} = \frac{\partial v}{\partial r}\frac{\partial r}{\partial x} + \frac{\partial v}{\partial s}\frac{\partial s}{\partial x}$$

$$= \frac{\partial v}{\partial r} + \frac{\partial v}{\partial s}$$

and

$$\frac{\partial^2 u}{\partial x^2} = \frac{\partial^2 v}{\partial r^2}\frac{\partial r}{\partial x} + \frac{\partial^2 v}{\partial s \partial r}\frac{\partial s}{\partial x} + \frac{\partial^2 v}{\partial r \partial s}\frac{\partial r}{\partial x} + \frac{\partial^2 v}{\partial s^2}\frac{\partial s}{\partial x}$$

$$= \frac{\partial^2 v}{\partial r^2} + 2\frac{\partial^2 v}{\partial r \partial s} + \frac{\partial^2 v}{\partial s^2}.$$

One can show in a similar manner that

$$\frac{\partial^2 u}{\partial t^2} = c^2\frac{\partial^2 v}{\partial r^2} - 2c^2\frac{\partial^2 v}{\partial r \partial s} + c^2\frac{\partial^2 v}{\partial s^2}.$$

Hence, equation (8.28) is equivalent to

$$\frac{\partial^2 v}{\partial r \partial s} = 0,$$

and thus, its solutions are given by

$$v(r, s) = p(r) + q(s),$$

where p and q are arbitrary functions of class C^2 (see Problem 8.11). Hence,

$$u(t, x) = p(x - ct) + q(x + ct)$$

are the solutions of equation (8.28). On the other hand, it follows from (8.29) that

$$p(x) + q(x) = f(x) \quad \text{and} \quad -cp'(x) + cq'(x) = g(x),$$

that is,

$$p(x) + q(x) = f(x) \quad \text{and} \quad -p(x) + q(x) = \frac{1}{c}G(x)$$

for some primitive G of g in the interval $[0, l]$. Therefore,

$$q(x) = \frac{1}{2}\left(f(x) + \frac{1}{c}G(x)\right) \quad \text{and} \quad p(x) = \frac{1}{2}\left(f(x) - \frac{1}{c}G(x)\right)$$

for every $x \in [0, l]$. Moreover, it follows from (8.30) that

$$p(-ct) + q(ct) = p(l - ct) + q(l + ct) = 0, \quad t \geq 0. \tag{8.31}$$

Therefore,

$$q(t) = -p(-t), \quad t \geq 0,$$

and the functions p and q are determined in the interval $[-l, l]$. Finally, it follows from (8.31) that

$$p(l - ct) = -q(l + ct) = p(-l - ct), \quad t \geq 0,$$

and thus, the functions p and q are $2l$-periodic. Therefore, they are determined in $\mathbb{R}$, and we obtain the solution

$$u(t,x) = \frac{1}{2}\left(f(x-ct) - \frac{1}{c}G(x-ct)\right) + \frac{1}{2}\left(f(x+ct) + \frac{1}{c}G(x+ct)\right)$$

$$= \frac{1}{2}\left(f(x-ct) + f(x+ct)\right) + \frac{1}{2c}\int_{x-ct}^{x+ct} g(s)\,ds.$$

Example 8.12

One can also find (formal) solutions of equation (8.28) by the method of separation of variables. We first observe that any linear combination of solutions of equation (8.28) satisfying condition (8.30) is still a solution of this equation and satisfies condition (8.30). Substituting $u(t,x) = T(t)X(x)$ in (8.28), we obtain

$$T''X = c^2 T X'',$$

and thus,

$$\frac{T''}{c^2 T} = \frac{X''}{X} = -\lambda$$

for some constant $\lambda \in \mathbb{R}$ (whenever $TX \neq 0$). We then obtain the equations

$$T'' + \lambda c^2 T = 0 \quad \text{and} \quad X'' + \lambda X = 0. \tag{8.32}$$

It follows from (8.30) that

$$T(t)X(0) = T(t)X(l) = 0, \quad t > 0,$$

which is equivalent to

$$X(0) = X(l) = 0$$

(although we do not yet know explicitly the function T, it follows from $TX \neq 0$ that $T(t) \neq 0$ for some t). By Proposition 7.2, the nonzero solutions of the problem

$$X'' + \lambda X = 0, \quad X(0) = X(l) = 0$$

are given by

$$X(x) = a\sin\left(\frac{n\pi x}{l}\right), \quad \text{with } a \neq 0,$$

for $\lambda = n^2\pi^2/l^2$ with $n \in \mathbb{N}$. The first equation in (8.32) then takes the form

$$T'' + \frac{n^2\pi^2 c^2}{l^2}T = 0,$$

and thus, its solutions are given by

$$T(t) = a_n \cos\left(\frac{n\pi ct}{l}\right) + b_n \sin\left(\frac{n\pi ct}{l}\right),$$

with $a_n, b_n \in \mathbb{R}$. Now we look for (formal) solutions of the form

$$u(t,x) = \sum_{n=1}^{\infty} \left[a_n \cos\left(\frac{n\pi ct}{l}\right) + b_n \sin\left(\frac{n\pi ct}{l}\right) \right] \sin\left(\frac{n\pi x}{l}\right).$$

We have

$$u(0,x) = \sum_{n=1}^{\infty} a_n \sin\left(\frac{n\pi x}{l}\right),$$

and assuming that the derivative $\partial u/\partial t$ is well defined and that it can be computed term by term, we also have

$$\frac{\partial u}{\partial t}(0,x) = \sum_{n=1}^{\infty} \frac{n\pi c}{l} b_n \sin\left(\frac{n\pi x}{l}\right).$$

Therefore, one can then take

$$a_n = \frac{2}{l} \int_0^l f(x) \cos\left(\frac{n\pi x}{l}\right) dx, \quad n \in \mathbb{N},$$

and

$$b_n = \frac{2}{n\pi c} \int_0^l g(x) \sin\left(\frac{n\pi x}{l}\right) dx, \quad n \in \mathbb{N}.$$

Example 8.13

Now we use the method of separation of variables to solve the equation

$$\frac{\partial^2 u}{\partial t^2} = \frac{\partial^2 u}{\partial x^2} + u + \frac{\partial u}{\partial t}, \tag{8.33}$$

for $t \geq 0$ and $x \in [0, \pi]$, with condition (8.30) satisfied for $l = \pi$. Substituting the function $u(t,x) = T(t)X(x)$ in equation (8.33), we obtain

$$T''X = TX'' + TX + T'X,$$

and thus,

$$\frac{T''}{T} = \frac{X''}{X} + 1 + \frac{T'}{T}.$$

whenever $TX \neq 0$. Therefore, there exists $\lambda \in \mathbb{R}$ such that

$$\frac{T'' - T'}{T} = \frac{X''}{X} + 1 = -\lambda,$$

and we obtain the equations

$$T'' - T' + \lambda T = 0 \quad \text{and} \quad X'' + (\lambda + 1)X = 0.$$

By Proposition 7.2, we have the nonzero solutions given by

$$X(x) = b\sin(nx), \quad \text{with } b \neq 0,$$

for $\lambda = n^2 - 1$ with $n \in \mathbb{N}$. It then remains to solve the equation

$$T'' - T' + \left(n^2 - 1\right)T = 0.$$

Since the roots of the polynomial $a^2 - a + n^2 - 1$ are $a = (1 \pm \sqrt{5 - 4n^2})/2$, we obtain the solutions

$$u(t, x) = \left(a_1 e^t + b_1\right)\sin x$$
$$+ \sum_{n=2}^{\infty}\left[a_n e^{t/2}\cos\left(\frac{\sqrt{4n^2 - 5}\,t}{2}\right) + b_n e^{t/2}\sin\left(\frac{\sqrt{4n^2 - 5}\,t}{2}\right)\right]\sin(nx),$$

with $a_n, b_n \in \mathbb{R}$ for each $n \in \mathbb{N}$.

8.4 Solved Problems and Exercises

Problem 8.1

Find a solution of the equation

$$\frac{\partial u}{\partial t} = \frac{\partial^2 u}{\partial x^2}, \quad t \geq 0, \ x \in [0, \pi],$$

with the conditions

$$u(t, 0) = u(t, \pi) = 0, \quad t > 0,$$

and

$$u(0, x) = x, \quad x \in (0, \pi).$$

Solution

By Proposition 8.1, the solution is given by

$$u(t,x) = \sum_{n=1}^{\infty} c_n e^{-n^2 t} \sin(nx),$$

where

$$c_n = \frac{2}{\pi} \int_0^\pi x \sin(nx)\, dx$$

$$= \frac{2}{\pi} \left(-\frac{x \cos(nx)}{n} + \frac{\sin(nx)}{n^2} \right) \Big|_{x=0}^{x=\pi} = \frac{2(-1)^{n+1}}{n}$$

for each $n \in \mathbb{N}$. We thus obtain

$$u(t,x) = \sum_{n=1}^{\infty} \frac{2(-1)^{n+1}}{n} e^{-n^2 t} \sin(nx).$$

Problem 8.2

Find a solution of the equation

$$\frac{\partial u}{\partial t} = 2 \frac{\partial^2 u}{\partial x^2}, \quad t \geq 0,\ x \in [0,1],$$

with the conditions

$$\frac{\partial u}{\partial x}(t,0) = \frac{\partial u}{\partial x}(t,1) = 0, \quad t > 0,$$

and

$$u(0,x) = e^x, \quad x \in (0,1).$$

Solution

By Example 8.4 with $l = 1$, we look for (formal) solutions of the form

$$u(t,x) = \sum_{n=0}^{\infty} c_n e^{-2n^2 \pi^2 t} \cos(n\pi x).$$

Taking $t = 0$, we obtain

$$u(0,x) = \sum_{n=0}^{\infty} c_n \cos(n\pi x) = \frac{2c_0}{2} + \sum_{n=1}^{\infty} c_n \cos(n\pi x).$$

Hence, by Theorem 7.29, we take

$$c_0 = \int_0^1 e^x \, dx = e - 1$$

and

$$c_n = 2 \int_0^1 e^x \cos(n\pi x) \, dx$$

$$= \frac{2e^x}{1 + n^2 \pi^2} \left(\cos(n\pi x) + n\pi \sin(n\pi x) \right) \Big|_{x=0}^{x=1}$$

$$= \frac{2[e(-1)^n - 1]}{1 + n^2 \pi^2}$$

for $n \in \mathbb{N}$. We thus obtain the solution

$$u(t, x) = e - 1 + \sum_{n=1}^{\infty} \frac{2[e(-1)^n - 1]}{1 + n^2 \pi^2} e^{-2n^2 \pi^2 t} \cos(n\pi x).$$

Problem 8.3

Find a solution of the equation

$$\frac{\partial u}{\partial t} = \frac{\partial^2 u}{\partial x^2} + u, \quad t \geq 0, \ x \in [0, \pi],$$

with the conditions

$$u(t, 0) = u(t, \pi) = 0, \quad t > 0,$$

and

$$u(0, x) = \cos x, \quad x \in (0, \pi). \tag{8.34}$$

Solution

By Example 8.5, we look for (formal) solutions of the form

$$u(t, x) = \sum_{n=1}^{\infty} c_n e^{(1 - n^2)t} \sin(nx).$$

In order that condition (8.34) is satisfied, we take

$$c_1 = \frac{2}{\pi} \int_0^{\pi} \cos x \sin x \, dx = 0$$

and

$$c_n = \frac{2}{\pi} \int_0^\pi \cos x \sin(nx)\, dx$$

$$= \frac{2n(1+(-1)^n)}{(n^2-1)\pi} = \begin{cases} 0 & \text{if } n \text{ is odd,} \\ 4n/[(n^2-1)\pi] & \text{if } n \text{ is even} \end{cases}$$

for $n > 1$. We thus obtain the solution

$$u(t,x) = \sum_{n=1}^\infty \frac{8n}{(4n^2-1)\pi} e^{(1-4n^2)t} \sin(2nx).$$

Problem 8.4

Find a solution of the equation

$$\frac{\partial u}{\partial t} = \frac{\partial^2 u}{\partial x^2} + 2u, \quad t \geq 0, \ x \in [0,\pi],$$

with the conditions

$$u(t,0) = u(t,\pi) = 0, \quad t > 0,$$

and

$$u(0,x) = 1, \quad x \in (0,\pi).$$

Solution

Proceeding in a similar manner to that in Example 8.5, we find the (formal) solutions

$$u(t,x) = \sum_{n=1}^\infty c_n e^{(2-n^2)t} \sin(nx).$$

Taking $t = 0$, we obtain

$$u(0,x) = \sum_{n=1}^\infty c_n \sin(nx) = 1.$$

Hence, by Example 7.31, we take

$$c_n = \frac{2}{n\pi}[1-(-1)^n] = \begin{cases} 0, & \text{if } n \text{ is even,} \\ 4/(n\pi) & \text{if } n \text{ is odd,} \end{cases}$$

which yields the solution

$$u(t,x) = \sum_{n=0}^{\infty} \frac{4}{(2n+1)\pi} e^{[2-(2n+1)^2]t} \sin\big((2n+1)x\big).$$

Problem 8.5

Find a solution of the equation

$$\frac{\partial u}{\partial t} = \frac{\partial^2 u}{\partial x^2} + \frac{\partial u}{\partial x}, \quad t \geq 0, \ x \in [0,1], \tag{8.35}$$

with the conditions

$$u(t,0) = u(t,1) = 0, \quad t > 0, \tag{8.36}$$

and

$$u(0,x) = e^{-x/2}\sin(\pi x), \quad x \in (0,1).$$

Solution

We first look for nonzero solutions of the form $u(t,x) = T(t)X(x)$. Substituting $u(t,x)$ in (8.35), we obtain

$$T'X = TX'' + TX',$$

and hence,

$$\frac{T'}{T} = \frac{X'' + X'}{X} = -\lambda$$

for some constant $\lambda \in \mathbb{R}$ (whenever $TX \neq 0$). This yields the equations

$$T' = -\lambda T \quad \text{and} \quad X'' + X' + \lambda X = 0.$$

The solutions of the first equation are given by

$$T(t) = ce^{-\lambda t}, \quad \text{with } c \neq 0.$$

On the other hand, it follows from (8.36) that $X(0) = X(1) = 0$. Thus, we must solve the problem

$$X'' + X' + \lambda X = 0, \quad X(0) = X(1) = 0.$$

Since the polynomial $v^2 + v + \lambda$ has the roots $(-1 \pm \sqrt{1-4\lambda})/2$, we consider three cases.

1. When $\lambda = 1/4$, we obtain the equation

$$X'' + X' + \lambda X = (D + 1/2)^2 X = 0,$$

whose solutions are given by

$$X(x) = ae^{-x/2} + bxe^{-x/2}, \quad \text{with } a, b \in \mathbb{R}.$$

It follows from $X(0) = 0$ that $a = 0$. Hence, $X(1) = be^{-1/2} = 0$, which yields $b = 0$ and $X(x) = 0$.

2. When $\lambda < 1/4$, we obtain the equation

$$X'' + X' + \lambda X = \left(D + \frac{1 + \sqrt{1 - 4\lambda}}{2}\right)\left(D + \frac{1 - \sqrt{1 - 4\lambda}}{2}\right)X = 0,$$

whose solutions are given by

$$X(x) = ae^{-(1+\sqrt{1-4\lambda})x/2} + be^{-(1-\sqrt{1-4\lambda})x/2},$$

with $a, b \in \mathbb{R}$. It follows from $X(0) = 0$ that $a + b = 0$. Hence,

$$X(1) = a\left(e^{-(1+\sqrt{1-4\lambda})/2} - e^{-(1-\sqrt{1-4\lambda})/2}\right).$$

Since $\lambda < 1/4$, we have

$$e^{-(1+\sqrt{1-4\lambda})/2} \neq e^{-(1-\sqrt{1-4\lambda})/2},$$

and it follows from $X(1) = 0$ that $a = 0$. Hence, $b = 0$ and $X(x) = 0$.

3. When $\lambda > 1/4$, we obtain the equation

$$X'' + X' + \lambda X = \left(D + \frac{1 + i\sqrt{4\lambda - 1}}{2}\right)\left(D + \frac{1 - i\sqrt{4\lambda - 1}}{2}\right)X = 0,$$

whose solutions are given by

$$X(x) = ae^{-x/2}\cos\left(\frac{\sqrt{4\lambda - 1}x}{2}\right) + be^{-x/2}\sin\left(\frac{\sqrt{4\lambda - 1}x}{2}\right),$$

with $a, b \in \mathbb{R}$. It follows from $X(0) = 0$ that $a = 0$. Hence,

$$X(1) = be^{-1/2}\sin\left(\frac{\sqrt{4\lambda - 1}}{2}\right) = 0,$$

and thus, $b = 0$ (which would give $X = 0$), or

$$\sin\left(\frac{\sqrt{4\lambda - 1}}{2}\right) = 0. \tag{8.37}$$

It follows from (8.37) that

$$\frac{\sqrt{4\lambda - 1}}{2} = n\pi, \quad \text{that is, } \lambda = n^2\pi^2 + \frac{1}{4},$$

for $n \in \mathbb{N}$. We thus obtain the solutions

$$X(x) = be^{-x/2}\sin(n\pi x), \quad n \in \mathbb{N}.$$

Now we observe that any linear combination of solutions of equation (8.35) satisfying condition (8.36) is still a solution of this equation and satisfies condition (8.36). We then look for (formal) solutions of the form

$$u(t,x) = \sum_{n=1}^{\infty} c_n e^{-(n^2\pi^2+1/4)t} e^{-x/2}\sin(n\pi x).$$

Taking $t = 0$, we obtain

$$u(0,x) = \sum_{n=1}^{\infty} c_n e^{-x/2}\sin(n\pi x) = e^{-x/2}\sin(n\pi x),$$

and hence, we must take

$$c_n = \begin{cases} 1 & \text{if } n = 1, \\ 0 & \text{if } n \neq 1. \end{cases}$$

Therefore, a solution is given by

$$u(t,x) = e^{-(\pi^2+1/4)t} e^{-x/2}\sin(\pi x).$$

Problem 8.6

Find a solution of the equation

$$\frac{\partial u}{\partial t} = \frac{\partial^2 u}{\partial x^2} + 4\frac{\partial u}{\partial x}, \quad t \geq 0, \ x \in [0, \pi],$$

with the conditions

$$u(t,0) = u(t,\pi) = 0, \quad t > 0,$$

and

$$u(0,x) = 5e^{-2x}\sin(8x), \quad x \in (0, \pi).$$

Solution

Proceeding in a similar manner to that in Problem 8.5, for a solution of the form $u(t, x) = T(t)X(x)$, we obtain the equations

$$T' = -\lambda T \quad \text{and} \quad X'' + 4X' + \lambda X = 0,$$

for some constant $\lambda \in \mathbb{R}$. In this case, the polynomial $v^2 + 4v + \lambda$ has the roots $-2 \pm \sqrt{4 - \lambda}$. One can easily verify that the problem

$$X'' + 4X' + \lambda X = 0 \quad \text{with } X(0) = X(\pi) = 0$$

has nonzero solutions if and only if $\lambda = n^2 + 4$ for some $n \in \mathbb{N}$, in which case

$$X(x) = be^{-2x} \sin(nx),$$

with $b \neq 0$. We then look for (formal) solutions of the form

$$u(t, x) = \sum_{n=1}^{\infty} c_n e^{-(n^2 + 4)t - 2x} \sin(nx).$$

Taking $t = 0$, we obtain

$$u(0, x) = \sum_{n=1}^{\infty} c_n e^{-2x} \sin(nx) = 5e^{-2x} \sin(8x),$$

and thus, we must take

$$c_n = \begin{cases} 5 & \text{if } n = 8, \\ 0 & \text{if } n \neq 8. \end{cases}$$

Hence, a solution is given by

$$u(t, x) = 5e^{-68t - 2x} \sin(8x).$$

Problem 8.7

Find a solution of the equation

$$\frac{\partial^2 u}{\partial x^2} + \frac{\partial^2 u}{\partial y^2} = 0, \quad (x, y) \in [0, a] \times [0, b], \tag{8.38}$$

with the conditions

$$u(x, 0) = u(x, b) = 0, \quad x \in (0, a), \tag{8.39}$$

and

$$u(0,y) = 0, \quad u(a,y) = 1, \quad y \in (0,b). \tag{8.40}$$

Solution

Substituting $u(x,y) = X(x)Y(y)$ in (8.38), we obtain

$$X''Y + XY'' = 0,$$

and hence,

$$\frac{X''}{X} = -\frac{Y''}{Y} = \lambda$$

for some constant $\lambda \in \mathbb{R}$. This yields the equations

$$X'' - \lambda X = 0 \quad \text{and} \quad Y'' + \lambda Y = 0.$$

On the other hand, it follows from (8.39) and (8.40) that

$$Y(0) = Y(b) = 0 \quad \text{and} \quad X(0) = 0.$$

By Proposition 7.2, the problem

$$Y'' + \lambda Y = 0, \quad Y(0) = Y(b) = 0$$

has nonzero solutions if and only if $\lambda = n^2\pi^2/b^2$ for some $n \in \mathbb{N}$, in which case

$$Y(y) = \sin\left(\frac{n\pi y}{b}\right)$$

is a solution. Now we solve the problem

$$X'' - \lambda X = 0, \quad X(0) = 0.$$

Since $\lambda = n^2\pi^2/b^2 > 0$, we have

$$X(x) = ce^{\sqrt{\lambda}x} + de^{-\sqrt{\lambda}x}, \quad \text{with } c,d \in \mathbb{R}.$$

It follows from $X(0) = 0$ that $c + d = 0$, and thus,

$$X(x) = c\left(e^{\sqrt{\lambda}x} - e^{-\sqrt{\lambda}x}\right) = 2c\sinh(\sqrt{\lambda}x).$$

We then look for (formal) solutions of equation (8.38) of the form

$$u(x,y) = \sum_{n=1}^{\infty} c_n \sinh\left(\frac{n\pi x}{b}\right) \sin\left(\frac{n\pi y}{b}\right). \tag{8.41}$$

Taking $x = a$, we obtain

$$u(a,y) = \sum_{n=1}^{\infty} c_n \sinh\left(\frac{n\pi a}{b}\right) \sin\left(\frac{n\pi y}{b}\right) = 1,$$

and thus, we must take constants c_n such that

$$c_n \sinh\left(\frac{n\pi a}{b}\right) = \frac{2}{b}\int_0^b \sin\left(\frac{n\pi y}{b}\right) dy$$

$$= \frac{2}{n\pi}(1 - (-1)^n)$$

for each $n \in \mathbb{N}$. Substituting these constants in (8.41), we obtain the solution

$$u(x,y) = \sum_{n=1}^{\infty} \frac{2(1 - (-1)^n)}{n\pi \sinh(n\pi a/b)} \sinh\left(\frac{n\pi x}{b}\right) \sin\left(\frac{n\pi y}{b}\right)$$

$$= \sum_{n=0}^{\infty} d_n \sinh\left(\frac{(2n+1)\pi x}{b}\right) \sin\left(\frac{(2n+1)\pi y}{b}\right),$$

where

$$d_n = \frac{4}{(2n+1)\pi \sinh((2n+1)\pi a/b)}.$$

Problem 8.8

Find a solution of equation (8.38), with the conditions

$$\frac{\partial u}{\partial y}(x,0) = \frac{\partial u}{\partial y}(x,b) = 0, \quad x \in (0,a),$$

and

$$\frac{\partial u}{\partial x}(0,y) = 0, \quad \frac{\partial u}{\partial x}(a,y) = \cos\left(\frac{5\pi y}{b}\right), \quad y \in (0,b).$$

Solution

By Example 8.10, since the condition

$$\int_0^b \cos\left(\frac{5\pi y}{b}\right) dy = 0$$

is satisfied, we look for (formal) solutions of equation (8.38) of the form

$$u(x,y) = \sum_{n=0}^{\infty} c_n \cosh\left(\frac{n\pi x}{b}\right) \cos\left(\frac{n\pi y}{b}\right)$$

$$= c_0 + \sum_{n=1}^{\infty} c_n \cosh\left(\frac{n\pi x}{b}\right) \cos\left(\frac{n\pi y}{b}\right). \tag{8.42}$$

Taking (formally) derivatives term by term with respect to x, we obtain

$$\frac{\partial u}{\partial x}(a,y) = \sum_{n=1}^{\infty} \frac{c_n n\pi}{b} \sinh\left(\frac{n\pi a}{b}\right) \cos\left(\frac{n\pi y}{b}\right) = \cos\left(\frac{5\pi y}{b}\right).$$

Hence, we must take $c_n = 0$ for $n \in \mathbb{N} \setminus \{5\}$, and

$$c_5 = \frac{b}{5\pi \sinh(5\pi a/b)}.$$

Substituting these constants in (8.42), we obtain the solution

$$u(x,y) = c_0 + \frac{b}{5\pi \sinh(5\pi a/b)} \cosh\left(\frac{5\pi x}{b}\right) \cos\left(\frac{5\pi y}{b}\right),$$

where the constant c_0 is arbitrary.

Problem 8.9

Find a solution of the equation

$$\frac{\partial^2 u}{\partial t^2} = c^2 \frac{\partial^2 u}{\partial x^2}, \quad t \geq 0, \; x \in [0, \pi],$$

with the conditions

$$u(t,0) = u(t,\pi) = 0, \quad t > 0,$$

and

$$u(0,x) = \sin(2x), \quad \frac{\partial u}{\partial t}(0,x) = \sin x, \quad x \in (0, \pi). \tag{8.43}$$

Solution

By Example 8.12 with $l = \pi$, we look for (formal) solutions of the form

$$u(t,x) = \sum_{n=1}^{\infty} \left[a_n \cos(nct) + b_n \sin(nct)\right] \sin(nx). \tag{8.44}$$

We have

$$u(0, x) = \sum_{n=1}^{\infty} a_n \sin(nx).$$

Taking (formally) derivatives in (8.44) term by term with respect to t, we obtain

$$\frac{\partial u}{\partial t}(0, x) = \sum_{n=1}^{\infty} ncb_n \sin(nx).$$

Hence, in order that condition (8.43) is satisfied, we take

$$a_n = \frac{2}{\pi} \int_0^{\pi} \sin(2x) \sin(nx)\, dx = \begin{cases} 1 & \text{if } n = 2, \\ 0 & \text{if } n \neq 2, \end{cases}$$

and

$$b_n = \frac{2}{\pi nc} \int_0^{\pi} \sin x \sin(nx)\, dx = \begin{cases} 1/c & \text{if } n = 1, \\ 0 & \text{if } n \neq 1. \end{cases}$$

We thus obtain the solution

$$u(t, x) = \cos(2ct) \sin(2x) + \frac{1}{c} \sin(ct) \sin x.$$

Problem 8.10

Find a solution of the equation

$$\frac{\partial u}{\partial t} = \frac{\partial u}{\partial x}, \quad t, x \in \mathbb{R}, \tag{8.45}$$

satisfying the condition $u(0, x) = e^x + e^{-x}$.

Solution

We first look for nonzero solutions of the form $u(t, x) = T(t)X(x)$. Substituting $u(t, x)$ in (8.45), we obtain $T'X = TX'$, and thus,

$$\frac{T'}{T} = \frac{X'}{X} = \lambda$$

for some constant $\lambda \in \mathbb{R}$ (whenever $TX \neq 0$). This yields the equations

$$T' = \lambda T \quad \text{and} \quad X' = \lambda X,$$

whose solutions are given respectively by

$$T(t) = ce^{\lambda t} \quad \text{and} \quad X(x) = de^{\lambda x},$$

with $c, d \in \mathbb{R}$. Hence, equation (8.45) has the solutions

$$u(t, x) = ae^{\lambda(t+x)},$$

with $a \in \mathbb{R}$. Since any linear combination of solutions of equation (8.45) is still a solution of this equation, one can consider solutions of the form

$$u(t, x) = \sum_{n=1}^{N} a_n e^{\lambda_n (t+x)},$$

with $a_n, \lambda_n \in \mathbb{R}$ for each $n = 1, \ldots, N$. Taking $t = 0$, we obtain

$$u(0, x) = \sum_{n=1}^{N} a_n e^{\lambda_n x} = e^x + e^{-x},$$

and thus, one can take $N \geq 2$ and

$$a_n = \begin{cases} 1 & \text{if } n = 1, 2, \\ 0 & \text{if } n > 2, \end{cases}$$

with $\lambda_1 = 1$ and $\lambda_2 = -1$. Thus, a solution of the problem is

$$u(t, x) = e^{t+x} + e^{-(t+x)}.$$

Problem 8.11

Find all functions $u = u(t, x)$ of class C^2 satisfying

$$\frac{\partial^2 u}{\partial x \partial t} = 0. \tag{8.46}$$

Solution

We write equation (8.46) in the form

$$\frac{\partial}{\partial x}\left(\frac{\partial u}{\partial t}\right) = 0.$$

This implies that $\partial u / \partial t = f(t)$, where f is an arbitrary function of class C^1 (since u is of class C^2). Hence,

$$u(t, x) = X(x) + T(t), \tag{8.47}$$

where X is an arbitrary function of class C^2, and where T is a primitive of f (which thus is also of class C^2). Conversely, if u is of the form (8.47), where X and T are arbitrary functions of class C^2, then u is of class C^2 and

$$\frac{\partial^2 u}{\partial x \partial t} = \frac{\partial}{\partial x}\left(\frac{\partial u}{\partial t}\right) = \frac{\partial}{\partial x} T'(t) = 0.$$

Thus, the desired functions are those in (8.47), with X and T of class C^2.

Problem 8.12

Use the method of separation of variables to solve the equation

$$\frac{\partial^2 u}{\partial x^2} + \frac{\partial^2 u}{\partial y^2} + \frac{\partial^2 u}{\partial z^2} = 0. \tag{8.48}$$

Solution

Substituting the function

$$u(x, y, z) = X(x)Y(y)Z(z) \tag{8.49}$$

in equation (8.48), we obtain

$$X''YZ + XY''Z + XYZ'' = 0.$$

For nonzero solutions, dividing by XYZ, we obtain

$$\frac{X''}{X} + \frac{Y''}{Y} + \frac{Z''}{Z} = 0,$$

and thus,

$$\frac{X''}{X} = -\left(\frac{Y''}{Y} + \frac{Z''}{Z}\right).$$

Since the left-hand side does not depend either on y or on z, and the right-hand side does not depend on x, there exists $\lambda \in \mathbb{R}$ such that

$$\frac{X''}{X} = -\left(\frac{Y''}{Y} + \frac{Z''}{Z}\right) = -\lambda.$$

Therefore,

$$X'' + \lambda X = 0 \quad \text{and} \quad \frac{Y''}{Y} + \frac{Z''}{Z} = \lambda.$$

Now we write the second equation in the form

$$\frac{Y''}{Y} = -\frac{Z''}{Z} + \lambda.$$

Since the left-hand side does not depend on z and the right-hand side does not depend on y, there exists $\mu \in \mathbb{R}$ such that

$$\frac{Y''}{Y} = -\frac{Z''}{Z} + \lambda = -\mu,$$

and thus,

$$Y'' + \mu Y = 0 \quad \text{and} \quad Z'' - (\lambda + \mu)Z = 0.$$

By solving the equations

$$X'' + \lambda X = 0, \quad Y'' + \mu Y = 0 \quad \text{and} \quad Z'' - (\lambda + \mu)Z = 0, \tag{8.50}$$

we obtain a solution of equation (8.48) of the form (8.49). Since linear combination of solutions of equation (8.48) are still solutions of this equation, one can consider (formal) solutions of the form

$$u(x, y, z) = \sum_{n=1}^{\infty} c_n u_n(x, y, z), \tag{8.51}$$

where

$$u_n(x, y, z) = X_n(x) Y_n(y) Z_n(z)$$

for some functions X_n, Y_n and Z_n satisfying (8.50).

Problem 8.13

Solve equation (8.48), for $t \geq 0$ and $x, y, z \in [0, \pi]$, with the conditions

$$u(0, y, z) = u(\pi, y, z) = 0, \quad y, z \in (0, \pi), \tag{8.52}$$

and

$$u(x, 0, z) = u(x, \pi, z) = 0, \quad x, z \in (0, \pi). \tag{8.53}$$

Solution

Proceeding as in Problem 8.12, we first consider solutions of the form (8.49), which corresponds to solving the equations in (8.50). Substituting the function

$u(x, y, z) = X(x)Y(y)Z(z)$ in the conditions (8.52) and (8.53), we obtain

$$X(0) = X(\pi) = 0 \quad \text{and} \quad Y(0) = Y(\pi) = 0.$$

By Proposition 7.2, the problems

$$X'' + \lambda X = 0 \quad \text{with } X(0) = X(\pi) = 0,$$

and

$$Y'' + \mu Y = 0 \quad \text{with } Y(0) = Y(\pi) = 0$$

have nonzero solutions if and only if

$$\lambda = n^2 \quad \text{and} \quad \mu = m^2$$

for some $n, m \in \mathbb{N}$. The solutions are given respectively by

$$X(x) = a \sin(nx) \quad \text{and} \quad Y(y) = b \sin(my),$$

with $a, b \neq 0$. The third equation in (8.50) then takes the form

$$Z'' - \left(n^2 + m^2 \right) Z = 0,$$

and has the solutions

$$Z(z) = c_{nm} e^{\sqrt{n^2+m^2}\, z} + d_{nm} e^{-\sqrt{n^2+m^2}\, z},$$

with $c_{nm}, d_{nm} \in \mathbb{R}$ for each $n, m \in \mathbb{N}$. One can then consider (formal) solutions of the form (8.51), that is,

$$u(x, y, z) = \sum_{n=1}^{\infty} \sum_{m=1}^{\infty} \sin(nx) \sin(my) \left(c_{nm} e^{\sqrt{n^2+m^2}\, z} + d_{nm} e^{-\sqrt{n^2+m^2}\, z} \right).$$

Problem 8.14

Discuss whether the method of separation of variables can be used to solve the equation

$$t \frac{\partial^2 u}{\partial x^2} + u \frac{\partial u}{\partial t} = 0. \tag{8.54}$$

Solution

We first look for solutions of the form $u(t, x) = T(t)X(x)$. It follows from (8.54) that

$$tTX'' + TXT'X = 0, \tag{8.55}$$

and thus, for solutions with $TT'X'' \neq 0$, we have

$$\frac{t}{T'} = -\frac{X^2}{X''} = -\lambda$$

for some constant $\lambda \neq 0$. Therefore,

$$T' = -\frac{t}{\lambda} \quad \text{and} \quad X'' = \frac{1}{\lambda}X^2.$$

The solutions of the first equation are given by

$$T(t) = -\frac{1}{2\lambda}t^2 + a, \quad \text{with } a \in \mathbb{R}.$$

The second equation has for example the solution $X(x) = 1/x^2$ for $\lambda = 1/6$. Hence, a family of solutions of equation (8.54) is

$$u(t,x) = \frac{a - 3t^2}{x^2}, \quad \text{with } a \in \mathbb{R}.$$

On the other hand, in general, a linear combination of solutions of equation (8.54) is not a solution of this equation. For example, if u is a solution, then

$$t\frac{\partial^2 (2u)}{\partial x^2} + (2u)\frac{\partial (2u)}{\partial t} = 2t\frac{\partial^2 u}{\partial x^2} + 4u\frac{\partial u}{\partial t}$$

$$= 2\left(t\frac{\partial^2 u}{\partial x^2} + u\frac{\partial u}{\partial t}\right) + 2u\frac{\partial u}{\partial t} = 2u\frac{\partial u}{\partial t}.$$

Hence, $2u$ is also a solution if and only if

$$\frac{\partial (u^2)}{\partial t} = 2u\frac{\partial u}{\partial t} = 0,$$

which shows that u^2, and thus also u, must be independent of t. In other words, we must have $u(t,x) = X(x)$. It then follows from (8.54), or from (8.55), that $tX'' = 0$, and $X(x) = bx + c$, with $b, c \in \mathbb{R}$. In particular, it is impossible to use linear combinations to generate new solutions from the solutions that are already known, namely

$$\frac{a - 3t^2}{x^2} \quad \text{and} \quad bx + c,$$

with $a, b, c \in \mathbb{R}$. In this sense, the method of separation of variables cannot be used to solve equation (8.54).

EXERCISES

8.1. Find a solution of the equation:

(a) $\dfrac{\partial u}{\partial t} = \dfrac{\partial^2 u}{\partial x^2}$, $t \geq 0$, $x \in [0,1]$,

with the conditions

$$u(t,0) = u(t,1) = 0, \quad t > 0,$$

and

$$u(0,x) = x + \sin(\pi x), \quad x \in (0,1);$$

(b) $\dfrac{\partial u}{\partial t} = 4\dfrac{\partial^2 u}{\partial x^2}$, $t \geq 0$, $x \in [0,2]$,

with the conditions

$$u(t,0) = u(t,2) = 0, \quad t > 0,$$

and

$$u(0,x) = x^2, \quad x \in (0,2);$$

(c) $\dfrac{\partial u}{\partial t} = \dfrac{\partial^2 u}{\partial x^2} - 2\dfrac{\partial u}{\partial x}$, $t \geq 0$, $x \in [0,1]$,

with the conditions

$$u(t,0) = u(t,1) = 0, \quad t > 0,$$

and

$$u(0,x) = e^x \sin(\pi x), \quad x \in (0,1);$$

(d) $\dfrac{\partial^2 u}{\partial x^2} + \dfrac{\partial^2 u}{\partial y^2} = 0$, $(x,y) \in [0,1] \times [0,2]$,

with the conditions

$$u(x,0) = u(x,2) = 0, \quad x \in (0,1),$$

and

$$u(0,y) = 0, \quad u(1,y) = y^2, \quad y \in (0,2);$$

(e) $\dfrac{\partial^2 u}{\partial x^2} + \dfrac{\partial^2 u}{\partial y^2} = 0$, $(x,y) \in [0,3] \times [0,4]$, with the conditions

$$\frac{\partial u}{\partial y}(x,0) = \frac{\partial u}{\partial y}(x,4) = 0, \quad x \in (0,3),$$

and

$$\frac{\partial u}{\partial x}(0,y) = 0, \quad \frac{\partial u}{\partial x}(3,y) = \cos\left(\frac{3\pi y}{4}\right), \quad y \in (0,4);$$

(f) $\dfrac{\partial^2 u}{\partial t^2} = 2\dfrac{\partial^2 u}{\partial x^2}, \quad t \geq 0, \ x \in [0,1],$

with the conditions

$$u(t,0) = u(t,1) = 0, \quad t > 0,$$

and

$$u(0,x) = \sin(\pi x), \quad \frac{\partial u}{\partial t}(0,x) = 2\sin(4\pi x), \quad x \in (0,1).$$

8.2. Find a solution of equation (8.45) with $u(0,x) = e^{3x} + 4e^{-x}$.

8.3. Write the Laplacian

$$\Delta = \frac{\partial^2}{\partial x^2} + \frac{\partial^2}{\partial y^2}$$

in polar coordinates (r,θ).

8.4. Show that the functions $u(t,x) = v(x+ct)$, with v of class C^1, satisfy the equation

$$\frac{\partial u}{\partial t} = c\frac{\partial u}{\partial x}.$$

8.5. Show that all solutions of the equation

$$\frac{\partial u}{\partial t} = c\frac{\partial u}{\partial x}$$

are of the form $u(t,x) = v(x+ct)$.

8.6. Find all solutions of the equation

$$\frac{\partial^2 u}{\partial x \partial t} = u$$

of the form $u(t,x) = T(t)X(x)$.

8.7. Discuss whether the method of separation of variables can be used to solve the equation:

(a) $\dfrac{\partial u}{\partial t} = 4t^2 \dfrac{\partial u}{\partial x}$;

(b) $\dfrac{\partial^2 u}{\partial t^2} = 4x \dfrac{\partial^2 u}{\partial x^2}$.

Index

L. Barreira, C. Valls, *Complex Analysis and Differential Equations*,   413
Springer Undergraduate Mathematics Series,
DOI 10.1007/978-1-4471-4008-5, © Springer-Verlag London 2012